LTE SMALL CELL OPTIMIZATION

LTE SMALL CELL OPTIMIZATION

3GPP EVOLUTION TO RELEASE 13

Edited by

Harri Holma
Antti Toskala
Jussi Reunanen

A catalogue record for this book is available from the British Library.

ISBN: 9781118912577

Set in 10/12pt Times by Aptara Inc., New Delhi, India

Cover image: © RossHelen/istockphoto

1   2016

Contents

Preface

We have witnessed a fast growth in mobile broadband capabilities during the last 10 years in terms of data rates, service availability, number of customers and data volumes. The launch of the first LTE network in December 2009 further boosted the growth of data rates and capacities. LTE turned out to be a success because of efficient performance and global economics of scale. The first LTE-Advanced network started in 2013, increasing the data rate to 300 Mbps by 2014, 450 Mbps in 2015 and soon to 1 Gbps. The number of LTE networks had grown globally to more than 460 by end 2015.

This book focuses on those solutions improving the practical LTE performance: small cells and network optimization. The small cells are driven by the need to increase network capacity and practical user data rates. The small cell deployment creates a number of new challenges for practical deployment ranging from interference management to low-cost products, site solutions and optimization. The network optimization targets to squeeze everything out of the LTE radio in terms of coverage, capacity and end-user performance.

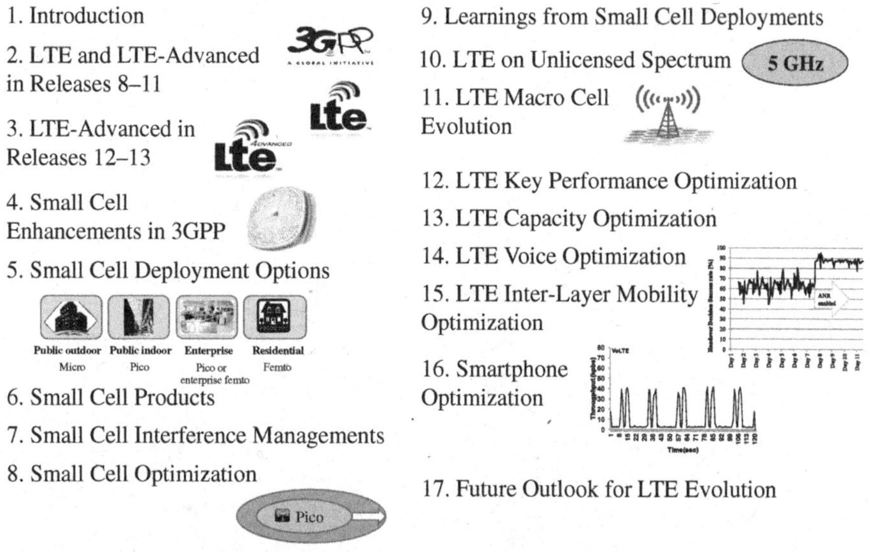

1. Introduction

2. LTE and LTE-Advanced in Releases 8–11

3. LTE-Advanced in Releases 12–13

4. Small Cell Enhancements in 3GPP

5. Small Cell Deployment Options

Public outdoor — Micro
Public indoor — Pico
Enterprise — Pico or enterprise femto
Residential — Femto

6. Small Cell Products

7. Small Cell Interference Managements

8. Small Cell Optimization

9. Learnings from Small Cell Deployments

10. LTE on Unlicensed Spectrum — 5 GHz

11. LTE Macro Cell Evolution

12. LTE Key Performance Optimization

13. LTE Capacity Optimization

14. LTE Voice Optimization

15. LTE Inter-Layer Mobility Optimization

16. Smartphone Optimization

17. Future Outlook for LTE Evolution

Figure P.1   Contents of the book

Smartphones, tablets and laptops are the main use cases for LTE networks currently, but LTE radio will be the foundation for many new applications in the future. Internet of things, public safety, device-to-device communication, broadcast services and vehicle communication are a few examples that can take benefit of future LTE radio.

The contents of the book are summarized in Figure P.1. Chapters 1–3 provide an introduction to LTE in 3GPP Releases 8–13. The small cell-specific topics are discussed in Chapters 4–10 including 3GPP features, network architecture, products, interference management, optimization, practical learnings and unlicensed spectrum. The LTE optimization is presented in Chapters 11–16 including 3GPP evolution, performance, voice, inter-layer and smartphone optimization. Chapter 17 illustrates the outlook for further LTE evolution.

Acknowledgements

The editors would like to acknowledge the hard work of the contributors from Nokia Networks, T-Mobile USA and Videotron Canada: Rajeev Agrawal, Anand Bedekar, Mihai Enescu, Amitava Ghosh, Tero Henttonen, Wang Hua, Suresh Kalyanasundaram, Jari Lindholm, Timo Lunttila, Riku Luostari, Bishwarup Mondal, Laurent Noël, Brian Olsen, Klaus Pedersen, Karri Ranta-aho, Claudio Rosa, Jari Salo, Rafael Sanchez-Mejias, Mikko Simanainen, Beatriz Soret and Benny Vejlgaard.

The editors also would like to thank the following colleagues for their valuable comments and contributions: Petri Aalto, Yin-tat Peter Chiu, Bong Youl (Brian) Cho, Jeongho (Jackson) Cho, Jinho (Jared) Cho, Anthony Ho, Richa Gupta, Kari Hooli, Kyeongtaek Kim, Kimmo Kivilahti, Ekawit Komolpun, Wai Wah (Endy) Kong, Dinesh Kumar, Karri Kuoppamäki, Andrew Lai, Franck Laigle, Mads Lauridsen, Hyungyoup (Henry) Lee, Jasin (Jason) Lee, Sami Lehesaari, Jun Liu, Jarkko Lohtaja, Yi-Nan (Evan) Lu, Mark McDiarmid, Luis Maestro, Deshan Miao, Marko Monkkonen, Balamurali Natarajan, Nuttavut Sae-Jong, Shuji Sato, Changsong Sun, Wangkeun (David) Sun, Kirsi Teravainen, Jukka Virtanen, Eugene Visotsky and Veli Voipio.

The editors appreciate the fast and smooth editing process provided by Wiley publisher and especially Tiina Wigley and Mark Hammond.

The editors are grateful to their families, as well as the families of all the authors, for their patience during the late night writing and weekend editing sessions.

The editors and authors welcome any comments and suggestions for improvements or changes that could be implemented in forthcoming editions of this book. The feedback may be addressed to: harri.holma@nokia.com, antti.toskala@nokia.com and jussi.reunanen@nokia.com

List of Abbreviations

3D	Three Dimensional
3GPP	Third Generation Partnership Project
AAS	Active Antenna System
ABS	Almost Blank Subframe
AC	Alternating Current
ACK	Acknowledgement
AIR	Antenna Integrated Radio
AM	Acknowledge Mode
AMR	Adaptive Multirate
ANDSF	Access Network Discovery and Selection Function
ANR	Automatic Neighbour Relations
APP	Applications
APT	Average Power Tracking
ARFCN	Absolute Radio Frequency Channel Number
ARQ	Automatic Repeat Request
AS	Application Server
ASA	Authorized Shared Access
AWS	Advanced Wireless Spectrum
BBU	Baseband Unit
BCCH	Broadcast Channel
BLER	Block Error Rate
BSIC	Base Station Identity Code
BSR	Buffer Status Report
BTS	Base Station
C-RNTI	Cell Radio Network Temporary Identifier
CA	Carrier Aggregation
CAPEX	Capital Expenditure
CAT	Category
CC	Component Carrier
CCA	Clear Channel Assessment
CCE	Control Channel Element
CDF	Cumulative Density Function
CDMA	Code Division Multiple Access
cDRX	Connected Discontinuous Reception

CoMP	Coordinated Multipoint
CPRI	Common Public Radio Interface
CN	Core Network
CPICH	Common Pilot Channel
CPU	Central Processing Unit
CQI	Channel Quality Indicator
CRC	Cyclic Redundancy Check
CRAN	Centralized Radio Access Network
CRS	Common Reference Signals
CRS-IC	Common Reference Signal interference cancellation
CS	Circuit Switched
CS	Cell Selection
CSCF	Call Session Control Function
CSFB	Circuit Switched Fallback
CSG	Closed Subscriber Group
CSI	Channel State Information
CSI-RS	Channel State Information Reference Signals
CSMO	Circuit Switched Mobile Originated
CSMT	Circuit Switched Mobile Terminated
CSSR	Call Setup Success Rate
CWIC	Code Word Interference Cancellation
CWDM	Coarse Wavelength Division Multiplexing
D2D	Device-to-Device
DAS	Distributed Antenna System
DC	Direct Current
DC	Dual Connectivity
DCCH	Dedicated Control Channel
DCH	Dedicated Channel
DCI	Downlink Control Information
DCR	Drop Call Rate
DFS	Dynamic Frequency Selection
DMCR	Deferred Measurement Control Reading
DMRS	Demodulation Reference Signals
DMTC	Discovery Measurement Timing Configuration
DPS	Dynamic Point Selection
DRB	Data Radio Bearer
DRS	Discovery Reference Signals
DRX	Discontinuous Reception
DSL	Digital Subscriber Line
DTX	Discontinuous Transmission
DU	Digital Unit
ECGI	E-UTRAN Cell Global Identifier
eCoMP	Enhanced Coordinated Multipoint
EDPCCH	Enhanced Downlink Physical Control Channel
EFR	Enhanced Full Rate
eICIC	Enhanced Inter-Cell Interference Coordination

eMBMS	Enhanced Multimedia Broadcast Multicast Solution
EPA	Enhanced Pedestrian A
EPC	Evolved Packet Core
EPRE	Energy Per Resource Element
eRAB	Enhanced Radio Access Bearer
ESR	Extended Service Request
ET	Envelope Tracking
EVM	Error Vector Magnitude
EVS	Enhanced Voice Services
FACH	Forward Access Channel
FD-LTE	Frequency Division Long Term Evolution
FDD	Frequency Division Duplex
feICIC	Further Enhanced Inter-Cell Interference Coordination
FFT	Fast Fourier Transformation
FSS	Frequency Selective Scheduling
FTP	File Transfer Protocol
GBR	Guaranteed Bit Rate
GCID	Global Cell Identity
GERAN	GSM EDGE Radio Access Network
GPON	Gigabit Passive Optical Network
GPS	Global Positioning System
GS	Gain Switching
GSM	Global System for Mobile Communications
HARQ	Hybrid Automatic Repeat Request
HD	High Definition
HetNet	Heterogeneous Network
HFC	Hybrid Fibre Coaxial
HO	Handover
HOF	Handover Failure
HPM	High-Performance Mobile
HSPA	High-Speed Packet Access
HSDPA	High-Speed Downlink Packet Access
HSUPA	High-Speed Uplink Packet Access
HTTP	Hypertext Transfer Protocol
IAS	Integrated Antenna System
IC	Interference Cancellation
ICIC	Inter-Cell Interference Coordination
IRC	Interference Rejection Combining
IEEE	Institute of Electrical and Electronics Engineers
IM	Instant Messaging
IMEI	International Mobile Station Equipment Identity
IMPEX	Implementation Expenditure
IMS	Internet Protocol Multimedia Subsystem
IMT	International Mobile Telecommunication
IoT	Internet-of-Things
IQ	In-phase and Quadrature

IRC	Interference Rejection Combining
IRU	Indoor Radio Unit
ISD	Inter Site Distance
IT	Information Technology
ITU-R	International Telecommunications Union – Radiocommunications Sector
JP	Joint Processing
JT	Joint Transmission
KPI	Key Performance Indicator
LAA	Licensed Assisted Access
LAN	Local Area Network
LAU	Location Area Update
LBT	Listen-Before-Talk
LMMSE-IRC	Linear Minimum Mean Squared Error Interference Rejection Combining
LOS	Line of Sight
LP	Low Power
LTE	Long-Term Evolution
LTE-A	LTE-Advanced
LU	Location Update
MAC	Medium Access Control
MBSFN	Multicast Broadcast Single Frequency Network
MDT	Minimization of Drive Testing
MeNB	Macro eNodeB
MeNodeB	Master eNodeB
M2M	Machine-to-Machine
MCL	Minimum Coupling Loss
MCS	Modulation and Coding Scheme
ML	Maximum Likelihood
MLB	Mobility Load Balancing
MIMO	Multiple Input Multiple Output
MME	Mobility Management Entity
MMSE	Minimum Mean Square Error
MOS	Mean Opinion Score
MRC	Maximal Ratio Combining
MRO	Mobility Robustness Optimization
MSS	Mobile Switching centre Server
MTC	Machine Type Communications
MTC	Mobile Terminating Call
MTRF	Mobile Terminating Roaming Forwarding
MTRR	Mobile Terminating Roaming Retry
M2M	Machine-to-Machine
NAICS	Network Assisted Interference Cancellation and Suppression
NAS	Non-access Stratum
NB	Narrowband
NLOS	Non-line of Sight
NOMA	Non-orthogonal Multiple Access
OBSAI	Open Base Station Architecture Initiative

O&M	Operations and Maintenance
OFDM	Orthogonal Frequency Division Multiplexing
OFDMA	Orthogonal Frequency Division Multiple Access
OPEX	Operating Expenditure
OS	Operating System
OSG	Open Subscriber Group
OTT	Over the Top
PA	Power Amplifier
PBCH	Physical Broadcast Channel
PCell	Primary Cell
PCFICH	Physical Control Format Indicator Channel
PCH	Paging Channel
PCI	Physical Cell Identity
PDCCH	Physical Downlink Control Channel
PDCP	Packet Data Convergence Protocol
PDF	Power Density Function
PDSCH	Physical Downlink Shared Channel
PDU	Protocol Data Unit
PESQ	Perceptual Evaluation of Speech
PHR	Power Headroom Report
PWG	Packet Data Network Gateway
PLMN	Public Land Mobile Network
PMI	Precoding Matrix Indicator
PoE	Power over Ethernet
POLQA	Perceptual Objective Listening Quality
PON	Passive Optical Network
PRB	Physical Resource Block
PSCCH	Physical Sidelink Control Channel
PSBCH	Physical Sidelink Broadcast Channel
PSCCH	Physical Sidelink Control Channel
PSD	Power Spectral Density
PSDCH	Physical Sidelink Discovery Channel
PSM	Power-Saving Mode
PSSCH	Physical Sidelink-Shared Channel
PUCCH	Physical Uplink-Control Channel
PUSCH	Physical Uplink-Shared Channel
pRRU	Pico Remote Radio Unit
QAM	Quadrature Amplitude Modulation
QCI	Quality of Service Class Identifier
QoS	Quality of Service
RA-RNTI	Random Access Radio Network Temporary Identifier
RACH	Random Access Channel
RAN	Radio Access Network
RAO	Random Access Opportunity
RAT	Radio Access Technology
RB	Radio Bearer

RCS	Rich Call Services
RE	Range Extension
RET	Remote Electrical Tilting
RF	Radio Frequency
RHUB	Remote Radio Unit Hub
RI	Rank Indicator
RIM	Radio Information Management
RLC	Radio Link Control
RNC	Radio Network Controller
RRH	Remote Radio Head
RAT	Radio Access Technology
RLF	Radio Link Failure
RLM	Radio Link Monitoring
RNC	Radio Network Controller
RNTP	Relative Narrow-Band Transmit Power
ROHC	Robust Header Compression
RoT	Rise over Thermal
RRC	Radio Resource Control
RRM	Radio Resource Management
RS	Reference Signal
RSCP	Received Signal Code Power
RSRP	Reference Signal Received Power
RSRQ	Reference Signal Received Quality
RSSI	Received Signal Strength Indication
RX	Reception
SC-FDMA	Single Carrier Frequency Division Multiple Access
SDU	Service Data Unit
S-GW	Serving Gateway
S-TMSI	SAE Temporary Mobile Subscriber Identity
S1AP	S1 Application Protocol
SeNB	Small eNodeB
SeNodeB	Secondary eNodeB
SCell	Secondary Cell
SCS	Short Control Signalling
SFN	Single Frequency Network
SGW	Serving Gateway
SI-RNTI	System Information Radio Network Temporary Identifier
SIB	System Information Block
SINR	Signal to Interference and Noise Ratio
SIP	Session Initiation Protocol
SISO	Single Input Single Output
SLIC	Symbol Level Interference Cancellation
SMS	Short Message Service
SNR	Signal-to-Noise Ratio
SON	Self-Organizing Network
SPS	Semi Persistent Scheduling

SR	Scheduling Request
SRB	Signalling Radio Bearer
SRS	Sounding Reference Signals
SRVCC	Single Radio Voice Call Continuity
SU	Single User
SWB	Super Wideband
TA	Timing Advance
TAU	Tracking Area Update
TBS	Transport Block Size
TC	Test Case
TCO	Total Cost of Ownership
TCP	Transmission Control Protocol
TDD	Time Division Duplex
TD-LTE	Time Division Long-Term Evolution
TETRA	Terrestrial Trunked Radio
TM	Transmission Mode
TPC	Transmission Power Control
TTI	Transmission Time Interval
TTT	Time to Trigger
TX	Transmission
UCI	Uplink Control Information
UDN	Ultra Dense Network
UDP	User Datagram Protocol
UE	User Equipment
UPS	Uninterruptible Power Supply
USB	Universal Serial Bus
USIM	Universal Subscriber Identity Module
UTRA	Universal Terrestrial Radio Access
UTRAN	Universal Terrestrial Radio Access Network
VAD	Voice Activity Detection
VDSL	Very High Bit Rate Digital Subscriber Line
VoLTE	Voice over LTE
V2X	Vehicle to Infrastructure
V2V	Vehicle to Vehicle
WB	Wideband
WCDMA	Wideband Code Division Multiple Access
WDM	Wavelength Division Multiplexing
Wi-Fi	Wireless Fidelity
WLAN	Wireless Local Area Network
X2AP	X2 Application Protocol

1

Introduction

Harri Holma

1.1 Introduction

Mobile broadband technology has experienced an incredibly fast evolution during the last 10 years. The first High-Speed Downlink Packet Access (HSDPA) network was launched 2005 enabling the high-speed mobile broadband and the first iPhone was launched 2007 creating the massive need for the mobile broadband. The data rates have increased more than 100-fold and the data volumes by more than 1000-fold during the last 10 years. The HSDPA started with 3.6 Mbps while the latest Long-Term Evolution (LTE)-Advanced networks deliver user data rates of 300 Mbps during 2014 and 450 Mbps during 2015. But still, we have just seen the first part of the mobile broadband era – the fast evolution continues forward. Also, the number of mobile broadband subscribers is increasing rapidly with affordable new smartphones providing the internet access for the next billions of users.

This chapter shortly introduces the status of LTE networks globally, the traffic growth, LTE in Third Generation Partnership Project (3GPP) and the spectrum aspects. The chapter also discusses the importance of the small cell deployments, the network optimization and the LTE.

LTE Small Cell Optimization: 3GPP Evolution to Release 13, First Edition.
Edited by Harri Holma, Antti Toskala and Jussi Reunanen.
© 2016 John Wiley & Sons, Ltd. Published 2016 by John Wiley & Sons, Ltd.

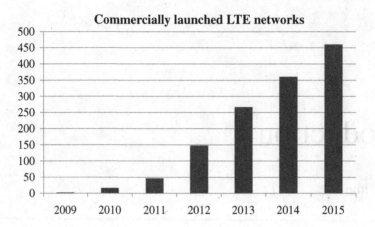

Figure 1.1   Commercially launched LTE networks [1]

1.2   LTE Global Deployments and Devices

The first commercial LTE network was opened by Teliasonera in Sweden in December 2009 marking the new era of high-speed mobile communications. The number of commercial LTE networks has already increased to 460 in more than 140 countries by end 2015. The fast growth of launches has happened during 2012–2015. It is expected that more than 500 operators in more than 150 countries will soon have commercial LTE network running. The number of launched networks is shown in Figure 1.1.

The very first LTE devices supported 100 Mbps in the form factor of Universal Serial Bus (USB) modem. Soon LTE capability was introduced into high end and mid-priced smartphones with the bit rate up to 150 Mbps. By 2015, LTE radio is found in most smartphones excluding the very low end segment at 50-USD. The data rate capability is increased to 300 Mbps, and 450 Mbps, in the latest devices. An example drive test throughput is shown in Figure 1.2 illustrating that it is possible to get very close to 300 Mbps in the good channel conditions in

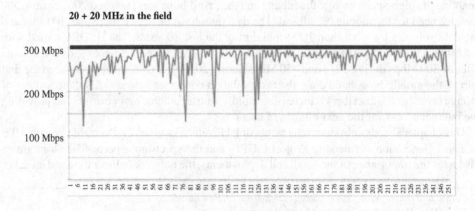

Figure 1.2   Drive test data rate with Category 6 LTE device

Figure 1.3 Example Category 6 LTE device supporting 300 Mbps

the field with Category 6 devices. An example of such a device is shown in Figure 1.3. Also, the support for Voice over LTE (VoLTE) is included, which allows to use LTE network not only for data connections but also for voice connections.

1.3 Mobile Data Traffic Growth

The mobile data traffic has grown rapidly during the last few years driven by the new smartphones, large displays, higher data rates and higher number of mobile broadband subscribers. The mobile data growth for 2-year period is illustrated in Figure 1.4. These data are collected from more than 100 major operators globally. The absolute data volume in this graph is more than million terabytes, that is, exabytes, per year. The data traffic has grown by a factor of 3.6× during the 2-year period, which corresponds to 90% annual growth. The fast growth of mobile data is expected to continue. The data growth is one of the reasons why more LTE networks are required, more spectra are needed, radio optimization is necessary and why small cells will be deployed. All this data growth must happen without increase in the operator revenues. Therefore, the cost per bit must decrease and the network efficiency must increase correspondingly.

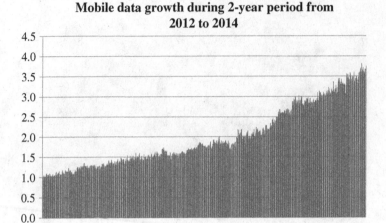

Figure 1.4   Mobile data growth during 2-year period

1.4   LTE Technology Evolution

The LTE technology has been standardized by Third Generation Partnership Project (3GPP). The LTE was introduced in 3GPP Release 8. The specifications were completed and the backwards compatibility started in March 2009. Release 8 enabled peak rate of 150 Mbps with 2×2 MIMO, low latency, flat network architecture and the support for 4-antenna base station transmission and reception. Release 8 enabled in theory also 300 Mbps with 4×4 MIMO but the practical devices so far have two antennas limiting the data rate to 150 Mbps. Release 9 was a relatively small update on top of Release 8. Release 9 was completed 1 year after Release 8 and the first deployments started during 2011. Release 9 brought enhanced Multimedia Broadcast Multicast Solution (eMBMS) also known as LTE-Broadcast, emergency call support for VoLTE, femto base station handovers and first set of Self-Organizing Network (SON) functionalities. Release 10 provided a major step in terms of data rates and capacity with Carrier Aggregation (CA), higher order MIMO up to eight antennas in downlink and four antennas in uplink. The support for Heterogeneous Network (HetNet) was included in Release 10 with the feature enhanced Inter-Cell Interference Coordination (eICIC). Release 10 was completed in June 2011 and the first commercial carrier aggregation network started in June 2013. Release 10 is also known as LTE-Advanced. Release 11 enhanced LTE-Advanced with Coordinated Multipoint (CoMP) transmission and reception, with further enhanced ICIC (feICIC), advanced UE interference cancellation and carrier aggregation improvements. First Release 11 commercial implementation was available during 2015. Release 12 was completed in 3GPP in March 2015 and the deployments are expected by 2017. Release 12 includes dual connectivity between macro and small cells, enhanced CoMP (eCoMP), machine-to-machine (M2M) optimization and device-to-device (D2D) communication. Release 13 work started during second half of 2014 and is expected to be completed during 2016. Release 13 brings Licensed Assisted Access (LAA) also known as LTE for Unlicensed bands (LTE-U), Authorized Shared Access (ASA), 3-dimensional (3D) beamforming and D2D enhancements.

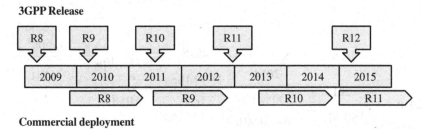

Figure 1.5 3GPP Release availability dates

The timing of 3GPP releases in standardization and in first commercial networks is shown in Figure 1.5. The main contents of each release are summarized in Figure 1.6.

The radio data rate has increased very fast with LTE and LTE-Advanced and 150 Mbps was available with Release 8 commercially during 2010 when using continuous 20 MHz spectrum allocation and 2×2 MIMO. The carrier aggregation with 10 + 10 MHz during 2013 enabled 150 Mbps also for those operators having just 10 MHz continuous spectrum. The aggregation of 20 + 20 MHz during 2014 pushed the peak rate to 300 Mbps and three-carrier aggregation (3CA) further increased the data rate up to 450 Mbps during 2015. The evolution is expected to continue rapidly in the near future with commercial devices supporting 1 Gbps with 100 MHz of total bandwidth. The evolution of device data rate capabilities is illustrated in Figure 1.7. The terminals used to be the limiting factor in terms of data rates but now the limiting factor is rather the amount of spectrum that can be allocated for the connection. Those operators having more spectrum resources available will have an advantage in terms of data rate capabilities. The carrier aggregation aspects are discussed in Reference [2].

1.5 LTE Spectrum

The LTE-Advanced needs lot of spectrum to deliver high capacity and data rates. The typical spectrum resources in European or some Asian markets are shown in Figure 1.8. All the

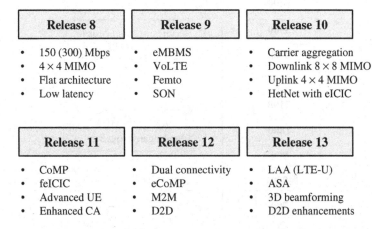

Figure 1.6 Main contents of each release

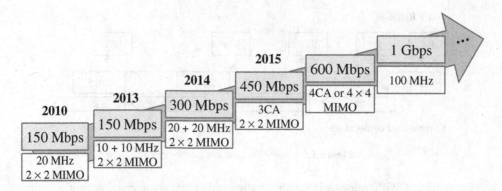

Figure 1.7 Peak data rate evolution in commercial devices

spectrum between 700 and 2600 MHz will be aggregated together. The carrier aggregation of the multiple spectrum blocks together helps in terms of network traffic management and load balancing in addition to providing more capacity and higher data rates. The higher frequency at 3.5 GHz can also be aggregated in the macro cells especially if the site density is high in the urban areas. Another option is to use 3.5 GHz band in the small cell layer. The unlicensed band at 5 GHz can be utilized in the small cells for more spectrum [3]. In the best case, more than 1 GHz of licensed spectrum is available for the mobile operators and more when considering the unlicensed spectrum.

1.6 Small Cell Deployments

The small cell deployments are driven by the need for improved network quality, higher data rates and more capacity. So far most of the base stations have been high power macro base

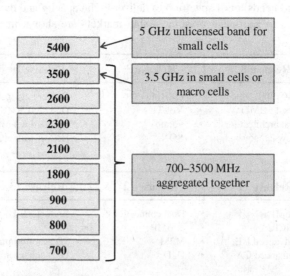

Figure 1.8 Example spectrum resources for LTE usage

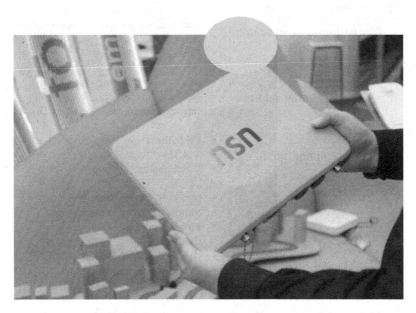

Figure 1.9   The LTE micro cell base station with 5 kg, 5 L and 5+5 W

stations while some small cells are being deployed in the advanced LTE networks. The small cell rollouts have come simpler than earlier because the new products are compact and include SON features for automatic configuration and optimization. An example micro base station product is shown in Figure 1.9: Nokia FlexiZone micro with 5+5 W of output power in 5 kg and 5 L outdoor capable package. The lower power indoor products are even smaller and more compact.

The small cell deployment is different from the traditional macro cells and requires new competences and tools. For example, the location of the cell is important for the maximal benefit of the small cell. If the small cell is located close to the hotspot users, it can offload lot of macro cell traffic. But if the small cell location is less optimal, it collects hardly any traffic and increases interference. The interference management is also critical between macro and small cells in the co-channel deployment.

The 3GPP brings a number of enhancements for HetNet cases in LTE-Advanced for interference management, for multi-cell reception and for higher order modulation.

1.7   Network Optimization

The LTE networks need optimization in order to deliver the full potential of LTE technology. The Key Performance Indicators (KPIs) like success rates and drop rates must reach sufficient levels, the data rates should match the expectations for 4G technology and the coverage must fulfil the operator promises. These basic optimizations not only include traditional RF and antenna optimization but also LTE-specific parameter and planning solutions like cell identity planning, link adaptation and power control parameterization and scheduler optimization. The hotspot areas and especially mass events call for more advanced optimization to deliver stable

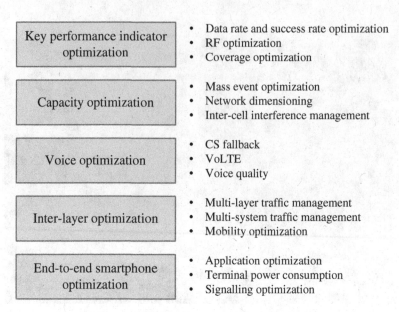

Figure 1.10   Key areas of LTE network optimization

network operator during the extreme loading when tens of thousands of customers are in the same location like sports stadiums. The interference management and control channels are the critical areas in those venues.

The initial LTE networks did not support voice but CS voice in the legacy 2G and 3G networks was used instead. The CS fallback requires some optimization while VoLTE needs further optimization steps to deliver the same or higher quality voice than with CS voice.

Both 2G and 3G networks typically had two frequency bands in use for each operator while LTE networks can have more than six LTE bands. Such complex spectrum allocation requires careful planning how the multiple frequencies are used together. The interworking to the underlying 3G network is also needed for both voice and data until LTE deployment reaches full coverage and VoLTE is used by all customers.

The end-user experience is impacted also by the terminal power consumption, by the application performance and by the device operating system. The end-to-end application optimization needs to consider all these aspects to deliver an attractive user experience to the customer's smartphone or other device.

The key areas of LTE network optimization are summarized in Figure 1.10 and will be explained in detail in this book.

1.8   LTE Evolution Beyond Release 13

The 3GPP is busy working with Release 13 during 2015 and 2016, but it will not be the end of LTE technology. The target is to utilize the massive ecosystem of the LTE radio also for other use cases and applications. Figure 1.11 illustrates new services that can be supported by the LTE radio and the corresponding technical solution. There are more than 7 billion

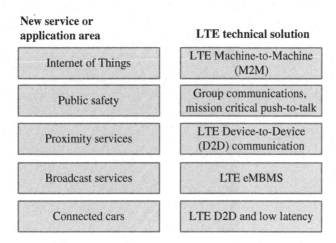

Figure 1.11 New application areas for LTE radio

mobile connections globally during 2016 but a lot more connected devices are expected in the future when many objects will have an internet connection – that is called Internet of Things (IoT). The LTE M2M optimization is targeting on that application area. The LTE can also be utilized for the public safety networks, like police and firemen. When the public safety applications utilize LTE technology, the performance will be far better than in the traditional public safety radios and the communication can even run on top of commercial networks making the solution also more cost efficient. The public safety applications need direct D2D communication. The same D2D functionality can also be utilized for many new proximity services between devices. The LTE broadcast service could eventually replace terrestrial TV by using eMBMS. Other new application areas can be connected cars for traffic safety and for entertainment or backhauling the Wi-Fi access points in the airplanes. The new service capabilities are illustrated in Reference [4].

1.9 Summary

The mobile broadband technology has experienced an impressive evolution during the last 10 years in terms of data usage, data rates and connected smartphones, tablet and computers. The data volumes have increased more than by a factor of 1000× and the data rates have increased by more than a factor of 100×. The further evolution continues in 3GPP and in commercial LTE networks. A number of optimization steps are required to take full benefit of LTE technology capabilities including network optimization and small cells. This book focuses on these topics: 3GPP evolution, LTE small cells and LTE network optimization.

References

[1] GSA – The Global Mobile Suppliers Association, July 2015.
[2] 'LTE-Advanced Carrier Aggregation Optimization', Nokia white paper, 2014.
[3] 'LTE for Unlicensed Spectrum', Nokia white paper, 2014.
[4] 'LTE-Advanced Evolution in Releases 12 – 14', Nokia white paper, 2015.

2

LTE and LTE Advanced in Releases 8–11

Antti Toskala

2.1 Introduction

This chapter presents the principles of first LTE version in Release 8 as well as the first development step with LTE Advanced in Releases 10 and 11. First the fundamental principles of LTE in Release 8 are covered, including the key channel structures as well as the key physical layer procedures. Then the enhancements in Releases 10 and 11 are introduced including the principles of carrier aggregation, enhanced multi-antenna operation as well as Co-operative Multipoint (CoMP) transmission principles. This chapter is concluded with the UE capabilities until Release 11.

2.2 Releases 8 and 9 LTE

Release 8 work was started in the end of 2004 and was finalized towards the end of 2008 with the introduction of the first full set of LTE specifications. The first LTE Release 8 was already a high-performance radio system that could deliver high capacity and data rates beyond 100 Mbps, as has been proven in the field as well. While the multiple access principle was new

---

LTE Small Cell Optimization: 3GPP Evolution to Release 13, First Edition.
Edited by Harri Holma, Antti Toskala and Jussi Reunanen.
© 2016 John Wiley & Sons, Ltd. Published 2016 by John Wiley & Sons, Ltd.

Table 2.1   LTE Releases 8 and 9 physical layer parameters for different bandwidths

Bandwidth	1.4 MHz	3.0 MHz	5 MHz	10 MHz	15 MHz	20 MHz
Sub-frame duration			1 ms			
Sub-carrier spacing			15 kHz			
FFT length	128	256	512	1024	1526	2048
Sub-carriers	72	180	300	600	900	1200
Symbols per slot			7 with short CP and 6 with long CP			
Cyclic prefix			5.21 or 4.69 µs with short CP and 16.67 µs with long CP			

compared to earlier system in 3GPP, there were several advanced technologies that had been taken into use from the development work done for the High-Speed Packet Access (HSPA) [1] technology over the past decade, including base station-based scheduling, link adaptation, physical layer retransmission and many other advances in the field of mobile communications. With LTE, these advances were combined with the advanced design of the multiple access technology enabling both efficient terminal power consumption with Single Carried FDMA (SC-FDMA) for uplink and low-complexity terminal receiver with Orthogonal Frequency Division Multiple Access (OFDMA) for downlink. The improvements in Release 9 did not change the peak data rates or key physical layer principles or the achievable data rates. Release 9 improvements mostly focused on enabling voice service, including necessary elements for the emergency calls such as prioritization and position location.

2.2.1   Releases 8 and 9 Physical Layer

The LTE Releases 8 and 9 support six different system bandwidths ranging from 1.4 to 20 MHz as shown in Table 2.1. The design has been done such that the timing is constant for all bandwidths, only the number of sub-carrier to receive by the UE changes as the function of the bandwidth. The designs for the synchronization signals as well as for the Physical Broadcast Channel (PBCH) are such that they are included in the six centre 180 kHz physical resource blocks regardless of the actual system bandwidth.

The downlink resource allocation is done with the Physical Resource Blocks (PRB), with the minimum resource allocation being one resource block. A resource block is 1 ms in time domain and 12 sub-carriers, each of them 15 kHz, thus 180 kHz in frequency domain. The number of OFDMA symbols fitting to a sub-frame depends on the cyclic prefix length chosen. With the short cyclic prefix, there is room for 14 symbols per each sub-frame while longer (extended) cyclic prefix reduces the number of symbols to 12. The downlink PRB structure is shown in Figure 2.1. The eNodeB scheduler will allocate the resources every 1 ms among the devices needing to the scheduled.

The downlink frame structure is using 1 ms sub-frame with 1, 2 or 3 symbols used for downlink control signalling on Physical Downlink Control Channel (PDCCH) and rest of the sub-frame left for user data carried on Physical Downlink Shared Channel (PDSCH), as shown in Figure 2.2.

With the Time-Division Duplex (TDD) operation, there are some differences in the frame structure as shown in Figure 2.2. There is the special sub-frame when the transmission direction changes between uplink and downlink. In the first versions of TDD

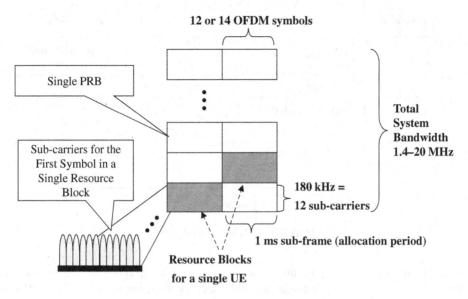

Figure 2.1   Downlink 1 ms allocation structure

(TD-LTE), the uplink/downlink configuration is static parameter, which also allows preventing uplink/downlink interference between different cells. The more dynamic allocation of resources is addressed in Release 12 as discussed in Chapter 4. The TDD cells can be parameterized based on the average traffic asymmetry by choosing the suitable uplink/downlink configuration. In the structure in Figure 2.3, the first sub-frame is always allocated for downlink direction while the first sub-frame after the special sub-frame is allocated for the uplink direction. After that the chosen configuration determines when downlink allocation starts

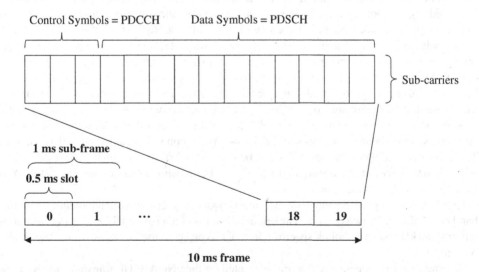

Figure 2.2   Downlink frame structure

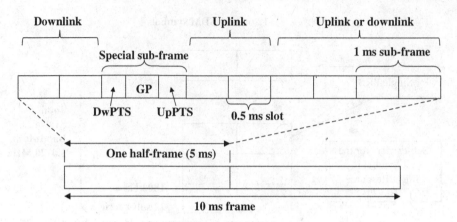

Figure 2.3 LTE TDD frame structure

again. The use of maximum asymmetry to boost downlink capacity needs to be done with care as then the resulting uplink capacity and especially range are respectively more limited if a UE with power limitation at cell edge can only transmit once per 10 ms frame. In general, 3GPP has minimized the differences between feature-driven development (FDD) and TDD to facilitate easy implementation of both eNodeBs and UEs that may support both modes of operation. There are already UEs in the market that support both FDD and TDD modes of operation, something which never really happened with 3G

The uplink resource allocation is also handled by the eNodeB, with the physical layer signalling indicating every millisecond which devices are allowed to transmit user data in the uplink and in which part of the uplink spectrum. The multiple access with SC-FDMA is based on the transmitter principle as shown in Figure 2.4, with the QAM modulator transmitting one symbol at the time, and not parallel in time domain like in the downlink with OFDMA. The resulting symbol duration is shorter than in the downlink direction and for that reason the cyclic prefix is not added after each symbol but after a block of symbols. As the block of symbols in the uplink has the same duration as a single OFDMA symbol in downlink, the relative overhead is the same from the cyclic prefix also in the uplink. The resulting impact on the eNodeB receiver is that there is now inter-symbol interference between the symbols in the block of transmitted symbols and thus more advanced receiver is needed in the eNodeB side. The Minimum Mean Square Error (MMSE) receiver shown in Figure 2.4 is only an example with the actual receiver solution being left for an eNodeB vendor, similar to the many other algorithms in the eNodeB, such as scheduler solution for both uplink and downlink directions. The sub-carrier mapping in Figure 2.4 can be used to determine in which part of the uplink transmission spectrum the transmitted signal will be transmitted, based on the instructions from the eNodeB (uplink resource grants).

Example of the uplink resource allocation is shown in Figure 2.5, with the uplink frequency band being shared by two devices. The uplink allocation for a single UE is always continuous n times 180 kHz piece of uplink spectrum in order to preserve the low peak-to-average uplink waveform.

Depending on the frequency resource allocated by the eNodeB, UE transmits always minimum of 180 kHz, with the example of the minimum allocation shown in Figure 2.5. When

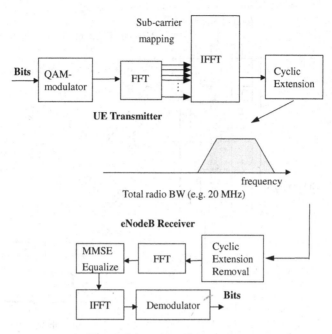

Figure 2.4   LTE uplink principle

more bandwidth is allocated, the duration of a single symbol in the uplink direction becomes shorter due to the increased symbol transmission rate. The generation of the uplink slot content shown in Figure 2.6 includes the use of reference signal sequence which is known in the eNodeB receiver as well and is used by the eNodeB receiver to facilitate the channel estimation in the uplink direction. The uplink structure includes pre-determined locations for physical layer control information, such as ACK/NACK feedback for packets received in the

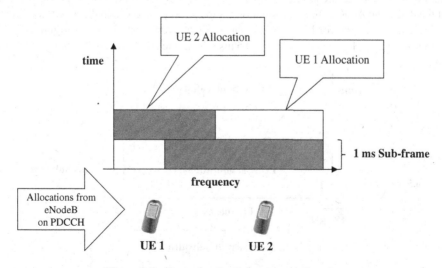

Figure 2.5   Example of uplink resource allocation

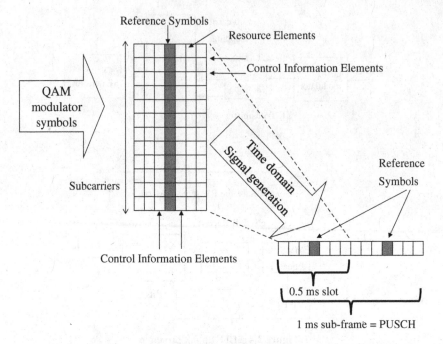

Figure 2.6 Generation of uplink slot contents for transmission

downlink direction or Channel Quality Information (CQI) for providing feedback to enable the frequency domain scheduling in the downlink direction.

The Physical Uplink Control Channel (PUCCH) can be used when there is no uplink allocation available for Physical Uplink Shared Channel (PUSCH). In case UE has earlier run out of data, PUCCH can be used to send scheduling request when more arrives to the buffer if there are no resource allocations existing. Other information to be sent on PUCCH included CQI feedback and ACK/NACK information for the packets received in the downlink direction. The frequency resource for PUCCH is located at band edges, as shown in Figure 2.7. A UE transmitting PUCCH will send first one 0.5 ms slot at one band edge and other slot then with

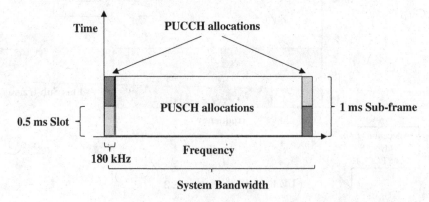

Figure 2.7 Uplink PUCCH allocations

other band edge. Same PUCCH resource can be assigned to multiple devices as UEs have their own code sequence thus allowing multiple simultaneous PUCCH transmissions using the same resources.

The LTE physical layer included already from Release 8 full use of Multiple Input Multiple Output (MIMO) transmission operation, with the support for two antenna reception being mandatory for the typical UE classes. Similarly, the first LTE networks also supported the dual-stream transmission from the start of the LTE deployments. Release 8 includes the support for up to four antenna MIMO operation, but going beyond two antennas is foreseen to be implemented based on Release 10 as discussed later in this chapter.

2.2.2   LTE Architecture

The design principle with LTE architecture was to enable well scalable solution, without too many layers. The architecture from radio point of view concluded to have all radio protocols in the eNodeB, thus resulting to the single element radio access network. The S1 interface towards core network was divided to two parts, S1-U for the user plane data and S1-MME for control plane information between the Mobility Management Entity (MME) and eNodeB. None of the user data goes through MME, thus MME processing load is depending only on the amount of users and their frequency of signalling events such as connection setups (attach), paging or mobility events like handovers. The MME does not make any decisions on handovers, as those are done by the eNodeB, but the MME keeps track where the UE is going in the networks and guides the Evolved Packet Core (EPC) gateways to set up the data connection to the correct eNodeB. In the case of idle mode, the MME keeps track of the location of the UE and knows which tracking areas (TA) to be used when there is a need to page a UE. The radio-related architecture is shown in Figure 2.8, with the X2-interface also visible between the eNodeBs. X2-interface is used mainly for radio resource management purposes, like sensing the handover command between eNodeBs, but also temporary for the purposes of data forwarding before core network tunnels have been updated to route the data for the new eNodeB.

2.2.3   LTE Radio Protocols

The Non-Access Stratum (NAS) Signalling is sent via eNodeB from UE to MME. This includes authentication when UE connects to the network. All the radio protocol layers are located in the eNodeB with the following key functionalities:

- Radio Resource Control (RRC) layer is responsible for the related signalling between UE and eNodeB, all the messages such as configuring the connection, measurements reports and RRC reconfiguration (handover command) are covered by the RRC layer. Compared to HSPA, LTE had less RRC states with only RRC idle and RRC connected states being defined, as shown in Figure 2.9.
- Packet Data Convergence Protocol (PDCP) handles the security (ciphering and integrity protection) as well as IP header compression operation. The PDCP compared to HSPA (WCDMA) includes now the security functionality and is thus used for control plane signalling as well (with the exception of the broadcasted control information).

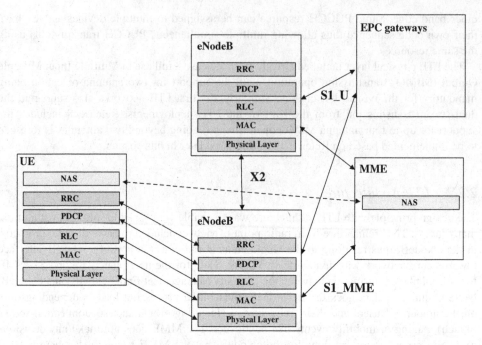

Figure 2.8   LTE radio architecture

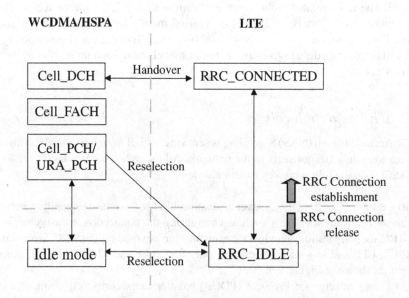

Figure 2.9   LTE RRC states

- Radio Link Control (RLC) layer covers the retransmission handling if the physical layer retransmission operation fails. For some services, RLC retransmissions are not used due to tight delay budget, such as with Voice over LTE, VoLTE. In such a case, RLC is operated without retransmissions, but RLC layer still handles the packet re-ordering functionality. This is needed since the physical layer retransmissions may end up changing the order of packets being provided from the physical layer and MAC layer to RLC.
- Medium Access Control (MAC) covers functions such as control of discontinuous reception and transmission, timing advance signalling, scheduling and retransmission handling and UE buffer and transmission power headroom reporting.

Release 9 was rather small step in terms of radio capabilities for LTE, as there was no increase in peak data rate, rather bringing smaller enabler such a emergency call prioritization or position location-related measurement, both important for the introduction of the VoLTE functionality in LTE, as elaborated in more details in Reference 2. From the physical layer point of view, a new Transmission Mode 8 (TM8) was added to enable the use of multi-stream MIMO operation when using beamforming operation. This was especially relevant for those LTE TDD deployments where eight-port beamforming antenna is being used.

2.3 LTE Advanced in Releases 10 and 11

Release 10 introduced the first LTE-Advanced release, bringing clear increase for the data capabilities of LTE networks. Both the average achievable data rates and the peak data rate were increased. The key technologies introduced with Releases 10 and 11 LTE Advanced are as follows:

> Carrier aggregation, enabling the UE to receive and transmit signals using more than one carrier in downlink and uplink direction.

> Evolved MIMO operation, with more efficient reference signal structures and support for use of up to eight transmit and eight receiver antennas in the downlink direction.

> Improved support for heterogeneous network operation, especially for the co-operation from multiple transmitters or receivers with CoMP, as introduced in Release 11.

There were also further enhancements in Releases 10 and 11 than the ones related to the LTE Advanced. Release 13 work on low-cost MTC is utilizing the Release 11 Enhanced PDCCH, which allows to have also the downlink signalling send with part of the downlink spectrum only and not like in Release 8 using the full system bandwidth. Enhanced PDCCH (EPDCCH) is covered more in Chapter 3 in connection with LTE-M introduction.

2.3.1 Carrier Aggregation

The introduction of carrier aggregation (CA) in Release 10 [3] is the most successful LTE-Advanced feature and has been deployed already by several operators in the field. The principle

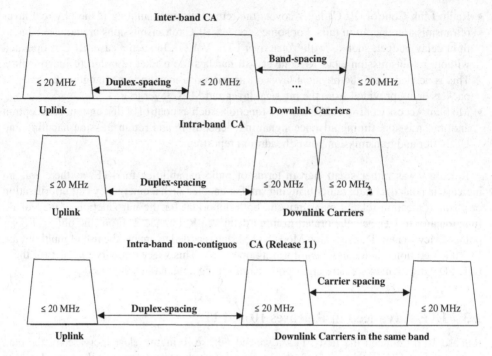

Figure 2.10   Different types of carrier aggregation

itself is relatively straightforward, with the UE now being able to receive up to five carriers (with downlink carrier aggregation) or able to transmit with up to five carriers in the uplink direction (with uplink carrier aggregation).

There are three different types of carrier aggregation as illustrated in Figure 2.10.

- Inter-band carrier aggregation, when the UE received two carriers that are from different frequency bands. For example, 850 and 1800 MHz bands are aggregated in Korea, whereas in Europe, 1800 MHz is aggregated together with 2600 MHz band.
- Intra-band carrier aggregation is usable when one operator has more than 20 MHz bandwidth available from the same frequency band.
- Intra-band carrier aggregation was extended for intra-band on-contiguous case in Release 11. This is needed in some countries where an operator has allocation in one band but the allocation for some reason is not continuous but there is another operator or other systems in between.

From the radio protocol point of view, carrier aggregation mostly impacting the MAC layer as layers above MAC do not see a difference from Release 8-based operation, with the exception of higher peak user data rates being enabled. The MAC layer scheduler in the eNodeB needs to consider the CQI feedback from UE to determine which carrier is best suited, in addition to network load situation and other factors considered in the eNodeB scheduler. The data are then preferably scheduled over the best-suited physical resource blocks from two

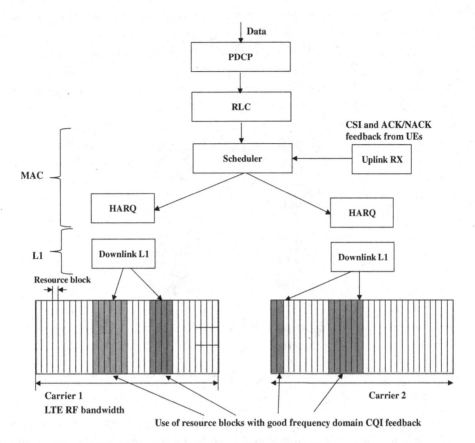

Figure 2.11   Frequency domain scheduling with downlink carrier aggregation

or more carriers. This provides extra frequency diversity and enables to increase the system capacity in addition to the peak data rates. Once the packet is placed on a particular carrier, the physical layer retransmissions are then taken care of by that particular carrier.

The carrier aggregation in Release 10 has direct impact on the peak data rate. With the first phase carrier aggregation, the approach in 2013 was to reach maximum data rate of 150 Mbps, as in Release 8, with more fragmented spectrum availability when a single frequency band could not offer 20 MHz for a single operator and single radio technology. The following steps enabled then going beyond that, with first up to 300 Mbps in 2014 and during 2015 up to 450 Mbps downlink peak data rate as shown in Figure 2.12. The end user will benefit from the improved peak data rates and also the average data rate increases, depending on the load of the network. With fewer users, there is bigger impact than with very large number of users, since in the latter case the eNodeB scheduler anyway can choose from large pool of users which to schedule. Then the capacity increase from the extra frequency domain diversity is smaller. There can be intermediate steps in data rate as well. An European operator which typically would have spectrum in 1800 and 2600 MHz, would typically have no more than 10 with 800 MHz band, thus such an operator would likely utilize a total of 50 MHz of spectrum, resulting to a peak data rate of 375 Mbps when having UE capable for three-carrier

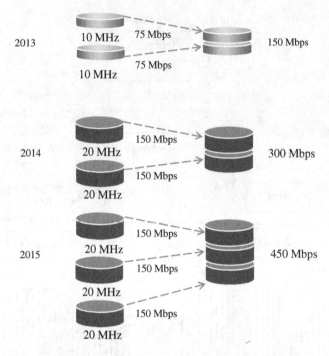

Figure 2.12   Carrier aggregation capability development in the commercial networks and UEs

aggregation in the downlink direction. In the markets where total of 60 MHz bandwidth is available for three-carrier aggregation, downlink peak data rates up to 450 Mbps were enabled in 2015 together with the availability of corresponding UEs.

From the performance point of view, the carrier aggregation gives benefits in the full cell area. The cell edge area receiving transmission from the two or more separate frequencies allows to improve the cell edge performance, while also allowing to maximize the use of higher frequency band; since for the cases when uplink link budget is not sufficient on higher frequency band, the uplink connection can be covered by a lower frequency band while downlink is received from both. For range-limited scenarios, in general, single uplink is more preferable since the transmission power can be concentrated on a single frequency band alone. Figure 2.13 shows example performance with two-carrier aggregation.

There is continuously increasing number of CA band combinations defined in 3GPP. The work on UE RF requirements was first started with limitation to two downlink only carriers. Depending on the bands to be aggregated together, there is some loss of the sensitivity due to the extra switch added in the receiver chain, to enable connecting the other receiver chain for other frequency band. There are 80 band combinations for two carriers aggregation defined in 3GPP in Release 12 until end of 2014, as given in [4]. There exists already several three-band combinations and during 2015 (Release 13) more cases are being added, including downlink band combinations with up to four different bands. The band combinations given in specifications [4] define for each band combination the possible relaxations in the reference sensitivity as well as transmit power allowed in the UE implementation. Also the supported

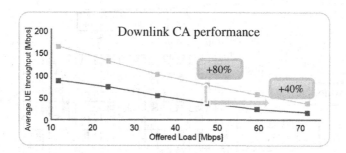

Figure 2.13   Carrier aggregation compared to Release 8 operation without aggregation

bandwidths that can be used with carrier aggregation in connection with a specific band combination are specified in [4].

2.3.2   Multiple Input Multiple Output Enhancements

The MIMO operation was enhanced to make it more efficient when increasing the number of antennas. The following improvements were done:

- Introduction of Channel State Information Reference Signals (CSI-RS). The UE will measure these from many antennas that are in use (configured) and with the CSI-RS, there is no need to continuously send from too many antennas if the conditions are not sufficient or there are no data to schedule for a UE with capability for more than 2 MIMO streams. The CSI-RS does not occur in every sub-frame, and since the rate of those is parameterized to be up to 80 ms, the impact on legacy devices is minimal.
- Demodulation Reference Signals (DM-RS), also known as User-Specific Reference Signals (URS). These end together with the actual data, which allows UE-specific transmitter operation when the pre-coder is applied to the DM-RS as well in addition to the actual data. With DM-RS/URS also beamforming operation is enabled.
- Enabling up to eight-stream MIMO operation in the downlink direction and respectively up to four antenna MIMO transmission in the uplink direction.

These improvements enable efficient use of more than two transmit antennas as supported by the UEs in the field as of today. With the CSI-RS, an LTE-Advanced UE supporting four-stream MIMO can measure and provide feedback for the eNodeB for the rank selection and pre-coder to be used. Only when the transmission for the four-stream capably UE is done, the network then sends the DM-RS together with the data from all four transmit antennas. This is illustrated in Figure 2.14.

2.3.3   HetNet Enhanced Inter-cell Interference Coordination

Part of the work for LTE Advanced in Releases 10 and 11 was boosting the performance with heterogenous networks operation using enhanced Inter-cell Interference Coordination (eICIC). The work focused on the co-channel case, where the both macro and small cell are

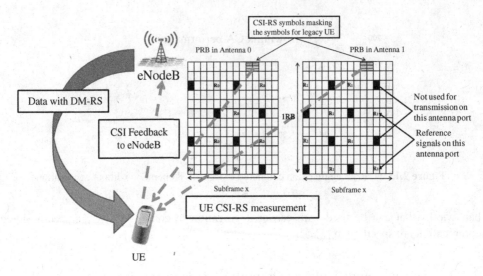

Figure 2.14 Release 10 MIMO operation with CSI-RS

on the same frequency. The dedicated frequency case was addressed with dual connectivity in Release 12 as covered in Chapter 4. The key for getting most of the small cells installed in the same carrier as the macro cell is to capture as much traffic as possible via the small cells. This can be achieved with good planning but always it is not trivial to predict exactly where the traffic hotspots are. The improvements done in LTE Advanced enable cell range extension, which allows the UE connect to a small cell even if the signal from the macro cell would be stronger. The solution introduced the following three methods for enabling such operation and capturing thus more traffic for the small cells:

- Introduction of Almost Blank Sub-frames (ABS). The purpose of ABS is to allow small cell to communicate during those frames with UEs that are actually having stronger signal from macro cell than from the small cell. The capacity is improved when multiple small cells can use the same ABS resources even if one macro cell has reduced total capacity. Use of ABS is illustrated in Figure 2.15.
- Use of offset values for selection between macro cell and small cell
- Advanced receivers to cancel the interference from the macro cell CRS.

The ABSs are not fully empty as they retain the minimum reference symbol configuration to support measurements of also legacy devices. If everything would be removed, a UE making handover and measurements could suddenly provide wrong results if some of the sub-frames would be fully empty.

When a UE is connected to small cell, while macro cell is stronger, it is very beneficial if the UE is able to cancel the interference from the macro cell transmission during the ABSs. As the source of interference is known and also it can be identified what reference signals are transmitted (by decoding the PCI), then the UE does not need to try blindly detecting the interfering symbols but it can take advantage of the knowledge. Cancelling the interference from any random cell would be clearly more demanding.

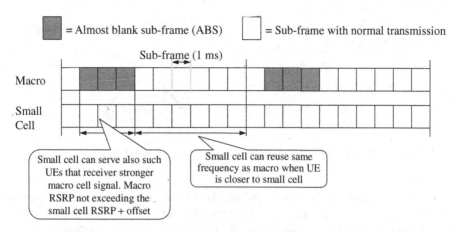

Figure 2.15 Use of almost blank sub-frame and offset parameter for range extension

The offset that can be used had limits as well, since the interference level from the macro cell will after some point in time start to limit the UE receiver performance, thus a UE cannot have tens of decibels stronger macro cell signal and still maintain good reception of the small cell on the same frequency. The higher the offset, the more users are served by the small cells instead of the macro cell. The study captured in Reference 5 concluded that going above 14 dB offset already started to reduce the system performance, while one also has to consider how much macro cell sub-frames are designated as ABS, since that will also reduce the data rate of those users not able to connect to the small cells, with further performance analysis presented in Chapter 7. With the Release 12 dual connectivity, as described in Chapter 4, there is of course no such dependency as macro and small cells operate on different frequencies.

2.3.4 Coordinated Multipoint Transmission

As part of Release 11 work, the support was introduced for the operation of Coordinated Multipoint (CoMP) Transmission, which enables the eNodeBs to operate more jointly. In Releases 8–10, the UE was communicating with one eNodeB at the time, while with Release 11 CoMP, the possibility to provide feedback for more than one eNodeB was introduced and similarly the possibility for receiving data from even more multiple eNodeBs. Likely in the uplink direction, CoMP could be achieved as well by eNodeB implementation where the multiple receivers in different cells/sectors are receiving the transmission from a single UE.

There are different types of CoMP operation possible in Release 11, with all of them relying on fast connectivity between eNodeBs. Would there be more delay (as was studied in Release 12), performance loss would be easily resulting, as was the case with the studies for eNodeB scheduler coordination. The following CoMP modes can be identified:

- Coordinated Scheduling and Beamforming (CS/CB) requires the least amount of information exchange between eNodeBs since UEs receive a data transmission from their respective serving cell only, but adjacent cells perform coordinated scheduling and/or transmitter precoding to avoid or reduce the inter-cell interference.

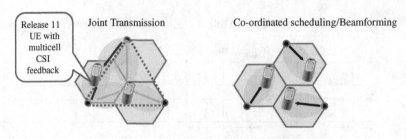

Figure 2.16 Different downlink CoMP schemes in Release 11

- Joint Processing (JP) CoMP aims to provide the UE data for multiple transmission points (eNodeBs) so more than just the serving cell can take part of the transmission for a UE. This requires large bandwidth, very low latency backhaul, in practice, fibre connectivity, between eNodeB sites. The possible modes are Dynamic Cell Selection (DCS) where the data would be sent always from one eNodeB only (based on the UE feedback) or Joint Transmission (JT) CoMP where multiple cells jointly and coherently transmit to one or multiple terminals on the same time and frequency resources. Use of the same time and frequency resources is important element of CoMP as then the transmission from neighbouring cell is not interference but rather part of the own signal transmission.

In the uplink direction, CoMP could be applied even for Release 8 UEs, as with the uplink CoMP the eNodeB is simply receiving also other UEs than the ones having that eNodeB as their serving cell. Since there are no extra transmissions in the uplink direction for enabling this, there is big gain potential in uplink direction. In downlink direction, the extra transmissions also create more interference, and thus the gain is not so obvious. The different CoMP methods are illustrated in Figure 2.16, with further details on the achievable CoMP performance gains covered in Chapter 7.

2.4   UE Capability in Releases 8–11

In Releases 8 and 9, there are five UE categories defined that reach in theory up to 300 Mbps with the support of four-stream MIMO operation. Currently, the devices in the field based

Table 2.2 UE categories introduced in Releases 8 and 9

UE category	Class 1	Class 2	Class 3	Class 4	Class 5
Peak rate DL/UL (Mbps)	10/5	50/25	100/50	150/50	300/75
RF bandwidth (MHz)	20	20	20	20	20
Modulation DL	64QAM	64QAM	64QAM	64QAM	64QAM
Modulation UL	16QAM	16QAM	16QAM	16QAM	64QAM
Rx diversity	Yes	Yes	Yes	Yes	Yes
BTS tx diversity	I-4 tx	I-4 tx	I-4 tx	I-4 tx	I-4 tx
MIMO DL	Optional	2×2	2×2	2×2	4×4

Table 2.3 UE categories introduced in Release 10 to enable first phase carrier aggregation

UE category	Class 6	Class 7	Class 8
Peak rate DL/UL (Mbps)	300/50	300/100	3000/1500
RF bandwidth (MHz)	40	40	100
Modulation DL	64QAM	64QAM	64QAM
Modulation UL	16QAM	16QAM	64QAM
Rx diversity	Yes	Yes	Yes
MIMO DL	2×2 (CA) or 4×4	2×2 (CA) or 4×4	8×8
MIMO UL	No	2×2	4×4

on Releases 8 and 9 are based on either categories 3 or 4. The first phase UEs in the market were typically able to reach in-line with category 3 the 100 Mbps downlink peak data rate (with FDD mode) and 50 Mbps uplink peak data rate when the network has 20 MHz spectrum for uplink and downlink carrier available. With category 4, the downlink peak data rate was increased to 150 Mbps. As discussed earlier, the two antenna/stream downlink MIMO capability is mandatory for all UEs, except the category 1 devices. With the introduction of LTE Advanced in Release 10, the additional UE categories 6, 7 and 8 were added. Currently, the most advanced devices in the market are Release 11 category 9 devices that can provide up to 450 Mbps peak data rate when network is able to have three downlink 20 MHz carriers aggregated with Release 10 carrier aggregation. Category 8 is rather theoretical one with all the possible features enabled. The latest categories in Release 11, categories 9 and 10, were added to the specifications as recently as in June 2014. While they are based on otherwise Release 10 principles, they are using three-carrier aggregation in downlink direction, thus enabling up to 450 Mbps when using three 20 MHz carriers and two-stream MIMO transmission. Category 9 devices, enabling 450 Mbps downlink peak data rate, have become available during 2015 in the field.

The use of carrier aggregation was in the market first realized with Class 4 type of devices, which could aggregate 10 + 10 MHz of downlink spectrum, before the Class 6 type of devices entered the market in 2014, allowing more than 150 Mbps which was possible with Release 8-based baseband capabilities.

Table 2.4 UE categories introduced in Release 11 to further enhance carrier aggregation capabilities

UE category	Class 9	Class 10	Class 11	Class 12
Peak rate DL/UL (Mbps)	450/50	450/100	600/50	600/100
RF bandwidth (MHz)	60	60	80	80
Modulation DL	64QAM	64QAM	64QAM	64QAM
Modulation UL	16QAM	64QAM	16QAM	64QAM
Rx diversity	Yes	Yes	Yes	Yes
MIMO DL	2×2 or 4×4	2×2 or 4×4	2×2 or 4×4	2×2 or 4×4
MIMO UL	No	2×2	No	2×2

The Release 12 new UE categories accommodate further higher data rates as well as new functionality such as 256-QAM downlink modulation. Release 12 UE categories are covered in Chapter 3.

2.5 Conclusions

In the previous sections, we have covered the LTE standards from the first LTE Release 8 standard to the LTE-Advanced Release 10 and 11 standards. These form the basis currently deployed in the field at the time of writing this book in mid 2015, with the Releases 12 and 13 features covered mostly in the Chapters 3 and 4 to follow in the marketplace in the next phase. While there are over 300 networks with LTE Release 8 rolled out to the market place, already more than 50 networks [6] have rolled-out LTE Advanced, mostly based on Release 10 downlink carrier aggregation.

References

[1] Holma, H. and Toskala, A., *HSPA*, John Wiley & Sons, Ltd (2014).
[2] Poikselkä, M., Holma, H., Hongisto, J., Kallio, J., and Toskala, A., *Voice Over LTE (VoLTE)*, John Wiley & Sons (2012).
[3] 3GPP Tdoc RP-RP-091440, 'Work Item Description: Carrier Aggregation for LTE', December 2009.
[4] 3GPP Technical Specification, TS 36.101, 'Evolved Universal Terrestrial Radio Access (E-UTRA); User Equipment (UE) Radio Transmission and Reception', Release 12, Version 12.6.0, December 2014.
[5] Holma, H. and Toskala, A., *LTE Advanced: 3GPP Solution for IMT-Advanced*, John Wiley & Sons (2012).
[6] www.gsa.com (accessed March 2015).

3

LTE-Advanced Evolution in Releases 12–13

Antti Toskala

3.1 Introduction

This chapter covers summary of Release 12 and 13 enhancements outside the big themes such as small cells, macro cell evolution or LTE-unlicensed. The topics covered in this chapter are Machine-Type Communications, enhanced CoMP with non-ideal backhaul, FDD–TDD carrier aggregation, WLAN interworking as well as device-to-device (D2D) operation intended both for public safety and commercial use cases.

3.2 Machine-Type Communications

One of the use cases for mobile communications is Machine-Type-Communication (MTC), also known as Machine-To-Machine (M2M) or as Internet of Things (IoT). A typical use case

---

LTE Small Cell Optimization: 3GPP Evolution to Release 13, First Edition.
Edited by Harri Holma, Antti Toskala and Jussi Reunanen.
© 2016 John Wiley & Sons, Ltd. Published 2016 by John Wiley & Sons, Ltd.

considered in 3GPP is the metering application where the data rates needed are relatively low. In such a case, the advanced capabilities of LTE are not likely to be needed, but even a modem based on GSM/GRPS could do the job. As GSM/GPRS modem prices are in the range of 5 USD or less, it offers an attractive approach to consider those. Also 3G modems based on HSDPA can be obtained at only slightly higher price point. The higher complexity of LTE modems causes them to be somewhat more expensive than 2G or 3G modems. If the price difference is too high, then a utility company installing a large number of devices will not want to invest too much extra cost per device just for the sake of LTE connectivity. For a single device, the extra cost of, for example, 5 dollars (if assuming 10 USD LTE modem price) or more is not sounding like a massive investment but when installing millions of devices then even such a relatively low difference starts to matter when the actual device itself seems not really to benefit from the 100 Mbps or even higher LTE data rates.

From an operator point of view, serving MTC type of devices is a relatively high value business, as often the number of bits transferred per subscription is low and often long-term contracts are made. However, if a utility company for their metering solution only equips the modems with 2G or 3G technology, one ends up being forced to commit in maintaining the 2G or 3G network for such a devices for next 10–15 years. If after next 5 years only MTC devices remain in the network, then the revenue may not be enough anymore to justify the maintenance cost for a 2G or 3G network of all the other traffic have move to 4G by then (or even getting started with 5G as discussed in Chapter 17).

The 3GPP conducted a study for LTE Release 12, with the results captured in [1], which concluded that making some simplifications, such as lower peak data rate and single RX chain the UE cost could be reduced. For Release 12, a new UE category was introduced which support only single antenna reception and reduced peak data rates of up to 1 Mbps. The new UE category is shown in Table 3.1 in comparison with category 1 and category 4, with the categories 3 and 4 being the current widely implemented UE categories, with the wider coverage of Release 8 UE categories provided in Chapter 2.

As can be seen from Table 3.1, the data rate capability of the category 0 UE is quite limited, offering only 1 Mbps peak data rate both in the uplink and downlink directions. This limitation is intended to ensure that this category would not be used for regular end user devices, such as low cost 4G-enabled tablets, since the maximum data rate of only 1 Mbps, clearly below of the low cost 3G-enabled chip sets. This was important for especially such markets where the operators are not able to control (at least not fully) what kinds of devices are introduced to their network. For example, a low cost 4G-enabled tablet would be otherwise tempted to operate with single receiver antenna if it could achieve also higher LTE data rates with such an implementation.

Table 3.1 Category 0 added in Release 12

	Rel-8 Cat-4	Rel-8 Cat-1	Rel-12 Cat-0
Downlink peak rate (Mbps)	150	10	1
Uplink peak rate (Mbps)	50	5	1
Max number of receiver antennas	2	1	1
Duplex mode	Full duplex	Full duplex	Half-duplex (opt)
UE receive bandwidth (MHz)	20	20	20
Maximum UE transmit power (dBm)	23	23	23

Table 3.2 Cost-saving potential with different complexity reduction approaches over Release 8 LTE

Technique	Approximate saving (%)
Single receiver antenna	14–18
Reduced maximum UE power	1–3
Half-duplex FDD	9–12
Reduced maximum UE bandwidth	6–10
Peak rate reduction	5–7

The 3GPP estimated the reductions achievable for different parts being as shown in Table 3.2 and presented in more details in [1].

The 3GPP has also specified Power Saving Mode (PSM) as part of Release 12. The PSM allows the UE stay registered with the network while sleeping longer periods of time. This then allows reducing the signalling when the UE is waking up and thus reduced duration of the needed on-time and thus reduced battery consumption, this is illustrated in Figure 3.1. If there is UE with mobility, then there could be need for Tracking Area Update (TAU) once reconnection to the network but still the amount of signalling is significantly reduced compared to the full registration procedure. This approach reduces the impact of the radio technology in use for the battery life of the device, especially for applications where there is constant need to transmit information to the network or neither need to be always available to respond to paging from network side or the relevant server controlling the M2M application. A metering device can then limit the needed set-up time and focus sending only the latest meter reading and quickly turn off the radio parts once the network has acknowledged the reception of the message. With the range-limited case, where all signalling is necessary with the maximum transmit power level or very the close to that, the improvement to the battery life can be quite significant.

The further work in 3GPP [2] for Release 13 is on-going to specify further reduced and cost optimized capability UE classes to enable even lower price point for LTE M2M modem implementation. Such a new class maintains 1 Mbps as the peak data rate (thus more than a 2G implementation using GSM) but instead to have other reductions compared to Release 12 category 0 UE such as smaller supported maximum RF bandwidth, down to 1.4 MHz,

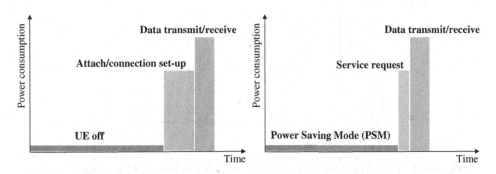

Figure 3.1 Release 12 power saving mode (PSM) operation

Figure 3.2   Example M2M modems HSPA (left) and LTE (right) (courtesy of Gemalto)

or even 200 kHz, instead of full 20 MHz. All devices abovementioned including Release 12 support the maximum 20 MHz bandwidth, as long as the 20 MHz bandwidth is part of the supported bandwidths in 3GPP specifications for the specific frequency band in use. The intention is to further narrow the gap with cost when compared to the 2G implementation based on GSM/GRPS technology or compared with the more capable HSPA+ modem (including also GSM capability), such as the one shown in Figure 3.2 with 80-pin connector.

An example of the narrowband operation is shown in Figure 3.3, with the MTC LTE UE supporting only 1.4 MHz being operated as part of the larger bandwidth. In such a case, the normal UE categories could be scheduled in any part of the 20 MHz carrier but the MTC UEs with limited RF bandwidth would be listening only to the part of the carrier where they are assigned. The 3GPP needs still also to solve how to handle the control signalling reception,

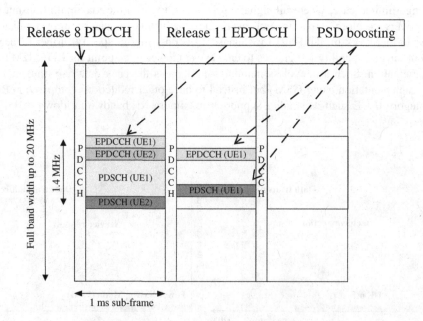

Figure 3.3   LTE-M operation using 1.4 MHz bandwidth with the 20 MHz system bandwidth

including reception of common control for all UEs (System Information Blocks, SIBs) in such a way that the existing devices do not get any significant performance impact.

The UE-specific physical layer downlink control signalling with LTE-M with narrow RF bandwidth is based on the Release 11 Enhanced DPCCH (EDPCCH), which allows sending the physical layer downlink control signalling without having to use the full system bandwidth as with the Release 8-based DPCCH as covered in Chapter 2.

Another dimension is also definition of a UE with smaller power class. Going to, for example, 20 dBm power class would allow more integrated implementation compared to the current 23 dBm power class, since smaller power amplifier is easier to integrate in the chip itself. With 200 kHz 23 dBm is needed to maintain the link budget.

Besides the cost, MTC work in 3GPP is also addressing the coverage aspects. This is driven by the fact that many metering applications are located in difficult conditions with large indoor penetration loss impacting the link budget. Such locations include basements and other underground facilities for gas, water and electricity metering applications. The work in Release 13 is targeting for 15 dB improvement for the link budget. Since latency is not that critical with metering applications, solutions impacting latency could be used to enhance the link budget. The following methods are being considered:

- Longer total transmission time, which could be achieved with TTI bundling for data channels and repetition signalling channels.
- Hybrid Automatic Repeat-reQuest (HARQ) could be used, enabling more transmission to be combined for data channels.
- Allocating more power in frequency domain for critical transmissions to an MTC UE with Power Spectral Density (PSD) boosting.

Further some of the channels could be decoded even with weaker signal conditions as of today assuming some of the performance requirements could be relaxed for the UE or eNodeB reception.

The expected UE categories for Release 13 foreseen to be as in Table 3.3, as presented in [3].

The work in 3GPP for Release 13 LTE MTC (often referred as LTE-M for 1.4 MHz case and NB-LTE or NB-IoT for 200 kHz case) is scheduled to be completed by the end of 2015 as visible in [4], while work for NB-LTE likely to finalize in 1H/2016.

Table 3.3 Expected Release 13 MTC UE categories

	Rel-8 Cat-4	Rel-13 (1.4 MHz LTE-M)	Rel-13 (200 kHz NB-LTE)
Downlink peak rate	150 Mbps	1 Mbps	128 kbps
Uplink peak rate	50 Mbps	1 Mbps	128 kbps
Max number of downlink spatial layers	2	1	1
Number of UE RF receiver chains	2	1	1
Duplex mode	Full duplex	Half duplex (opt)	Half duplex (opt)
UE RX/TX bandwidth	20 MHz	1.4 MHz	200 kHz
Maximum UE transmit power	23 dBm	~~20 dBm	23 dBm
Modem complexity	**100%**	**20%**	**<20%**

3.3   Enhanced CoMP

Release 11 defined the use of Co-operative Multipoint Transmission (CoMP) as covered in Chapter 2. The work in Release 11 focused for the case of ideal backhaul since non-ideal backhaul limits the gains achievable with CoMP to a large extend, making solutions relying on joint transmission rather impractical.

Release 12 work with Enhanced CoMP was focusing on the support of scheduler coordination with the non-ideal backhaul, not impacting the UE side at all. The study done in 3GPP concluded that significant gains are hard to find with such operation, as reported in [5] and also as shown in Figure 3.4, where only one result presented from different companies was showing any improvement in the total cell capacity and other results indicated rather loss of capacity. Thus, it was not considered reasonable to add new network elements but rather 3GPP agreed in introducing the following enhancements for the X2-interface application protocol (X2AP) to facilitate further scheduler coordination.

The study also found, as expected, that the backhaul latency has major impact for the gain potential. While with 5 ms backhaul latency, the median results from different companies, as shown in [5] gives some gain, with the 50 ms backhaul latency, there is rather loss than performance gain from the scheduler coordination.

The actual specification work was also divided into two different Releases due to the time limitations in Release 12 finalization. Release 12 enhanced CoMP was completed finally covering the following enhancements:

- CoMP hypotheses, which covers hypothetical resource allocation including benefit metric to reflect the benefit achieved if an eNodeB would follow the suggested allocation given by

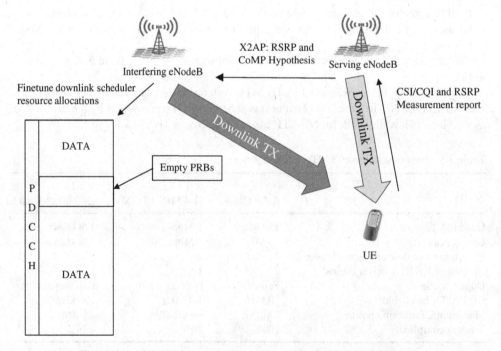

Figure 3.4   Scheduler coordination with enhanced CoMP in Release 12

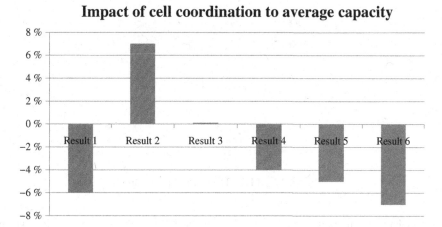

Figure 3.5 Cell average capacity performance with scheduler coordination from different companies as given in [5] with high resource utilization and nine cells coordinated

the hypotheses. The eNodeB is not mandated to follow the suggestion (and it may receive signalling from multiple eNodeBs suggesting conflicting allocations) but it is left for the eNodeB scheduler implementation how to reflect the hypothesis in the eNodeB scheduling and other radio resource management functionality.

- Reference Signal Receiver Power (RSRP) measurements as reported by the UE. The RSRP measurement result obtained from the UEs can be used similar to aid the scheduler and RRM operation and also to validate some of the hypothesis information.

Further the enhancements in Release 13 are expected to include, as described in [6], and work to be finalized by the end of 2015 in 3GPP:

- Sending UE CSI/CQI information over X2 interface between the eNodeBs. This allows the scheduler coordination operation to see also the frequency domain feedback as provided by the UEs in the neighbouring cell. Release 12 CoMP hypotheses can of course reflect this as desired to be taken into account by the actual serving eNodeB. If UE CSI/CQI information would be provided with frequent interval for large number of UEs over X2 interface, then care must be taken in order not to overload the X2 interface.
- Enhanced Relative Narrow-Band Transmit Power (RNTP), which in Release 8 provides more limited information of the transmitter power per each 180 kHz wide Physical Resource Blocks.

3.4 FDD–TDD Carrier Aggregation

Release 12 contains an important building block for bringing together FDD and TDD operation modes, namely FDD–TDD carrier aggregation. It allows to extend the carrier aggregation framework in such a way that PCell could be on an FDD band and SCell could be on a TDD band, or also other way around, as shown in Figure 3.6. In Release 10-based carrier aggregation, both bands have to be FDD bands or TDD bands. In most cases, the FDD band

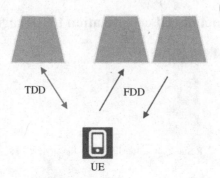

Figure 3.6   FDD–TDD carrier aggregation

is a lower band in frequency thus more likely to be used as PCell, thus in the first phase, the interest for practical combinations, such as 800 MHz with 2.6 GHz TDD naturally leads towards FDD PCell and TDD SCell.

From an operator point of view, the use of FDD–TDD carrier aggregation makes it easier to utilize both paired and unpaired bands. If an operator has already aggregated, for example, two FDD bands of 20 MHz each, the UE can reach 300 Mbps downlink peak data rates using carrier aggregation. If now there is only a single TDD band available (with 20 MHz), then using only TDD for the connection would mean smaller achievable peak data rate even if the actual network capacity would be increased. This would lead to such a situation in the small cell environment (assuming TDD used with small cells at higher frequency without carrier aggregation) that an operator would provide worse user experience with small cells than with macro cell environment, at least with low load conditions.

Most of the carrier aggregation principles are valid with FDD–TDD aggregation. Mainly some changes were needed especially for the handling of Ack/Nack signalling, especially if TDD band would be the PCell as then the uplink resources are not always available. From the practical deployment point of view, the most cases currently on the table are such that TDD band is clearly at higher frequency thus that becomes naturally the capacity extension band while coverage is ensured with the FDD band, as shown in Figure 3.7.

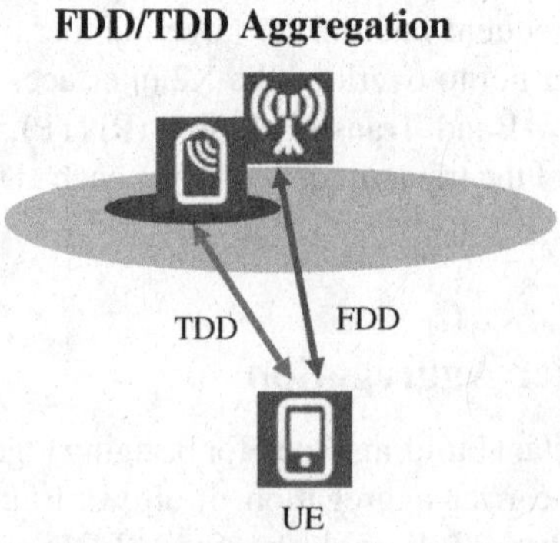

FDD/TDD Aggregation

Figure 3.7   Using FDD band as the coverage band with FDD–TDD CA

The 3GPP is working on further FDD–TDD band combinations beyond Release 12, with more being added to the work program for 2015. Even if a new band combination is added in Release 13 or later, it is added as Release-independent band combination, thus can be implemented on top of Release 12 without having to wait for any other Release 13 features.

- Band 8 (900 MHz) and Band 40 (2300 MHz). This was completed as part of the original TDD–FDD carrier aggregation work item [7] in June 2014 as the example band combination. In this case also, only one uplink is supported together with two downlinks (as with all cases covered in Release 12 timeline for FDD–TDD carrier aggregation).
- Band 1 (2100 MHz) and Band 41 (2600 MHz)
- Band 1 (2100 MHz) and Band 42 (3.5 GHz)
- Band 1 (2100 MHz) and Band 40 (2300 MHz)
- Band 3 (1800 MHz) and Band 38 (2600 MHz Europe)
- Band 3 (1800 MHz) and Band 40 (2300 MHz)
- Band 5 (850 MHz) and Band 40 (2300 MHz)
- Band 20 (800 MHz) and Band 40 (2300 MHz)

3.5   WLAN-Radio Interworking

As part of 3GPP Release 12 work was done to address the problem of a UE with good LTE connection selecting a poor performing WLAN network. The solutions are designed to operate together (not necessary requiring) with the Access Network Discovery and Selection Function (ANDSF) functionality specified on the core network side. The 3GPP did first a study [8] that identified the following three options to be considered:

- Solution 1 considers both ANDSF and RAN guidance and policies for WLAN off-loading control. In this approach, eNodeB provides RAN assistance information to the UE (broadcast or dedicated signalling) The UE would use the RAN assistance information, UE measurements and information provided by WLAN and policies that are obtained via the ANDSF to guide the WLAN selection.
- Solution 2 was based on providing the offloading rules from eNodeB via Radio Resource Control (RRC) signalling (dedicated or broadcasted), with the thresholds provided then guiding the WLAN selection.
- Solution 3 was based on the traffic steering for UEs in RRC CONNECTED state. The UEs would be controlled by the network using dedicated traffic steering commands, very similar to a handover operation (not just moving the actual connection but only part of the traffic). The solution would have been based on WLAN measurements from the UE done for the WLAN networks as defined by the eNodeB for UE to measure.

The 3GPP concluded to focus on the actual specification for combination of solutions 1 and 2, with the Release 12 work mostly completed by September 2014 for the work item started [9].

The radio-level functionality in Release 12 allows to define different parameters for controlling the selection of WLAN (and thus giving guidance of when to come back to LTE), as shown on Figure 3.8.

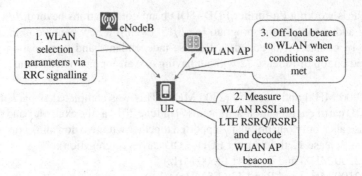

Figure 3.8   WLAN off-loading control from LTE eNodeB

The following parameters are defined in Release 12 3GPP specifications, with part of them actually obtained either from the WLAN AP signalling or from the UE WLAN measurements:

- The offload preference indicator can be used to configure the UE whether there is preference to use WLAN for off-loading data or not.
- There are parameters on the available backhaul capacity (given in Kbps) in the WLAN AP. The two different thresholds for Backhaul for Downlink Bandwidth allow defining higher value based on which WLAN should be used and lower value based on which LTE should be used instead. This value is provided by WLAN AP. Respectively the parameters can be set for the uplink backhaul bandwidth as well.
- The parameter 'thresholdBeaconRSSI' gives two different values to avoid selecting WLAN AP with too low signal strength (or rather move back to LTE in such a case) and when WLAN AP RSSI is considered strong enough to move from LTE to WLAN AP instead.
- The LTE RSRP and RSRQ parameters avoid offloading for a UE with good LTE signal quality (close to LTE eNodeB) and rather encourage off-loading for users at the cell edge area.
- The 't-SteeringWLAN' parameter defines the timer value during which the rules should be fulfilled. This avoids moving back and forth very rapidly between LTE and WLAN; for example, when the WLAN AP load varies.
- The LTE eNodeB can signal the parameters individually for each user, if there is a need to address specific situation for a particular device. This makes it possible to have, for example, longer timer values for some UEs or even restrict the off-loading for some UEs due to frequent bearer changes.

The Release 12 solution has some known shortcomings, and the 3GPP Release 13 is expected to address the further enhancements, for example, with the co-located WLAN AP with LTE eNodeB. The solutions on the table range from extending the Release 12 solution, defining the earlier rejected solution 3 (handover type operation) or go all the way to the radio-level aggregation where the data would be divided in the PDCP layer between LTE and WLAN based on the radio conditions and load conditions both on LTE- and WLAN-sided. The example of a possible aggregation framework is shown in Figure 3.8. The aggregation would move closer to the radio the approach currently available in the application layer where the same bearer would be divided between LTE and WLAN, but invisible to the radio layer.

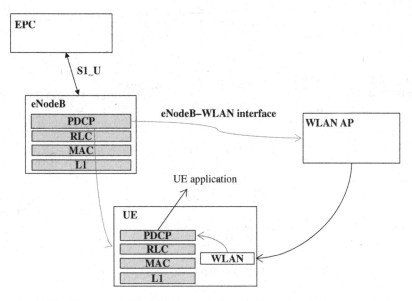

Figure 3.9   Proposed Release 13 approach for WLAN/LTE radio-level aggregation

Such an approach has the benefit of avoiding any changes to the WLAN infrastructure while the full aggregation would also require LTE eNodeB still to receive all the traffic going via WLAN AP as well, in contrast to full off-loading (and available only for the co-located case). To which extend such an aggregation would prove better than the application-level multipoint TCP operation depends on the environment and is yet proposed to be studied in 3GPP, with the study scheduled to be completed during 2015. The numbers given in [10] suggest on the average 15% improved throughput in the conditions where the situation changes dynamically and thus radio aggregation would be faster to react to changes than multipoint TCP approach would be. There is an interface needed for non-co-located cases, indicated in Figure 3.9 as well. In the UE side, the WLAN functionality needs to be also integrated tightly with the LTE functionality, most probably in the same baseband chip supporting both LTE and WLAN for ensuring low enough delay and sufficient communications between the LTE and WLAN transceivers.

The LTE radio-level aggregation is more dynamic from the reaction to the environment, thus somewhat better performance could be expected, the HTTP level or Multipoint level aggregation decisions are done higher up in the network, as shown in Figure 3.10. With the radio-level interworking, LTE needs to obtain more dynamic info from the Wi-Fi conditions, otherwise the performance is not any better than multipoint TCP. The peak data rate is the same in all the alternatives.

3.6   Device-to-Device Communication with LTE

In Release 12, the first wave of Device-to-Device (D2D) communication is being completed. The primary use case for Release 12 is the public safety operation. The existing narrowband public safety solutions, such as TETRA, are well suited for public safety voice communications but have a clear shortcomings due to limited data rate capabilities as a narrowband system. Also

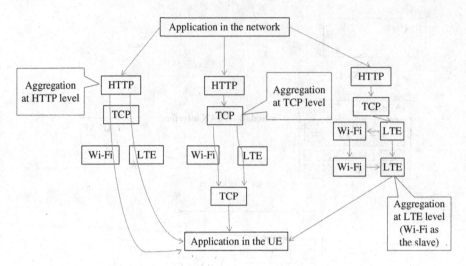

Figure 3.10   Aggregation with Wi-Fi on different levels

in some cases, the inter-operability is rather limited between different vendors, for example, considering the variety of systems (or vendor specific variants) in the market in the United States. The first wave of public safety solutions with LTE has the idea of an overlay LTE network with the voice being carrier by the legacy public safety radio while data services would be provided using LTE network. The evolution step, however, also enables mission critical voice with LTE with the specification of group calls as well as D2D operation, latter known as direct mode in TETRA side. The LTE as such is well suited for voice communication but requires some protocol enhancements to enable group call (which is being done using the eMBMS) and enabling the D2D operation new physical channels are needed for the sidelink connection between UEs.

The D2D operation may take place with two main different scenarios, as shown in Figure 3.11.

- In-coverage case, where the eNodeB coverage is available (in the special case only one of the devices would be in coverage area). In this case, network can provide guidance when the UEs with D2D capability could page/search for other D2D UEs, known as D2D discovery. This allows less resource consuming search procedure for the signals from other D2D UEs. It remains also important to allow physical resource sharing between the regular communications and D2D communications.

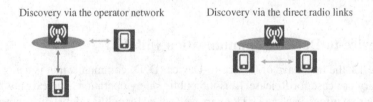

Figure 3.11   Discovery within network coverage and without network coverage

Figure 3.12 Sidelink with D2D communications

- This can be done as part of SIB broadcast or with dedicated RRC signalling.
- Out-of coverage case. In this case, no regular coverage is available, thus the devices need to manage on their own without timing reference. This is a critical requirement for the public safety operation.

There are further different communication modes (after the actual discovery has taken place), with the actual channels between two UEs called as 'sidelink' to differentiate from the existing uplink and downlink channels, as shown in Figure 3.12.

The sidelink operation has parameters like maximum transmit power that is configured by the network (pre-configured for out of coverage case). The UE needs now to transmit using ODFMA which in Release 8 LTE was not used in uplink due high Peak-To-Average (PAR) requirements. Use of OFDMA in the UE transmitter suggests respectively for the UE to use smaller transmission for sidelink operation than for normal uplink operation. The work for D2D Release 12 details is to be completed in March 2015 and then further work in Release 13 enhancements is to start, as covered in [11].

3.7 Single Cell Point to Multipoint Transmission

For Release 13 study was started on single cell point to multipoint operation. This kind of operation, as illustrated in Figure 3.13, could be envisaged for use cases such as public safety (for group call) and also approaches like provided traffic information for cars in a given area.

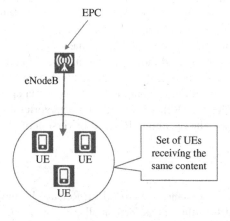

Figure 3.13 Single cell point to multipoint principle

Table 3.4   New Release 12 UE categories

UE category	Class 11	Class 12	Class 13	Class 14	Class 15
Peak rate DL/UL (Mbps)	600/50	600/100	390/50	390/100	3900/150
RF bandwidth (MHz)	80	80	40	40	100
Modulation DL	64QAM	64QAM	256QAM	256QAM	256QAM
Modulation UL	16QAM	16QAM	16QAM	16QAM	64QAM
Rx diversity	Yes	Yes	Yes	Yes	Yes
MIMO DL	2×2 (CA) or 4×4	2×2 (CA) or 4×4	2×2 (CA) or 4×4	2×2 (CA) or 4×4	8×8
MIMO UL	No	2×2	No	2×2	4×4

The study described in Reference 12 is scheduled to be finalized by end of 2015. In some cases, eMBMS could be used for the same purpose when the area of providing is large and cells are synchronized so Single Frequency Network (SFN) can be formed. This kind of operation has been suggested also for communications between vehicles and road-side infrastructures (V2X) or between different vehicles (V2V), but 3GPP has not yet concluded whether any specific work will be done on for this with LTE or not in Release 13.

3.8   Release 12 UE Capabilities

As part of Release 12 finalization, several additional UE categories were introduced, as covered in Table 3.4, in addition to the new MTC category 0 with 1 Mbps peak data rate. There is further work expected with the UE capabilities to introduce classes up to 800 Mbps and 1 Gbps during 2015 still for Release 12. For example, with the use of 256QAM modulation in the downlink direction, the peak data rate for a single 20 MHz carrier becomes 200 Mbps, thus a UE with support of five aggregated 20 MHz carriers can reach 1 Gbps downlink peak data rate. The 3GPP can then introduce even further capabilities based on Release 12 capabilities, as with more than 2 MIMO stream then more than 200 Mbps per carrier could be enabled. In the uplink direction, the use of 64QAM in the uplink is being enabled with the addition of the capability not dependent on a particular UE category, together with the development of the relevant UE RF requirements in Release 13 (since they were missing from the Release 8 due to lack of commercial interest to uplink 64QAM at the time).

With the work on-going to extend the maximum number of carrier to be aggregated up to 32 carriers in Release 13, the theoretical peak data rate in downlink direction could reach up to 32×200 Mbps = 6.4 Gbps with the assumption of two antenna operation, with eight-antenna operation even to 19.2 Gbps. In Release 13, new UE categories are expected to be added to specifications, especially to the upcoming Release 13 version of [13] during 2016.

3.9   Conclusions

In the previous sections, the enhancements for LTE Release 12 and the expected outlook for Release 13 was covered for other areas then small cells, which is captured in Chapter 4 or LTE-Unlicensed (In 3GPP term Licensed Assisted Access (LAA) for LTE being used) as explained in Chapter 10. Release 13 is going to reflect some of the identified enhancements need, including further work in WLAN interworking as well as further D2D to facilitate better also the non-public safety use case. Some of the Release 13 items are rather clear and small, like adding the additional X2 parameters for enhanced CoMP while with especially WLAN interworking 3GPP is still searching the direction which to consider in terms of level of integration with WLAN, ranging between moderate enhancements for WLAN selection parameters up to radio-level aggregation with WLAN. The work on new UE categories continues with the beyond 1 Gbps categories in Release 13 are to be added in later versions of [13] during 2016.

References

[1] 3GPP Technical Report, TR 36.888, 'Study on Provision of Low-Cost Machine-Type Communications (MTC) User Equipments (UEs) Based on LTE', v.12.0.0, June 2013.
[2] 3GPP Tdoc RP-141645, 'Further LTE Physical Layer Enhancements for MTC', September 2014.
[3] 3GPP Tdoc RP-141180, 'Motivation for New WI: Further LTE Physical Layer Enhancements for MTC', September 2014.
[4] 3GPP Tdoc RP-141660, 'New WI Proposal: Further LTE Physical Layer Enhancements for MTC', September 2014.
[5] 3GPP Technical Report, TR 36.874, 'Study on CoMP for LTE with Non-Ideal Backhaul', v.12.0.0, December 2013.
[6] 3GPP Tdoc RP-141032, 'New Work Item on Enhanced Signalling for Inter-eNB CoMP', June 2014.
[7] 3GPP Tdoc RP-131399, 'Updated WID: LTE TDD-FDD Joint Operation Including Carrier Aggregation', September 2013.
[8] 3GPP TR 37.834, 'Study on WLAN/3GPP Radio Interworking', v.12.0.0, December 2013
[9] 3GPP Tdoc RP-122101, 'LTE/WLAN Interworking', RP-132101, 2013.
[10] 3GPP Tdoc RP-141401, 'Motivation for E-UTRAN and WLAN Aggregation', September 2014.
[11] 3GPP Tdoc RP-142311, 'Enhanced LTE D2D', December 2014.
[12] 3GPP Tdoc RP-142205, 'Study on Support of Single-Cell Point-to-Multipoint Transmission in LTE', December 2014.
[13] 3GPP Technical Specification, TS 36.306, 'User Equipment (UE) Radio Access Capabilities', v.12.3.0, December 2014.

4

Small Cell Enhancements in Release 12/13

Antti Toskala, Timo Lunttila, Tero Henttonen and Jari Lindholm

4.1 Introduction

This chapter covers the small cell enhancements introduced in Release 12 in 3GPP as well as the items on-going for Release 13. First the architecture impacts from the dual connectivity architecture options specified for Release 12 are introduced, and then the dual connectivity principle is covered, followed with the more detailed impacts for the radio protocols as well as physical layer enhancements including 256 QAM and small cell on/off procedures. The resulting performance is introduced. Finally the chapter is concluded with the Release 13 enhancements, including uplink bearer split.

4.2 Small Cell and Dual Connectivity Principles

The 3GPP started the work for the small cell-related improvements following the workshop held in June 2012 on LTE-Advanced evolution, which concluded the need for evolution for

LTE Small Cell Optimization: 3GPP Evolution to Release 13, First Edition.
Edited by Harri Holma, Antti Toskala and Jussi Reunanen.
© 2016 John Wiley & Sons, Ltd. Published 2016 by John Wiley & Sons, Ltd.

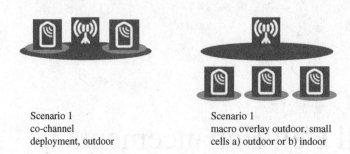

Scenario 1
co-channel
deployment, outdoor

Scenario 1
macro overlay outdoor, small
cells a) outdoor or b) indoor

Figure 4.1   Small cell co-channel and overlay scenarios

the small cell operation. The scenario of importance was especially the use of small cells with a dedicated small cell frequency, as shown in Figure 4.1 as earlier Releases contained already solution such as CoMP or eICIC that can be used with the co-channel scenario.

In Release 10-based intra-site carrier aggregation, as explained in Chapter 2, all the carriers are coming from the same co-located eNodeB. The carrier aggregation can be implemented also with radio frequency (RF) head behind dark fibre configuration with the baseband units/processes handling the aggregated cells just being in the same location.

An important extension to the carrier aggregation principle was adopted in Release 12 with the dual connectivity (also known as inter-site carrier aggregation), intended to be used between a macro and small cell, the connection is now between two eNodeBs (master eNodeB and secondary eNodeB) that are not co-located. The UE needs in this case to be able to provide physical layer feedback signal to both eNodeBs separately (as non-ideal backhaul is assumed) thus the UE has always connection to both macro eNodeB and small cell eNodeB. This allows aggregating the maximum data rate from the macro and small cells and enables the use of small cells more efficiently as the macro cell is used to maintain the connectivity even if the small cell layer does not offer necessary continuous coverage. Small cells may also due potentially higher frequency, have clearly smaller coverage area and may not necessary be always usable as primary cell.

4.3 Dual Connectivity Architecture Principle

Following the 3GPP study phase for small cell enhancements, several different architecture options were considered, as reported in 3GPP TR 36.842 [1], with two of the architecture options agreed to be standardized for Release 12 as shown in Figure 4.2.

The selected dual connectivity architecture options have the following main characteristics:

- In Architecture 1A, the user plane data (i.e. S1-U) is split at the serving gateway (S-GW) between the master eNode (MeNodeB) and the secondary eNodeB (SeNodeB). This means the user plane protocol stacks are identical in both eNodeBs.
- Architecture 1A was motivated with the potential limitations in the backhaul connectivity between the macro and small cells. That is why a single bearer is only provided either via the small cell or via the macro cell.

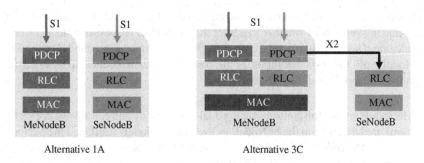

Figure 4.2 Dual connectivity downlink user plane architectures

- In the case of only a single service being active, there is no possibility aggregate of the data from both cells and respectively the dual connectivity will neither increase peak data rate nor provide the carrier aggregation benefits from dynamic scheduling and load sharing between two carriers.
- The overall security is managed by MeNodeB, but each eNodeB manages its own ciphering, so two different security keys are needed at the UE.
- In Architecture 3C, all the user plane data (i.e. S1-U) are routed via the master eNodeB. The PDCP protocol is only located in the MeNodeB, and only the lower layer protocols (i.e. RLC, MAC) are identical in both eNodeBs.
- The Architecture 3C provides all the carrier aggregation benefits for throughput increase and can be used when the backhaul between the eNodeBs is good.
- The security solution is unchanged since the PDCP is only located at MeNodeB and this architecture can be rolled out without any changes in the core network side.
- Since the user plane data for all bearers arrives first for the macro eNodeB which can then dynamically select between the small cell and macro cell. With this approach also the service interruptions are smaller compared when losing the connection with small cell in Architecture 1A.
- The UE also gets the increased peak rate even if only a single service is active. The gains as reported in [1] reach up to 90% increased user throughput at lower load levels.
- The 3GPP Release 12 allows the bearer split on the downlink direction only while the uplink user plane data for each bearer is provided via one cell only, and then forwarded to macro cell. This is foreseen to be extended after Release 12 to enable aggregation benefits in the uplink direction as well as discussed in later section.

4.4 Dual Connectivity Protocol Impacts

For the control plane side, the fundamental difference is that the cell serving as the master eNodeB (typically the macro cell) remains as the master of the correction and will send RRC signalling to the UE as well as be the receiving entity for the RRC signalling from the UE. The macro and small cell eNodeBs will enhance control information over the X2 interface. All the RRC measurement reports are provided for the macro eNodeB for processing and deciding on RRM actions, including possible handovers. The measurement reports may also be provided to the SeNodeB when SCells are requested to be added, to aid in the SeNodeB decision-making process of whether to accept the MeNodeB request or not.

To understand how the dual connectivity functions, one must understand how the MeNodeB and SeNodeB split the responsibilities, as the control of the dual connectivity is partially split between the MeNodeB and the SeNodeB:

- The MeNodeB retains full control over the dual connectivity and can always decide to release any configured SCell from the SeNodeB or even the SeNodeB itself, and the SeNodeB shall comply. The MeNodeB also controls the overall bearer structure, RRC connectivity and measurements at the UE side. Only MeNodeB can choose to request addition of SCells to the SeNodeB part.
- The SeNodeB retains control over its own resources and decides on its own radio configuration part: When MeNodeB requests dual connectivity, the SeNodeB is in control of its radio configuration and MeNodeB will not modify it. The SeNodeB can also neither request to start dual connectivity nor add an SCell, but can choose to reject such request from MeNodeB. But like MeNodeB, SeNodeB can at any time request to release an SCell or the dual connectivity itself, and MeNodeB will comply.

This division of responsibilities is depicted in Figure 4.3.

Release 12 bearer split defines how the bearer can be divided in downlink domain for UEs that support dual connectivity and necessary signalling is in place to support such a situation.

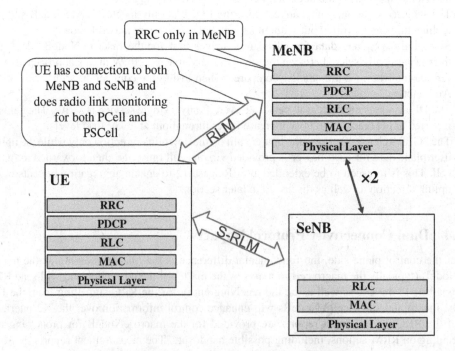

Figure 4.3   Division of control between MeNodeB and SeNodeB

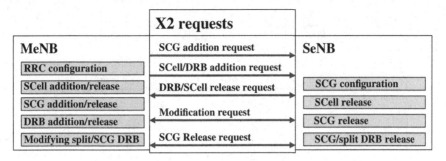

Figure 4.4   Dual connectivity control and user plane architecture for the bearer split

The layer 2 structure with dual connectivity and bearer split in use is shown in Figures 4.4 and 4.5. The PDCP layer does the split when aggregation from macro and small cells is possible.

4.5   Dual Connectivity Physical Layer Impacts and Radio Link Monitoring

In the LTE Releases 8–11 PUCCH has always been sent on the PCell, but in the case of dual connectivity, this is not possible any more. Because of delays in the backhaul, the operation where HARQ-ACK feedback related to downlink transmission from the SCell of SeNodeB would be transmitted to the PCell and then routed to the SeNodeB via backhaul cannot be done as the feedback would arrive to the SeNodeB too late.

Cells in the dual connectivity are divided into two groups as shown in figure 4.6: The cells that are controlled by MeNodeB are called as Master Cell Group (MCG) and uplink control information (UCI) related to transmissions in this cell group is sent via PUCCH on the PCell

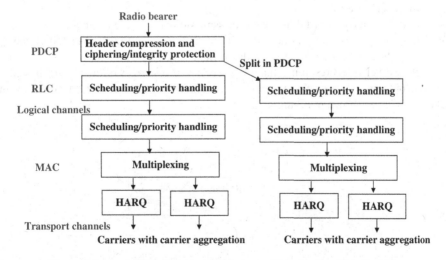

Figure 4.5   Layer 2 architecture with bearer split in the downlink direction

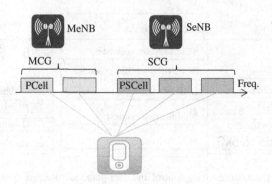

Figure 4.6   PUCCH on PCell and PSCell

or via PUSCH on one of the cells in the MCG. The cells that are controlled by SeNodeB are called as Secondary Cell Group (SCG) and PUCCH related to SCG transmissions is sent on the Primary SCell (PSCell). Because of this there is no need to transmit UCI in the backhaul.

As the role of PSCell being similar to that of the PCell, the UE also does Radio Link Monitoring (RLM) for the PSCell just like it does for the PCell, to avoid situations where UE could cause excess UL interference or be unable to be served by the SeNodeB. The detection of a Radio Link Failure on the PSCell is called Secondary Radio Link Failure (S-RLF). When S-RLF happens, the UE stops receiving and transmitting on the SeNodeB and informs MeNodeB via RRC message that the SeNodeB has failed. The UE still continues measuring the SeNodeB according to its configuration, but to recover use of the SeNodeB, the MeNodeB must restart the dual connectivity.

The UCI transmission rules defined in earlier releases are used within cell group to determine which physical channel (PUCCH or PUSCH) is used for UCI transmission and which PUCCH resource or PUSCH cell is used. Also periodic CSI dropping rules, handling of UCI combinations and HARQ-ACK timing and multiplexing are done by applying rules of previous release within cell group.

In dual connectivity, scheduling of UE transmissions is more challenging than in carrier aggregation, because scheduling decisions in the SeNodeB and in the MeNodeB cannot be instantaneously coordinated. Because of this it can easily happen that uplink grants sent from MeNodeB and SeNodeB result in situation where power resources of the UE are exceeded. In order to avoid this, it is possible to configure guaranteed power levels for transmissions in the MCG and SCG, as shown in Figure 4.7.

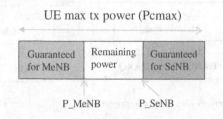

Figure 4.7   Guaranteed powers and remaining power

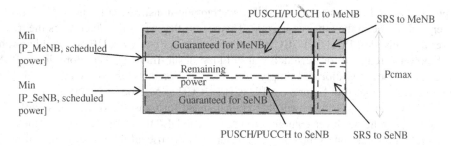

Figure 4.8 Power control mode 1

The guaranteed power level of a cell group is configured as the percentage of the maximum UE transmission power. For both MCG and SCG, guaranteed power level values can be selected from the following values: 0%, 5%, 10%, 15%, 20%, 30%, 37%, 44%, 50%, 56%, 63%, 70%, 80%, 90%, 95% and 100%. The configuration where UE power resources are simply divided between MCG and SCG (i.e. P_MeNB + P_SeNB = 100%) can be easily done. It is also possible to make a configuration where P_MeNB + P_SeNB < 100%. In this case, there is remaining power that is not dedicated to certain cell group and this remaining power can be dynamically allocated to MCG or SCG depending on scheduling. After the uplink power is allocated to MCG and SCG, Release 11 power scaling and channel dropping rules are applied within the cell group. It should be noted that guaranteed powers and remaining power allocation are applicable to PUCCH, PUSCH and SRS transmissions but PRACH can use all the UE power, if needed.

Dual connectivity can be used both in synchronous and asynchronous networks. Two power control modes have been defined for dual connectivity: PCM1 (power control mode1) is used in synchronous and PCM2 in asynchronous networks. All the dual connectivity capable UEs must support PCM1. If UE supports both power control modes, eNB configures the power control mode to be used by the UE.

In the Figure 4.8, power control mode 1 is depicted. The priority order for allocating the remaining power to MCG and/or SCG transmissions is based on UCI type. The priority order is the following: HARQ/ACK = SR > CSI > PUSCH without UCI > SRS. If same type of UCI transmission takes place simultaneously in MCG and SCG, MCG transmission is prioritized. If less than the guaranteed power is required for transmission to one of the cell groups, unused power can be used in the other cell group. The allocation of remaining power is done separately to the SRS symbol.

If network is not synchronous, eNodeB configures PCM2 for the UEs. In the asynchronous case, sub-frame boundaries are not aligned between transmissions to MeNodeB and SeNodeB, as can be seen in Figure 4.9. In the standard it has been assumed that UE is not fast enough

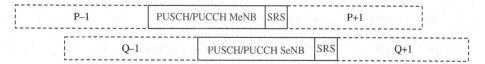

Figure 4.9 Asynchronous operation

to process the uplink grant that schedules overlapping transmission to the latter part of the current sub-frame in the other cell group. This means that when UE determines the transmit power for MeNodeB PUSCH/PUCCH in sub-frame P in Figure 4.9, UE does not know, how much power is required for SeNodeB transmission in sub-frame Q. Therefore, in the PCM2, the remaining power is allocated simply to the transmission that starts in the earlier sub-frame. So in the case depicted in Figure 4.9, PUSCH/PUCCH to MeNodeB in sub-frame P can use all the UE power except the power that is guaranteed for transmissions to SeNodeB. If UE knows based on higher layer configuration (e.g. TDD UL/DL configuration) that there is no overlapping transmission in the other cell group, UE can use all the power in one of the cell groups. The rule that earlier sub-frame can use the remaining power is applicable also in the case that only SRS is transmitted in the earlier sub-frame and the actual transmission in the other cell group in fact takes place prior to the SRS transmission.

In dual connectivity, both contention-based and non-contention-based random access procedures are supported in the SCG. Because it is difficult to coordinate PRACH transmissions to the MeNodeB and SeNodeB, it has been agreed that UEs have to support simultaneous PRACH transmissions to the MeNodeB and SeNodeB. In the PRACH power control, two principles have to be taken into account. The first thing is that PRACH transmission in the PCell is the highest priority channel of all the UE transmissions. The second thing is that, if UE has PRACH transmission that span over multiple sub-frames, the power of that transmission should not change in the middle of the transmission. In PCM1, if UE is power limited when sending PRACHs and the two PRACHs are intended to start at the same sub-frame, the priority order to allocate uplink power is such that PCell PRACH has the highest priority, other PRACHs have the second highest priority and the priority of other channels is lower than PRACHs. If there is a PRACH that has started already in the earlier sub-frame, that transmissions can continue without changes in the transmit power. In PCM2, the same priority order as in PCM1 is used to allocate uplink power. The PRACH transmissions typically take place only in subset of all UL sub-frames and UE may be able to determine the required power for PRACH transmission well before the actual transmission takes place. So in PCM2, UE should, in some cases, be able to prioritize the PCell PRACH over some lower priority transmission also in the case that PCell PRACH overlaps the latter part of some other transmission in the other cell group. If PRACH transmissions overlap and maximum UE transmission power is exceeded, it is up to UE implementation whether power of the lower priority PRACH is scaled down or PRACH is dropped. In this case, depending on physical layer indication to MAC, power ramping may not be done for the retransmission of the preamble.

When UE sends power headroom report to the eNodeB, it includes power headroom information of all the cells in the report. For the own cell group, that is, when UE reports Power Headroom (PH) of MCG transmissions to MeNodeB, or PH of SCG transmissions to SeNodeB, the reporting mechanism defined in Release 11 is used. When UE reports PH of SCG transmissions to the MeNodeB or PH of MCG to SeNodeB, type 2 power headroom of the PUCCH cell is always included in the report. In addition, depending on configuration power headroom in all the cells in the other cell group is based on reference format calculation (virtual power headroom) or it is calculated based on actual transmission in the cells, where PUSCH/PUCCH transmission takes place. In asynchronous dual connectivity, it is up to UE implementation, whether PHR of the cells in the non-scheduling eNodeB is calculated based on the first or the latter overlapping sub-frame.

The performance with dual connectivity is covered Chapter 8.

4.6   Other Small Cell Physical Layer Enhancement

Besides the dual connectivity, the main areas impacting the physical layer are the use of 256 Quadrature Amplitude Modulation (256QAM), small cell on/off procedure and over the air synchronization between eNodeBs, which are elaborated further on the following sections.

4.6.1   256QAM for LTE Downlink

- One of the new LTE features introduced in Rel-12 is the support for higher order modulation, that is, 256QAM, for downlink data transmission. The main motivation for introduction of 256QAM is the increase in peak data rates, achievable mainly in small cells. While 256QAM is, in principle, applicable also in macro cells, the required very high SINR of more than 20 dB is not often achieved in macro cell environments. Moreover, achieving high SINR also imposes stringent RF requirement for the eNodeB transmitter, which are easier to meet at lower output power levels, that is, in small cells. Similarly, for the UE, support for 256QAM requires better receivers from RF requirements point of view.
- From the physical layer point of view, 256QAM is a relatively simple feature. Changes in the specifications relate to extensions in signalling of modulation and coding scheme, transport block sizes as well as to channel quality indicator definition. In terms of peak data rates, the largest transport block size with 256QAM is about 30% larger than that of 64QAM. Correspondingly, new UE categories Cat11–Cat15 have been introduced to accommodate support for 256QAM with increased peak data rates as presented in Chapter 3. A UE with 256QAM support can achieve roughly 200 Mbps per 20 MHz downlink carrier peak data rate.

4.6.2   Small Cell ON/OFF Switching and Enhanced Discovery

Switching cells on and off is possible already based on earlier LTE releases and the SON-based schemes standardized in Release 9 have been successfully used especially when the network load is very low. With these schemes, cells in OFF (i.e. dormant) state do not transmit any signals and consequently UEs are not able to detect those cells. To return an OFF cell back to service, X2 signalling was standardized to allow an eNodeB to request a neighbouring eNodeB to switch on the OFF cell. This limits the achievable ON/OFF transition times to a timescale of hundreds of milliseconds, limiting the usefulness of these schemes to situations with very low network load such as during the night-time.

One of the features discussed intensively and adapted to LTE Rel-12 standards was the small cell on/off operation and associated cell discovery. The feature is used to, for example, reduce network energy consumption and interference during the times when the network load is low by allowing fast on/off switching of cells, with the energy saving potential analyzed more in the next section.

Rel-12 On/Off Mechanism

The work done in Rel-12 improves the efficiency of on/off mechanism by facilitating shorter transition times to/from eNodeB dormant (i.e. OFF) state. There are two mechanisms for achieving this:

Figure 4.10   Illustration of eNodeB on–off operation

1. **Transmission of DL Discovery Reference Signals** to facilitate fast discovery of dormant cells. UE RRM measurements of DRS are also introduced.
2. **Configuration of an OFF-cell as a deactivated SCell.** Using SCell activation/deactivation for on/off-switching reduces the transition times between the ON and dormant states.

The transmission of discovery signals for cell on/off is illustrated in Figure 4.10. First (during e.g. a low network load) the network decides to turn a cell off. The decision is then followed by an on-off transition period, during which the network releases the UEs in the cell to be turned OFF, using, for example, handover, connection release, redirecting RRC_IDLE mode UEs to different frequency layers etc. Once the network is satisfied that there are no longer UEs camping in the cell it can then make the final decision to turn the cell off and start a dormancy period. During the dormancy period, an eNodeB may transmit (e.g. periodically) DRS to allow for the UEs supporting the feature to discover and measure the dormant cell. Finally, at some point of time, the network can decide to turn the cell back on based on, for example, UE measurements.

Configuration of an OFF-cell as a deactivated SCell at UE side allows the on/off switching to be used efficiently. When discovery signals based measurements are configured for the carrier of a deactivated SCell, the UE cannot assume that the SCell transmits any signals other than discovery signals. This allows for the network to utilize SCell activation/deactivation for ON/OFF switching, reducing the ON/OFF transition times to the timescale of few tens of milliseconds.

Discovery Signal Transmission and Measurement Procedure

The discovery reference signals (DRS), whose structure is shown in Figure 4.2, consist of synchronization and references signals introduced already in LTE Rel-8: Primary Synchronization Signal (PSS), Secondary Synchronization Signal (SSS), and Cell-specific Reference Signal (CRS). Additionally, Channel State Information Reference Signals (CSI-RS) standardized in Rel-10 can also be configured as part of DRS. The PSS/SSS/CRS facilitate cell discovery and RRM measurements similar to the normal LTE operation while the CSI-RS allows for discovery of transmission points within the cell, enabling, for example, the so-called single-cell CoMP operation via RSRP measurements.

Unlike in prior LTE systems, where PSS/SSS/CRS are transmitted continuously, the Discovery Reference Signals are transmitted with a more sparse periodicity for the purpose of cell detection and RRM measurements. In Figure 4.11, one instance of DRS transmission is denoted as a DRS Occasion. A DRS Occasion has duration of 1–5 sub-frames and includes PSS/SSS and CRS corresponding to antenna port 0 in the same time/frequency locations as

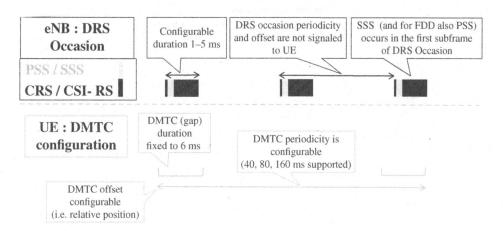

Figure 4.11   The structure of discovery deference signals and the related measurement timing

in ordinary LTE operation. Additionally, a DRS Occasion may comprise of transmission of several CSI-RS resources, each typically corresponding to a transmission point. In other words, DRS occasion can be seen as a snapshot of an ordinary LTE transmission on an unloaded carrier.

The UE performs discovery measurements according to eNodeB-given per-carrier Discovery Measurement Timing Configuration (DMTC). The DMTC indicates the time instances when the UE may assume DRS to be present for a carrier, a bit similar to measurement gap configuration used for inter-frequency RRM measurements. A DMTC occasion has a fixed duration of 6 ms and a configurable periodicity of 40, 80 or 160 ms. The network needs to ensure that the transmission times of DRS occasions of all cells on a given carrier frequency are aligned with the DMTC configuration in order to ensure those cells can be discovered. Hence, the network needs to be synchronized with the accuracy of approximately one sub-frame (or better) for the discovery procedure to work. Network may also not configure a UE to use all DRS occasions for the DMTC.

The DRS-based UE measurements differ slightly from legacy measurements.

- The UE may only assume presence of CRS/CSI-RS only during the DMTC.
- The DRS-based CRS (RSRP/RSRQ) measurements are done as in legacy LTE. This means the same measurement events (i.e. events A1–A6) are also applicable and all legacy options (e.g. cell-specific offsets) can be utilized.
- For the DRS-based CSI-RS measurements, only CSI-RS-based RSRP is supported. The network also explicitly configures the CSI-RS resources that the UE measures, and UE only triggers measurement reports for those CSI-RS resources.
- Two new measurement events have been defined for DRS-based CSI-RS measurements: The event C1 compares the measurement result of a CSI-RS resource against an absolute threshold value (similar to the existing event A4). The event C2 compares the measurement result of a CSI-RS resource against the measurement result of a pre-defined reference CSI-RS resource (similar to the existing event A3).

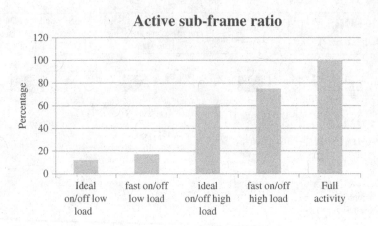

Figure 4.12   Power-saving gains (upper limit) with different on/off dynamics

4.6.3   Power Saving with Small Cell ON/OFF

The use of small cell ON/OFF functionality allows eNodeB to avoid having to transmit continuously during every 1 ms sub-frame. If the eNodeB transmitter has to be on continuously, then even with the low traffic situation lot of transmission power is consumed. During 1 ms time it is difficult to ramp down and up the eNodeB power amplifier. The use of small cells introduces more variance in the number of active users connected to a single cell. The evaluation, as originally presented in [2], considers the statistical small cell load and what could be achieved in terms power saving with on the other hand ideal on/off (1 ms basis) and on the other hand of on/off operation was enabled on 100 ms basis. As can be seen in Figure 4.11, 100 ms basis operation (assuming minimum 50 ms on period when there is traffic) is able to achieve most of the power-saving benefits. The results in Figure 4.12 represent an upper limit assuming no energy would be consumed when the transmitter is off.

4.6.4   Over the Air Synchronization Between eNodeBs

In the case there is desire to synchronize the network, typically external timing reference has been used, such as GPS. For the small cell deployments, the alternative approach of using timing measurements over the air, such that an eNodeB would consider one eNodeB as master in terms of timing (source selected for synchronization) and then measure and obtain the timing information from the transmission of the source eNodeB. Release 12 specifications contain signalling (over the S1-interface) which allows the SON functionality to help the over the air synchronization process by providing information of the muting pattern of the eNodeBs. By muting the detection of timing, information can be made more reliable as network planning does not normally try to ensure that eNodeBs can hear each other.

4.7   Release 13 Enhancements

With the Release 12 solutions as such concluded by the end of 2014, 3GPP is defining how to enhance further in Release 13 the small cells and other solutions, including carrier aggregation.

The discussion contains approaches like uplink bearer split [3] as Release 12 supports only downlink bearer split, but uplink direction would also benefit from such a possibility.

In the area of carrier aggregation, 3GPP is preparing for more spectrum availability by defining the possible number of aggregated carrier up to 32 carriers from the signalling point of view [4]. This is not really intended to go up to 32 different frequency bands but with solutions like LTE Licensed Assisted Access (LAA), as covered in Chapter 10, allow using more than four carriers on the unlicensed band. The considered 5 GHz band, in many cases, has in the order of few hundred megahertz of available spectrum. It is also expected that with new frequency bands like 3.5 GHz or the new bands from WRC-15 one operator likely have more than 20 MHz within a single frequency band. This work also covers introduction of PUCCH in SCell with carrier aggregation, as in Release 12 it is possible to send the same (or similar) L1 control information for SCell only in connection with dual connectivity but not in the Release 10-based carrier aggregation case.

4.8 Conclusions

In the previous sections, small cell-related enhancements have been covered in Release 12 and first set of relevant items started in Release 13 for LTE-Advanced further evolution. While in many markets, small cell are not yet critical to be implement due to low traffic volume in LTE, in some markets, there is more interest for small cell deployments. Release 12 solution forms good basis for small cell deployments, especially when downlink direction is foreseen to dominate in terms of data volumes. Release 13 solutions will build on these principles while provided obviously additional enhancements as well.

References

[1] 3GPP TR 36.842, 'Study on Small Cell Enhancements for E-UTRA and E-UTRAN; Higher Layer Aspects', v.12.0.0, December 2013.
[2] 3GPP Tdoc R1-133485, 'On Small Cell On/Off Switching', August 2013.
[3] 3GPP Tdoc RP-150156, 'New WI Proposal: Dual Connectivity Enhancements for LTE', March 2015.
[4] 3GPP Tdoc RP-142286, 'New WI Proposal: LTE Carrier Aggregation Enhancement Beyond 5 Carriers', December 2014.

5

Small Cell Deployment Options

Harri Holma and Benny Vejlgaard

5.1 Introduction

The number of small cells will increase during the next years driven by the need for more capacity and for better coverage. The small cell concept covers a number of different options from relatively high power outdoor micro cells to indoor pico cells and very low power femto cells. The small cell network architecture has a number of options depending on the use case and depending on the available transport infrastructure. The small deployments also need to consider the frequency usage between macro cells and small cells. This chapter discusses many of these small cell deployment options. Section 5.2 presents the main motivations for the small cell deployment and Section 5.3 illustrates the network architecture options. The frequency usage between macro cell and small cell is discussed in Section 5.4 and the solutions for finding the optimal location for the small cell is considered in Section 5.5. The indoor small cells are presented in Section 5.6. The cost aspects are presented in Section 5.7 and the chapter is summarized in Section 5.8.

LTE Small Cell Optimization: 3GPP Evolution to Release 13, First Edition.
Edited by Harri Holma, Antti Toskala and Jussi Reunanen.
© 2016 John Wiley & Sons, Ltd. Published 2016 by John Wiley & Sons, Ltd.

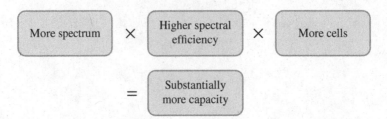

Figure 5.1   Three main components for increasing network capacity

5.2   Small Cell Motivation

The radio network capacity can be enhanced with more spectrum, with higher spectral efficiency and with more cells. The new frequency bands between 700 and 2700 MHz are used for increasing the macro cell capacity. Adding more spectrum is relatively simple for getting more capacity. The LTE-Advanced features increase the spectrum efficiency allowing more capacity from the same spectrum. These two solutions are used to squeeze out more capacity from the macro cell sites. When these solutions are exhausted, then small cells will be needed for the capacity reason. These three components for adding capacity are shown in Figure 5.1.

With fast growing data traffic, the need for small cells will be evident soon. The additional spectrum and the increased spectral efficiency will not be able to carry the rapidly increasing data traffic in the long term. The small cells will also be driven by the need for better coverage to deliver higher user data rates. At the same time, the small cell implementation has become simpler with small and compact products. Also 3GPP has developed a number of features that make the small cell deployment simpler. The availability of new wireless backhaul solutions has made the small cell rollout faster. The drivers and enablers for the small cells are shown in Figure 5.2.

5.3   Network Architecture Options

Before looking into details of the small cell products, we need to consider the network architecture options which will impact the choice of the small cell product. The three main architecture solutions are shown in Figure 5.3. The left-most option shows the case with a complete all-in-one small base station including all the protocol layers 1–3. The small cell is directly connected to the packet core with S1 interface. The interworking with the macro cell layer uses X2 interface. The right-most option shows low power radio frequency (RF) head which is connected to the macro cell baseband. The small cell is just RF without any layer 1–3 functions. The interface uses Common Public Radio Interface (CPRI) or Open Base Station Architecture Initiative (OBSAI) between RF head and macro cell. That interface requires high bandwidth and low latency. The core network connection goes only from the macro cell. The low power RF head appears as a new macro cell sector from the core network point of view. This option is referred to as Centralized Radio Access Network (CRAN) or Cloud RAN since the baseband processing can be done in the centralized location. The term 'cloud' is misleading since the latency requirements are tough and the baseband processing must be located relatively close to the RF head and not far-away in the cloud. The middle option is the latest addition to 3GPP specification in Release 12 to support dual connectivity where UE

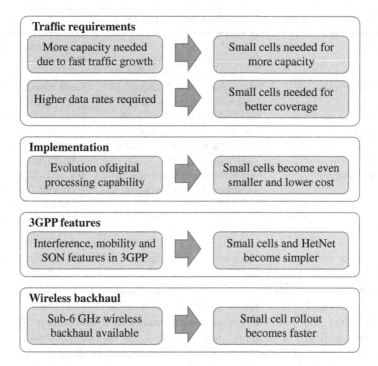

Figure 5.2 Drivers and enablers for small cell deployments

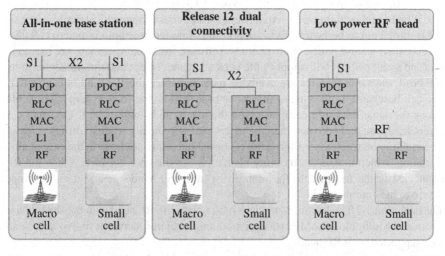

- Stand-alone BTS
- Low transport requirements

- Small BTS connected to macro
- Shared PDCP in macro BTS
- Inter-site carrier aggregation

- Small BTS = RF heada
- Heavy transport requirements = direct fiber

Figure 5.3 Small cell network architecture options

Table 5.1 Comparison of all-in-one small base station and radio head

	All-in-one base station	Radio frequency head (RF head)
Transport requirements	Low requirements. Transport can use radio, copper or fibre	Tough requirements. Direct fibre needed in practice
Feature parity between macro and small cells	Yes possible, but requires extra planning in development	Yes, comes naturally with the common baseband
Mobility between macro and small cells	Inter-site handovers visible also to the packet core	Intra-site handovers are not visible to the packet core
Coordinated Multipoint (CoMP)	Joint Processing not possible but Dynamic Point Selection is possible	Yes, all CoMP options possible between macro and small cells

has simultaneous connection to both macro and small cells. The low latency functionalities are located in the small cell including Layer 1, Media Access Control (MAC) and Radio Link Control (RLC) while Packet Data Convergence Protocol (PDCP) is located in the macro cell. The X2 interface between macro and small cells has clearly more relaxed requirements than CPRI/OBSAI interface. There is also other intermediate functionality split alternative between these options. For example, part of layer 1 can be integrated with RF, like Fast Fourier Transformation (FFT) while Turbo decoding can be centralized.

The RF head solution has a number of benefits since it is a straight forward extension of the macro cell. On the other hand, the interface between macro cell and RF heads needs very high capacity. These two options are compared in Table 5.1. The RF head requires high capacity for the CPRI/OBSAI interface typically beyond 1 Gbps, and very low latency. The all-in-one base station has relaxed transport requirements which make the practical deployment simpler.

It has turned out to be beneficial to support the same features both in macro cell and small cell. It is called feature parity. If the features, software loads and roadmaps are harmonized in macro and small cells, it will simplify the network management and optimization. In the case of RF head, the same features are available naturally since the RF head is connected to the macro cell baseband. In the case of all-in-one base station, the feature parity can be obtained by using the same hardware components in the macro and small cells. If the small cell uses different hardware and software than the macro cell, it typically leads to different features.

The mobility between macro and small is normal inter-site handover in the case of all-in-one base station, which makes the mobility visible also to the core network. When the RF head is connected to the macro cell, the mobility is intra-site handover and does not impact S1 interface to the core network.

The evolution to LTE-Advanced tends to be simpler with RF head with the shared baseband. Coordinated Multipoint (CoMP) transmission and reception between macro and small cells can be supported with RF head.

The transport requirements between base station and RF head have major differences. The transport with the base station is called backhaul since the transmission is behind the baseband. The transport with the RF head is called fronthaul since the transmission is in front of the baseband. An example case is illustrated in Figure 5.4 with 20 + 20 MHz LTE and 2×2 MIMO. The peak rate is 300 Mbps with 20 + 20 MHz carrier aggregation, which is also the backhaul

20 + 20 MHz FDD with 2 × 2 MIMO

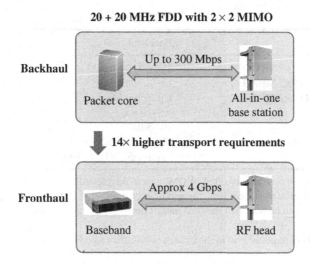

Figure 5.4   Transport requirements with small base station and RF head

requirement. It is also possible to use less backhaul, for example, 100 Mbps, if the small cell is mainly used for providing more coverage instead of high capacity or high peak rates. If multiple small cells share the same transport, then the total transport rate can be less than the sum of the peak rates because different cells are not typically fully loaded at the same time. That benefit is called trunking gain.

The fronthaul requirement with CPRI is over 4 Gbps with 20 + 20 MHz and 2×2 MIMO and increases a function of the bandwidth and the number of antennas. The data rate is independent of the cell loading or throughput requirements. The throughput requirement can be lowered by using CPRI compression but it comes at the expense of signal RF quality leading some of degradation of Error Vector Magnitude (EVM).

We will next illustrate how the fronthaul requirement increases as a function of bandwidth and antenna lines with OBSAI interface. The OBSAI line rate needs to be in multiples of 768 Mbps, which means 768, 1536, 3072 and 6144 Mbps. The master frame size in bytes is i*N_MG*(M_MG*19+K_MG) where i = 1, 2, 4, 8 that means multiples of 768 line rate. The master frame illustrating the sequence according to which LTE messages are inserted to the bus (parameter set M_MG = 21, N_MG = 1920, K_MG = 1, i = 1). We take into account that the sampling rate for 20 MHz LTE is 30.72 MHz and for OBSAI I and Q samples are mapped into 16 bits each. Single antenna 20 MHz stream requires 0.98304 Gbps. The corresponding OBSAI stream is using i*N_MG*(M_MG*19+K_MG) and the parameters above give 1.032192 Gbps. As the line rate for OBSAI is 768 Mbps this requires 1.536 Gbps OBSAI. The needed data rate for 2×2 MIMO with 20 MHz is 1.96608 Gbps, OBSAI frame rate 2.064384 Gbps and needed OBSAI interface 3.072 Gbps. The required fronthaul data rate grows with larger bandwidth in carrier aggregation and with more antennas. If we have 20 + 20 MHz carrier aggregation with four-antenna base station, the required frame rate is approximately 8 Gbps.

A comparison of backhaul and fronthaul requirements is presented in Table 5.2. The RF head requires also low latency in addition to the high data rate. The one-way delay must

Table 5.2   Comparison of backhaul and fronthaul requirement

	Backhaul	Fronthaul
Bandwidth 20 MHz 2×2 MIMO	Up to 150 Mbps*	2 Gbps[†]
Bandwidth 40 MHz 2×2 MIMO	Up to 300 Mbps*	4 Gbps[†]
Bandwidth 40 MHz 4×2 MIMO	Up to 300 Mbps*	8 Gbps[†]
Latency	<5–10 ms	<0.5 ms
Jitter	<1–5 ms	<8 ns
Practical transport options	Microwave radio, GPON, CWDM, VDSL, HFC	CWDM, NGPON2

GPON, gigabit passive optical network; CWDM, coarse wavelength division multiplexing; VDSL, very high bit rate digital subscriber line; HFC, hybrid fiber coaxial.
*Backhaul capacity can be less than the peak rate. Trunking gain can be used to connect multiple base stations.
[†]Fronthaul capacity is required also in the case of low traffic radio heads.

be clearly below 0.5 ms to run the low latency functions including retransmissions from the baseband. Therefore, the baseband processing cannot be faraway in the cloud but the physical location needs to be relatively close to the RF.

Also the delay jitter must be very low in the order of a few nanoseconds. The very high data rate means that it is difficult to use wireless backhaul for the RF head especially when LTE-Advanced increases bandwidth and number of antennas leading to higher fronthaul data rate requirements. The very low latency and jitter means that IP routing cannot be used in the fronthaul.

We have now learnt that RF heads have some benefits over all-in-one base stations but RF heads also require very fast and low latency transport network. Therefore, small cells with RF heads are practical in those networks with rich fibre infrastructure, like in Korea. The RF heads also need tight interaction between radio and transport teams to troubleshoot any potential issues.

5.4   Frequency Usage

The operators will aggregate all their spectrum between 700 and 2600 MHz together into one pool with carrier aggregation. That solution maximizes the macro cell data rates and capacity and is supported by the future terminal carrier aggregation capabilities. The 3.5 GHz band may be used in the macro cells especially in those areas where the macro cell density is high. If the macro cell grid is sparse, then 3.5 GHz may be used as the dedicated small cell frequency. The high frequencies, like 5 GHz unlicensed bands, will be used in the small cells. The small cells can share the same frequency with macro cells or use a dedicated frequency. The frequency usage aspects are considered in Table 5.3. The co-channel deployment maximizes the spectral efficiency but requires more careful inter-cell interference management. The dedicated frequency deployment is simple from the interference management point of view. More detailed interference management solutions are considered in Chapter 7.

Figure 5.5 shows an example dense urban deployment of pico cells using two different spectrum options: shared in-band spectrum and dedicated out-band spectrum. To reach an

Table 5.3 Frequency usage in small cells

	Co-channel with macro cells (in-band)	Dedicated small cell frequency (out-of-band)
Interference management between macro and small cells	Preferable clear dominance areas needed for small cells. Enhanced ICIC (eICIC) can be utilized.	Simple interference management
Mobility management between macro and small cells	Intra-frequency handover	Inter-frequency handovers needed which requires proper measurement triggering
Capacity	Maximizes the overall spectral efficiency	Beneficial for high number of small cells

outage of 5% only ~1/3 of the small cells would be needed to provide the same capacity with out-band small cells compared to in-band small cells. We see a breakeven of in-band versus out-of-band deployment of around two small cells per macro cell depending on the traffic load. An in-band solution is more attractive with a lower number of micro cells. Meanwhile out-of-band performs better with a high micro cell density. The in-band deployment increases network capacity and coverage and is recommended if the spectrum is limited and the macro networks are fully developed. The cost efficiency is lower than with out-band micros. The frequency usage is considered in Reference 1.

5.5 Selection of Small Cell Location

It is highly important to place the small cell optimally in the network. The small cell placement needs to consider the RF propagation and the user locations. The small cell should preferably

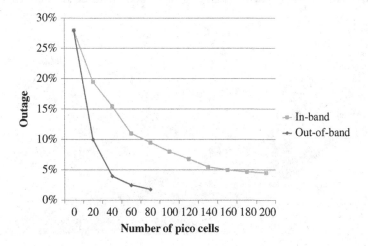

Figure 5.5 Co-channel (in-band) versus dedicated frequency (out-of-band) small cell

have a clear dominance area to minimize the interference between small cell and macro cell. The small cell should also be located close to the users. The geolocation concept can be used to find out the locations of the users and identify the traffic hotspots in the macro network. Geolocation can be applied for the network troubleshooting, optimization and for the selection of the new small cell locations. Geolocation estimates are based on a number of information elements collected from the network:

– Cell identity
– Distance measurement from the round trip time information
– Path loss from the signal level measurements
– Observed time difference of arrival
– GPS information from the UE

   The accuracy of geolocation estimates from round trip time is impacted by the non-line-of-sight propagation which makes the measured distance appear larger than the true distance between UE and the macro cell. Geolocation services can provide information about coverage holes, call drop locations, cell dominance areas, interference levels and voice and data traffic locations. The areas for the network optimization can be analyzed and the solutions can be implemented including parameter tuning, antenna optimization or adding a new small cell to provide better hotspot coverage and more capacity. An example output of geolocation calculations is shown in Figure 5.6. The advanced geolocation tools can also provide the traffic location information in three dimensions including the building floors as shown in Figure 5.7.
   Other sources of location information can be obtained, for example, from the social media. When people upload pictures to social media, the picture can also include the location information. By combining the information from large number of uploaded content from different media can give further information about the location of the traffic for the future small cell

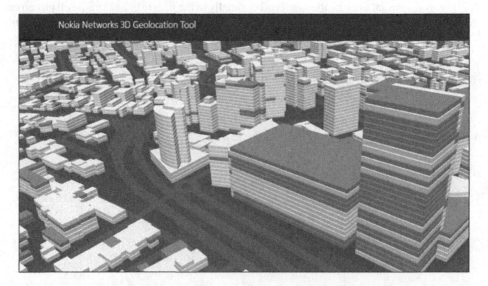

Figure 5.6   Geolocation solution identifies the location of traffic

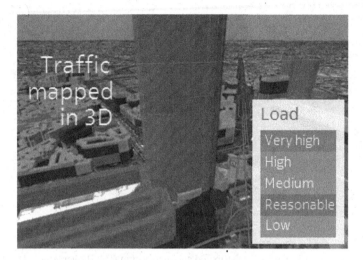

Figure 5.7   3D geolocation solution

location. Also 3GPP Minimization of Drive Testing (MDT) can provide useful information for the network upgrades.

5.6   Indoor Small Cells

Indoor cells provide a viable coverage and capacity solution in dense indoor traffic hotspots such as train stations, airports, shopping malls or enterprise buildings. The different indoor solutions are introduced in this section. The indoor solutions are presented in Reference 2 and the traffic steering is discussed in Reference 3.

5.6.1   Distributed Antenna Systems

Distributed Antenna System (DAS) is the traditional solution for indoor coverage in high rise buildings and in large public premises. The DAS uses passive distributed antenna system and a high power base station. The DAS concept is illustrated in Figure 5.8. DAS includes a large number of small indoor antennas, the coaxial cabling and splitters. A high power base station is connected to DAS. The setting up of DAS requires heavy work because the cable installation takes a lot effort. The DAS should preferably be installed during the building construction phase. Once installed, DAS has a number of benefits. DAS can provide consistent coverage because it uses several indoor antennas which help keeping indoor users connected to the indoor cells instead of having ping-pong between indoor and outdoor cells. The main benefit of DAS solution is that the same antenna network can be shared by multiple bands, multiple technologies and even by multiple operators. If the DAS has been designed for GSM and HSPA, it is typically simple to add LTE just by plugging in LTE base station as long as the antennas and splitters support the new LTE bands. It is also possible to share the DAS system by multiple operators which is an attractive solution for underground systems, shopping malls and other public premises.

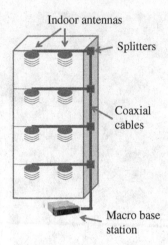

Indoor antennas

Splitters

Coaxial
cables

Macro base
station

Figure 5.8 Distributed antenna system (DAS) overview

The typical DAS installation has only single coaxial cable which makes the usage of Multiple Input Multiple Output (MIMO) difficult. If MIMO is required in DAS, then two parallel cables are needed in the installation phase. The lack of MIMO is not considered as a major problem since DAS installation provides good signal levels which enables reasonably high data rates even with single-stream transmission.

The capacity upgrade in DAS can be challenging. The number of sectors in DAS must be defined in the installation phase. Adding more frequencies is typically simple in DAS but adding more sectors can be difficult. Sector splitting can lead to major redesign requirements in the cable installation.

Distributed Antenna System (DAS) solutions can be classified as passive, active or hybrid systems.

- Passive DAS: In passive systems, the wireless signal from the RF source is distributed to the antennas for transmission without any amplification through a series of passive components.
- Active DAS: In active DAS, the RF signal from a source is converted to a digital signal for transmission over fibre optics or cable. It is fed to multiple remote units that convert the signal back to an RF signal for transmission through an antenna.
- Hybrid DAS: A hybrid DAS system is a combination of passive and active systems. In a hybrid DAS system, fibre optic or CAT5 cable is still used to connect the head end (master unit) to the remote units. However, passive DAS is used for distributing the RF to the antennas from the remote units.

5.6.2 Wi-Fi and Femto Cells

Wi-Fi can offload a large part of indoor traffic since a major part of global wireless data traffic are generated indoors and all smartphones and laptops are equipped with Wi-Fi connectivity. The indoor offload will connect users to the closest cell which reduces interference and transmission power and reducing battery consumption.

Wi-Fi is an important local area technology option for heterogeneous networks, complementing mobile technologies to enhance performance for the user as well as improving offload capacity. One of the criteria for Wi-Fi to become a successful part of the mobile network is technologies and procedures that enable efficient traffic steering between cellular and Wi-Fi, a seamless Wi-Fi/cellular access and therefore a better user experience.

The use of Wi-Fi technology is the preferred means of offloading data from macro cells for users at home or in the office. Smartphones should use Wi-Fi where possible. For public Wi-Fi deployment, careful selection is crucial for effective offload while providing the best user experience. Outdoor Wi-Fi deployment has limited potential where mature macro networks are already installed. It also requires careful planning to limit interference sources from the unlicensed spectrum. Furthermore, many DSL lines are limited to less than 10 Mbps, which is slower than a typical LTE macro cell. Therefore, the practical Wi-Fi data rates may be lower than LTE data rates from the macro cell because of Wi-Fi backhaul limitations.

Load-based traffic steering between macro, micro, pico clusters and Wi-Fi/femto layers will be needed in order to use the available spectrum efficiently. Furthermore, automatic authentication is needed for Wi-Fi offload to reach its full potential, because manual authentication will prevent some users from going through the registration process.

Figure 5.9 shows an example of indoor data offloading to Wi-Fi cells in a macro and micro overlaid network. The graph shows the percentage of users whose data rate is below 10 Mbps. These users are considered to be in outage. The number of users with below 10 Mbps is significantly reduced from 12% to 5% with 200 Wi-Fi cells in a 1 km^2 area. An alternative would be to deploy more indoor Wi-Fi cells and fewer outdoor cells as shown with an example of 1700 Wi-Fi and 100 micro cells. The split between outdoor and indoor cells depends on which one is the most cost-efficient solution. It has been shown that similar performance can be achieved by deploying indoor femto cells.

Indoor LTE femto cells tend to be simpler from the interference point of view compared to the outdoor small cells. In-band deployment is the default option due to operators having

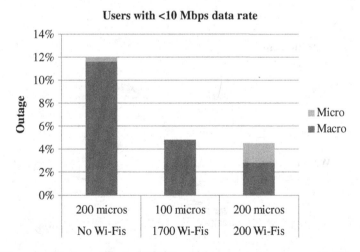

Figure 5.9 Example of indoor offload via Wi-Fi cells in a dense urban area with 20 macro sites and 200 micro cells

limited spectrum resources. Femto cells do not typically have as advanced schedulers as micro and pico cells and the necessary interfaces to coordinate with macro in order to reduce interference. This makes femto cells more suitable for indoor coverage. Femto cells' inability to manage interference means they are not very suited to a high density or large environment that requires very dense deployment or a very large number of cells.

The challenges of femto deployment become even more pronounced when a femto cell is configured with a Closed Subscriber Groups (CSG) identity. A user that is not part of the CSG group will connect to the macro network and experience or cause significant interference problems as normal mobility is overruled by the subscriber group admissions. The optimum performance will be achieved by configuring all femto cells as Open Subscriber Groups (OSG). However, femto cells provide excellent voice coverage extensions and the low transmission power and building attenuation isolate the femto cells very well from the macro cells.

5.6.3   Femto Cell Architecture

Femto cells are connected to operator's network by using the broadband line at home or in the office. The number of femto cells can be very high: there are operators where the number of 3G femto cells is 10–20 times more than the number of macro base stations. The network architecture and scalability need to be considered with such a high number of small cells. The two architecture options are shown in Figure 5.10. One option is to use femto gateway between femto cells and packet core network elements Mobility Management Entity (MME) and Serving Gateway (S-GW). The target of femto gateway is to hide the large number of femto cells from the core network. The femto gateway is always used with 3G femto cells and it can also be used with LTE femto cells. Another option is to connect femto cells directly to the packet core in the same way as marco cells. It is also possible to use femto gateway only for the control plane towards MME. We consider the pros and cons of the different architecture options below.

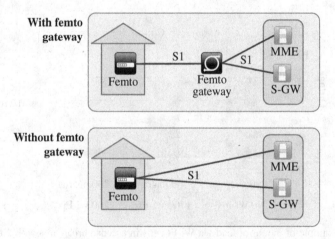

Figure 5.10   Femto architecture options

Femto cell architecture with femto gateway:

– Low scalability requirements to MME and S-GW. The packet core network does not connect
  to each individual femto cell but only to the femto gateway.
– Switching off and on femto cells will not cause signalling load to MME. If there is a large
  scale power failure in the housing area and the power comes back, then large number of
  femto cells will register simultaneously to the network.
– Less mobility signalling towards EPC. If there are multiple femto cells in the enterprise
  premises and UE is moving between those cells, the mobility is managed by the femto
  gateway and does not impact the packet core.
– Femtos do not need to support S1-Flex since the support is in the gateway. S1-Flex is used
  for supporting multiple core networks.
– Paging optimization can be done in femto gateway.
– Femto gateway can protect the packet core against attacks from femto cells.
– Similar architecture as in 3G. All 3G femto deployments have femto gateway. The same
  network element can be utilized also with LTE femtos.

Femto cell architecture without femto gateway:

– No single point of failure. If femto gateway fails, a large number of femto cells will stop
  working. Therefore, femto gateway needs to have high reliability.
– Less network elements without femto gateway. It is easier to upgrade new features to femto
  cells when femto gateway support is not required.
– Lower latency. Femto gateway adds some delay. The additional delay can be so low,
  1–2 ms, that it has only marginal impact to end-to-end latency.

Femto cells can utilize a concept called Local breakout where the data is routed directly
from the femto cell to the local network. The concept is shown in Figure 5.11. The local break
out has the benefit that the user can access local content and devices, for example, in-home
multimedia, data backups or enterprise intranet. Therefore, femto cells can also create new
services over LTE radio, which are not possible with the traditional network architecture. The

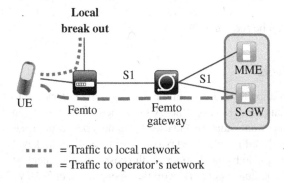

Figure 5.11   Local break out concept with femto cells

Table 5.4 Cost-efficient indoor offload recommendation in traffic hotspot areas

Offload technology	Recommendations
Wi-Fi	Deployment for capacity enhancements, especially in high public indoor traffic areas
Femto	Residential deployment of femto cells provides excellent coverage and capacity for voice and data
Indoor pico	Deployment providing coverage but focused on capacity in indoor public and private hot zones. High number of cells deployed with easy, low impact and fast scalability.
DAS	Suitable to provide cost-efficient coverage in large buildings and is operator neutral. Less cost-efficient for capacity-driven scenarios and small buildings than Wi-Fi.

latency can also be minimized improving the performance for the internet access for the local content. The benefit for the operator is that the local traffic is offloaded from the operator's core network.

Femto cells are designed for low cost implementation and for simple installation which makes it possible for any user to connect femto cell to his broadband connection. The simple installation requires that femto cells have a number of automatic configuration and optimization features for plug & play operation.

5.6.4 Recommendations

Indoor Wi-Fi deployment achieves the lowest cost, lowest energy consumption and the best network performance in a high-traffic urban environment. Indoor femto cell deployment requires a similar number access points to provide the same performance; however, it shares the spectrum with the micro cell layer and can cause interference and is not suitable for very dense, very large deployments.

Wi-Fi is a good supplement to an installed DAS system to help bring capacity for large indoor venues that require an operator neutral deployment. An LTE pico or pico cluster-type solution can be a good complement to an existing DAS system and add significant capacity or boost subscriber experience. Finally, deployment of indoor pico cells can reduce the scale of deployment to provide a cost-optimized solution. The summary of the indoor offload recommendation can be seen in Table 5.4.

5.7 Cost Aspects

Total cost of ownership (TCO) is one of the most important deciding factors when choosing a network deployment path. However, the TCO in each case depends on the operator's current installed base, its spectrum situation and user equipment penetration. The different deployment paths have been analyzed from a TCO perspective to outline the key TCO trends. The target of a TCO calculation is to bring together all the costs of a technical solution over its lifetime which in this case is complete network evolution scenarios over 5–10 years. These TCO values can then be compared to discover the best deployment options. For a fair comparison, it is assumed that the different network evolution scenarios perform in the same way and satisfy the same traffic requirements. A number of different deployment options are considered for the network evolution.

5.7.1 Macro Network Extension

Tilt optimization is a very cost-efficient method for signal-to-interference and noise ratio (SINR) optimization and thereby increases network capacity. Tilt optimization should always be pursued before any further optimizations. If spectrum is available, adding more carriers to already existing macro sites provides easy and low-cost capacity enhancements. The main cost is in CAPEX (Capital Expenditure) and IMPEX (Implementation Expenditure) – OPEX (Operating Expenditure) for the base station increases only slightly. The OPEX comes mainly from electricity, Operations and Maintenance (O&M), backhaul and software fees. However, dedicating spectrum to micro cells can provide an even bigger increase in capacity. Therefore, traffic growth and traffic hotspots play an important role in any site evolution strategy. Furthermore, reframing of spectrum is a cost-efficient way to increase both coverage and capacity. The most cost-efficient approach is to deploy the lower spectrum initially for coverage and deploy the higher spectrum later for macro or micro cells, depending on the traffic density and spectrum availability.

Sectorization in the vertical or horizontal plane provides a simple yet cost-efficient way to increase capacity in the macro network. The main portion of the cost is CAPEX and IMPEX (equipment, antennas and deployment) but OPEX is also raised owing to higher electricity costs, backhaul and additional site rent for new antennas. Six-sectorization is most efficient for uniform traffic distribution and may not be the best option in localized areas of high traffic or in very dense urban deployments where adaptive beamforming could be more attractive.

5.7.2 Outdoor Small Cells

Micro and pico cell deployment is a cost-efficient way of increasing network capacity and coverage. The realization of outdoor small cells by micro base stations means CAPEX for compact micro equipment, but OPEX is very significant for backhaul and site rental. The IMPEX for site acquisition and deployment are also relevant cost factors. Micro cells should be deployed in dedicated spectrum if available. However, for low to medium traffic-density areas or already-deployed macro spectrum, in-band deployment is the preferred solution. Alternatively, small cells can be realized by sharing (or pooling) baseband functions with macro cells and deploying the outdoor small cell RF as a low-power Remote Radio Head (RRH) with a dedicated fibre-based fronthaul transport. Low power RRH does not include any dedicated baseband, which can save CAPEX and ease operations and maintenance. However, this is offset by the requirement to have a dark fibre between the RRH and macro baseband module.

5.7.3 Outdoor Pico Cluster

For outdoor hot zones, pico cluster-based solution can provide a very economical approach compared to other traditional solutions and cell site splitting. A pico cluster solution helps to reduce TCO by simplifying backhaul, managing inter- and intra-layer interference to provide

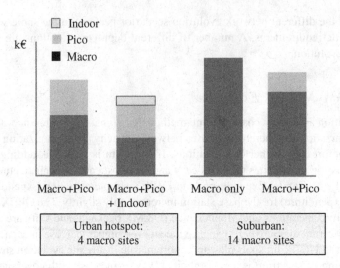

Figure 5.12   Total cost of ownership in different HetNet scenarios

higher performance and limit the amount spectrum planning. The pico cluster provides virtually unlimited scalability, limiting the impact to the core network with local break out and simplifying the operations management and installation.

5.7.4   Indoor Offloading

Wi-Fi is always a low-cost supplement to macro and micro cell deployments since the spectrum is freely available. However, the cost of Wi-Fi depends on the particular backhaul and site acquisition. Wi-Fi and small cells have very similar TCO performance, with similar CAPEX and almost identical installation and operational costs. Indoor small cells offer large benefits for residential and office installations, while public installations should be based on the traffic density and the available spectrum. The underlying assumption for residential and office scenarios is that backhaul at the deployment locations can be reused without incurring site costs. The cost in offices is assumed to be about four or five times higher than the cost in a residential home. Future multisystem pico solutions will provide a best of both worlds approach with Wi-Fi and cellular support and a very cost-effective and scalable solution for indoor coverage and capacity deployments.

Figure 5.12 illustrates the result of a case study about TCO for different network evolution paths. The study was done for urban hotspot and for suburban area. The result shows that the lowest cost in the hotspot area is provided by the combination of macro cells, outdoor pico cells and indoor solution. The lowest cost in suburban case was achieved with mostly macro cells combined with some outdoor pico cells.

5.8   Summary

The small cell deployments are driven by the need for more capacity and better coverage. The small cell rollouts are getting simpler with a number of new 3GPP features and the small base station implementation is more practical with compact products. The practical small cell deployments need to consider a number of options in terms of network architecture, frequency usage and finding the optimal location for the small cell. The network architecture can be selected based on the available transport network and based on the requirements for the radio features. The 3GPP specifications allow flexible deployment options for the small cells. The most cost-efficient solution is typically the combination of macro and small cells including Wi-Fi.

References

[1] 'Deployment Strategies for Heterogeneous Networks', Nokia white paper, 2014.
[2] 'Indoor Deployment Strategies', Nokia white paper, 2014.
[3] 'Business Aware Traffic Steering', Nokia white paper, 2014.

6

Small Cell Products

Harri Holma and Mikko Simanainen

6.1 Introduction

Small cells are needed in a number of application areas, which also call for a number of different product categories. This chapter presents examples of small cell products. Outdoor small cells, also called as micro cells, provide higher capacity in hot spot areas, better coverage and higher data rates. Micro cells need to have relatively higher transmission power to provide outdoor to indoor penetration and support a high density of users within the cell area. Operators can use public cells also in indoor premises like shopping malls, train stations or office complexes. These cases are typically called pico cells. The required output power is lower in the indoor installations. Also low-power femtocells can be used to provide coverage for small offices. Residential coverage can be organized with home femtocells. The small cell application areas are summarized in Figure 6.1.

This chapter is organized as follows. 3GPP base station categories are illustrated in Section 6.2. Micro and pico base station products are introduced in Sections 6.3 and 6.4, femtocells

LTE Small Cell Optimization: 3GPP Evolution to Release 13, First Edition.
Edited by Harri Holma, Antti Toskala and Jussi Reunanen.
© 2016 John Wiley & Sons, Ltd. Published 2016 by John Wiley & Sons, Ltd.

Public outdoor   Public indoor   Enterprise   Residential

Micro   Pico   Enterprise   Femto
                                    femto

Figure 6.1   Small cell application areas

in Section 6.5 and remote RF heads in Section 6.6. The distributed antenna systems (DASs) are illustrated in Section 6.7. The integration of Wi-Fi is discussed in Section 6.8 and backhaul products in Section 6.9. The chapter is summarized in Section 6.10.

6.2 3GPP Base Station Categories

3GPP has defined four base station categories with different RF requirements. The categories are defined with the maximum output power and the minimum coupling loss (MCL). MCL refers to the minimum path loss between the base station antenna and user equipment (UE) antenna. MCL indicates the type of base station installation. High MCL requires that UE cannot get too close to the base station antenna which is typical for the high-mounted macro cell antennas. Low MCL means that UE can get very close to the base station antenna. The categories are defined so that some RF specifications can be relaxed for the low-power base stations to enable lower cost implementation. Sensitivity requirements are relaxed because the link budget is anyway limited by the low power in downlink. Higher spurious emissions are allowed for the low-power products because there are no co-sited base stations with high sensitivity requirements. Some of the blocking requirements are higher for the low-power base stations because interfering UEs can get closer to the base stations. Frequency stability requirements are relaxed for the lower power base stations because there are no high vehicular speeds used in those cells.

The four categories are shown in Figure 6.2. The wide area category applies to all high-power base stations with more than 6 W of output power. The wide area category can also be called as macro base station. The required MCL is 70 dB. The medium range base station, or micro base station, can have up to 6 W of power. The local area base station is limited to 0.25 W and the home base station to 0.10 W power.

6.3 Micro Base Stations

Micro cell and pico cell products are targeted for installations in public areas. The installation is done by the operator and the small cell needs to work smoothly together with the macro cell. We consider some of the important characteristics of the small cell products below:

- The size and the weight of the base station should preferably be small to make the practical installation simple. The size of the complete macro base station is typically more than 50 kg

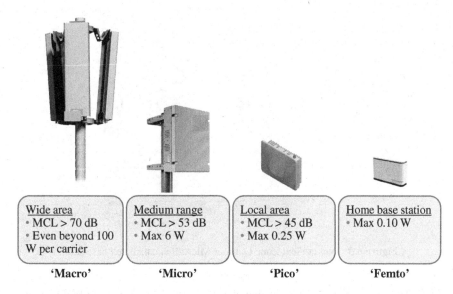

Wide area
- MCL > 70 dB
- Even beyond 100 W per carrier

'Macro'

Medium range
- MCL > 53 dB
- Max 6 W

'Micro'

Local area
- MCL > 45 dB
- Max 0.25 W

'Pico'

Home base station
- Max 0.10 W

'Femto'

Figure 6.2   3GPP base station categories

and 50 litres, while the size of a micro base station can be below 5 kg and 5 litres. It is easier to find a good location for the small cell installation if the product form factor is very small.

- The base station product should be weather proof to allow outdoor installation without any additional casing.
- An integrated antenna is simple from the deployment point of view. On the other hand, an antenna connector allows for using external antennas, which gives more flexibility especially for the outdoor installations.
- The output power needs to be sufficiently high especially in the outdoor installations to provide indoor coverage and dominance area in the presence of a co-channel macro cell. Several watts or even 10 W of power may be needed for the outdoor installations. The power level should also be adjustable so that lower power can be used to limit the interference and the coverage areas. The base station should preferably support an adjustable setting for the transmit power that the base station cannot exceed, in order to eliminate health risks caused by radio frequency emissions in places where users could get close to the base station.
- 2 × 2 MIMO support is needed since long-term evolution (LTE) devices also support 2 × 2.
- Power consumption should preferably be low to keep the base station size, the electricity bill and the running costs low.
- Small cells need to support a large number of simultaneous users to make it possible to use small cells for high-capacity hot spots, like sports stadiums or other mass events.
- Feature parity between small cell and macro cell is preferred. Feature parity means that the same features are supported by the small cells as by the macro cells. Feature parity helps in providing the same end user experience also in the small cells. Feature parity also helps in the operator network planning.
- A synchronization solution is needed in the small cell: frequency synchronization for Release 8 frequency division duplex (FDD) and time synchronization for time division duplex

Figure 6.3   Nokia Flexi Zone Outdoor Micro/Pico small cell base station

(TDD). Time synchronization is needed also for many LTE-Advanced FDD features. The synchronization can utilize satellite-based global positioning system (GPS) or backhaul-based IEEE 1588v2 or Synchronous Ethernet.

- Flexible backhaul options are needed to support different deployment options. The small base station should be able to take benefit of many different types of backhaul solution, such as microwave radio, optical or electrical Ethernet, or passive optical network (PON).
- Self-organizing network (SON) features are important in any new base stations, but especially important in the small cells since the number of small cells can be high and the cost of deployment needs to be low. SON includes self-configuration and self-optimization features.
- Future evolution to multi-band capability, carrier aggregation and LTE-Advanced features is preferred.

An example micro cell product is illustrated in Figure 6.3. The product is targeted for both outdoor and indoor installations. A more detailed list of features is shown in Table 6.1.

6.4   Pico Base Stations

The coverage challenge in in-building premises has accentuated in recent times due to the use of modern building materials that significantly attenuate the radio signals from outdoor cells.

While the micro base station, possibly powered-down, can be used in harsh indoor environments such as train and bus stations, more cost-optimized small cell solutions such as pico base stations are required for enterprise and office environment. At RF transmission levels of local area base station – 250 mW or less – health concerns due to radio frequency emissions are avoided.

Indoor pico base stations typically use office LAN cabling for backhaul, either existing or installed for the purpose. The LAN cabling should be of category CAT5e or better to allow

Table 6.1 Nokia Flexi Zone Outdoor Micro/Pico base station features

Weight	5 kg
Volume	5 litres
Antenna	Integrated omni and directional antenna, connectors for external antennas
Temperature range	−40°C to +55°C
Output power	Max 2 × 5 W
	Min 2 × 0.250 W
MIMO support	2 × 2 MIMO
Power consumption	<100 W typical
Max capacity	840 connected users
Feature parity with macro cell	Yes
Synchronization	GPS, IEEE 1588v2 or Synchronous Ethernet
Backhaul	Microwave radio, Ethernet over fibre or copper, GPON
SON	Yes

for transmission of high LTE data rates. It may be possible to also route the base station traffic through the company LAN switches and routers. Careful testing is, however, necessary as old LAN equipment impair the signal quality, for example, by causing excessive delay variation to synchronization packets.

If the existing LAN infrastructure can be reused, building an indoor system with pico base stations can be more cost efficient when compared to DAS. In addition, a much higher capacity is provided. Subscribers of multiple operators can be served by network-sharing arrangement between operators.

Power over Ethernet (PoE) is a cost-efficient and practical way of providing power to the pico base station, as dedicated power supply at the location of the base station is avoided. Since the power supply of office information technology (IT) infrastructure equipment is typically uninterruptible power supply (UPS)-protected, using PoE also protects the office pico base stations from interruptions of the mains supply.

There are different grades of PoE depending on the required power to be supplied to the powered device: PoE, PoE+ and PoE++. The first two have been standardized by IEEE, while PoE++ standard had not yet been published by the time of writing this text. The characteristics of different versions of PoE are presented in Table 6.2.

Table 6.2 Characteristics of Power over Ethernet versions

IEEE standard	Maximum power available to powered device (W)	LAN cable category	Maximum length of LAN cable (m)
802.3af (PoE)	12.95	CAT3 or higher	100
802.3at (PoE+)	25.5	CAT5 or higher	100
PoE++*	70	CAT5 or higher	100

* Not yet standardized, example values of a commercial product.

Figure 6.4   Nokia Flexi Zone Indoor Pico LTE base station

High-category LAN cables have a lower wire resistance than low-category cables and are therefore also better suited to PoE.

An example pico cell product is illustrated in Figure 6.4. Its main product specifications are shown in Table 6.3.

First dual-mode pico base stations, capable of simultaneously operating in WCDMA/HSPA and LTE mode, were announced in 2014. Products from several vendors were commercially available during 2015.

Table 6.3   Nokia Flexi Zone Indoor Pico LTE base station features

Weight	1.9 kg
Volume	2.8 litres
Temperature range	0°C to +40°C
Output power	Min 2 × 50 mW
	Max 2 × 250 mW
MIMO support	2 × 2 MIMO
Power consumption	<50 W
Max capacity	Up to 400 simultaneous users
Feature parity with macro cell	Yes
Option for integrated Wi-Fi	Yes
Synchronization	GPS, IEEE 1588v2 or Synchronous Ethernet
Backhaul	Ethernet over copper
Power feed	PoE++ or AC/DC adapter
SON	Yes

6.5 Femtocells

Femtocells are used for providing residential or enterprise indoor coverage. Small size and simple installation are important since femtocells are typically installed by the end user without any telecommunications or radio skills. Millions of WCDMA and CDMA femtocells have been installed in homes, and with the rapid proliferation of Wi-Fi-capable smartphones, 3G femtocells are nowadays primarily used to solve the voice coverage problem.

Femtocells are also deployed in small office premises requiring only one or a few access points. Other solutions are usually selected for larger indoor deployments. Femtocells have low output power and limited capacity in terms of supported users, which make them unsuitable for most outdoor installations.

Femtocells, especially residential femtocells, are typically deployed in an uncontrolled manner for cost reasons. Uncontrolled deployment means, for example, that, as the access point is installed by the customer, the exact position of a femtocell in a building or its coverage area relative to neighbouring macro or small cells is out of the control of the mobile network operator. This also means that only some basic network planning measures are applied. Therefore, femtocells, also known as femto access points, typically apply some relatively advanced SON features. Some important femto SON features are

- Automatic setup
- Downlink listen mode to identify the co-channel signal and the macro cells in the area
- Downlink power level setting based on the received signal level from the co-channel macro and small cells and based on the terminal reports
- Periodic monitoring of the surroundings during operation, to detect possible changes

Since residential femtocells are designed to automatically connect to the operator's network once plugged to the public Internet, it is necessary to ensure that the geographical location of the femto access point does not change from the one originally agreed upon between the operator and the customer. This is called location locking. Various methods of location locking have been designed, such as GPS coordinates, surrounding RF map, country-specific IP addressing range, DSL line identity and MAC address of the ADSL router initially assigned to the femto access point.

Another characteristic feature of femtocells is the possibility to restrict the mobile network service access to a closed subscriber group (CSG).

A 3G femto access point is connected to the core network via a femto gateway, located in mobile operator's premises. A femto gateway can typically have thousands of femto access points under its control. The telecom protocol used to connect the 3G femto access point to the gateway is the 3GPP-standardized Iuh protocol. For an LTE femto access point the gateway is optional and it can connect to the evolved packet core (EPC) directly over a standard S1 interface. An example 3G femto access point for office use is shown in Figure 6.5 and its specifications are presented in Table 6.4. An example femto gateway is shown in Figure 6.6. Femtocells are extensively discussed in [1].

In the case of small cells intended for office and enterprise use, the difference between a femto and a pico base station may be blurry. For example, a femto access point product may have a transmit power exceeding the value reserved for the femto category in 3GPP.

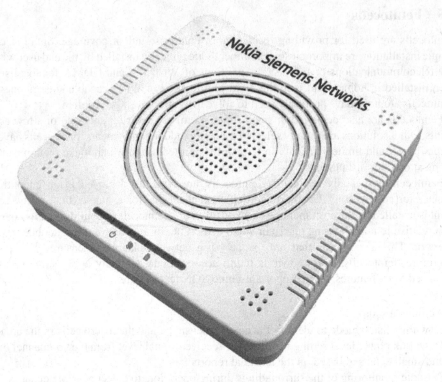

Figure 6.5   3G femto access point

6.6   Low-Power Remote Radio Heads

A low-power remote radio head (RRH) is a product comparable to a micro or pico base station, but one in which the digital baseband part and the RF part are separated. The baseband part typically is that of an ordinary macro base station. The connection between these two – called

Table 6.4   Specifications of a 3G femto access point

Weight	505 g
Volume	1.38 litres
Temperature range	0°C to +40°C
Output power	Max 100 or 250 mW
Peak data rate	HSDPA 21 Mbps and HSUPA 5.76 Mbps
Power consumption	<13.5 W
Max capacity	Up to 16 simultaneous users
Feature parity with macro cell	No
Option for integrated Wi-Fi	No
Synchronization	Network time protocol or over the air from macro downlink
Backhaul	Ethernet over copper
Power feed	PoE+ or AC/DC adapter
SON	Yes

Figure 6.6 Femto gateway

Figure 6.7 Nokia Flexi Metro remote radio head (RRH)

fronthaul – usually requires optical fibre due to the high bandwidth and tight delay constraints of the baseband signal.

An example low-power RRH product is illustrated in Figure 6.7. A more detailed list of features is shown in Table 6.5.

Table 6.5 Nokia Flexi Metro RRH specifications

Weight	5 kg
Volume	5 litres
Temperature range	–40°C to +55°C
Output power	Min 2 × 50 mW
	Max 2 × 5 W
Antennas	Integrated and external
MIMO support	2 × 2 MIMO
Power consumption	<100 W
Max capacity	840 connected users
Feature parity with macro cell	Yes
Synchronization	GPS, IEEE 1588v2 or SyncE
Fronthaul	OBSAI over fibre
SON	Yes

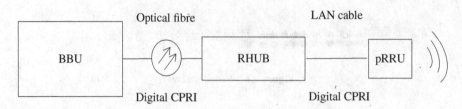

Figure 6.8   Lampsite architecture

6.6.1   *Alternative Remote Radio Head Designs for Indoor Use*

For indoor use, two alternative RRH designs are discussed in the following paragraphs. Both utilize the combination of optical fibre and LAN cabling. Optical fibre facilitates wide reach in large buildings or campuses, while the use of LAN cabling for the transmission and PoE in the last section is simple and cost efficient. Both systems lend themselves to medium to large indoor deployments, but may be too costly in smaller indoor environments compared to pico BTSs, for example.

The first, Huawei Lampsite, is a three-stage solution composed of a baseband unit (BBU), a remote radio unit hub (RHUB) and pico remote radio units (pRRUs). The BBU may be connected to several RHUBs and each RHUB may be connected to several pRRUs.

The Lampsite architecture is presented in Figure 6.8. The baseband signal is transported over an optical fibre to the remote radio hub. This node converts the digital optical signal into a digital electrical signal for transmission over a LAN cable to the pico RRH. The RHUB also provides PoE feed to the pRRUs.

The pRRUs can be equipped with various radio access technology (RAT)-specific modules to support multiple radio technologies simultaneously: WCDMA, LTE and Wi-Fi.

The second alternative, presented in Figure 6.9, is the Dot by Ericsson. It is also a three-stage solution composed of a digital unit (DU), an indoor radio unit (IRU) and a RRH called Dot.

The DU is the baseband part of a macro base station. The digital baseband signal (CPRI) is fed to IRU over an optical fibre. The IRU converts the digital optical baseband signal into an electrical analogue signal, which is then fed to the RRH over a LAN cable. The IRU also supports PoE functionality for powering the RRHs.

One benefit of the design is that the RRH can be made small and lightweight.

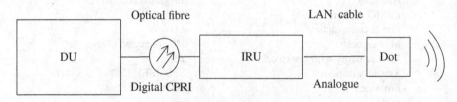

Figure 6.9   Dot architecture

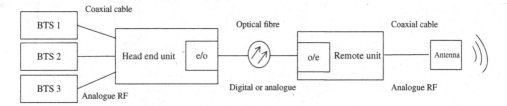

Figure 6.10 Active DAS architecture (e/o=electric to optical. o/e=optical to electric)

6.7 Distributed Antenna Systems

In a DAS the RF signal of one or more base stations is distributed to a number of remote antennas in order to cover a pre-defined indoor or (less often) an outdoor area. There are two main categories of DAS: passive DAS and active DAS.

In a passive DAS, a coaxial feeder cable is used for signal distribution along with passive RF components such as splitters and couplers. Being all passive, the system is reliable and has a moderate cost of materials. On the other hand, the use of coaxial cable brings two potential drawbacks: First, the system size is limited by the combined attenuation of the feeder itself and the required power splitting components. Second, installation of thick feeder is labour intensive and expensive.

An active DAS combines optical and coaxial transport in order to reduce cabling cost and to extend the reach of the system. A schematic of an active DAS architecture is presented in Figure 6.10. The RF signal from one or more base stations (or other sources such as repeater) is fed to a head end unit which conditions the RF signals to an appropriate power level and combines them for transmission over optical fibre. The signal is then either directly modulated to an analogue optical signal or, alternatively, converted to a digital signal before the electrical-to-optical conversion. Uplink and downlink signals can be carried over a single fibre by using wavelength division multiplexing (WDM). The remote unit performs optical-to-electrical, and optionally digital-to-analogue, conversion. In a typical office building installation, the optical part would be installed in the riser while thin coaxial cable would be used in the horizontal runs to the antennas.

Active DAS systems can incorporate multiple operators, RATs and frequency bands in the same system. They are typically used for building large neutral host systems in office buildings or public venues.

6.8 Wi-Fi Integration

Where the mobile broadband network is overloaded, WLAN, or Wi-Fi, provides an opportunity for relieving the congestion. Traffic steering between the cellular and Wi-Fi networks is possible through access network discovery and selection function (ANDSF). It is a 3GPP-defined entity in the core network that assists the terminal in discovering Wi-Fi networks and defining when and where to connect to Wi-Fi, as well as in setting priorities between Wi-Fi networks and the cellular network. ANDSF settings are sent over the air to the ANDSF client in the mobile terminal.

Figure 6.11   Ruckus ZoneFlex R700 Wi-Fi access point

Hotspot 2.0 is a Wi-Fi alliance-defined method for a (Hotspot 2.0 capable) terminal to select the most suitable Wi-Fi access point out of several candidates, based on information derived from the access point prior to attaching to it. Unlike ANDSF, Hotspot 2.0 can provide information of the quality of Wi-Fi networks.

802.11ac is the latest evolution of the 802.11 series of Wi-Fi standards. It enables improved bitrates in the 5 GHz band thanks to, for example, a wider RF bandwidth (up to 160 MHz), more spatial streams (up to 8) and a higher order modulation (up to 256 QAM).

An example Wi-Fi access point with 802.11ac capability is presented in Figure 6.11 and its main technical specifications are summarized in Table 6.6.

Table 6.6   Ruckus ZoneFlex R700 specifications

Weight	1 kg
Volume	2 litres
Temperature range	0–50°C
Wi-Fi standard	IEEE 802.11 a/b/g/n/ac; 2.4 and 5 GHz concurrent operation
Data rates on the physical layer	Up to 450 Mbps (2.4 GHz) Up to 1300 Mbps (5 GHz)
Max transmit power (country-specific rules apply)	29 dBm on 2.4 GHz; 27 dBm on 5 GHz
Antenna	Adaptive antenna array with >3000 unique antenna patterns; physical gain 3 dBi
Supported clients	500
Basic service set identifications	Up to 32 (2.4 GHz) Up to 8 (5 GHz)
Power consumption	7 W typical
Power input	PoE 802.3af or AC/DC adapter

Figure 6.12   DragonWave Harmony Radio Lite microwave radio for wireless backhaul

As shown earlier, some new 3GPP small cell products have an integrated Wi-Fi 802.11ac capability.

6.9   Wireless Backhaul Products

Wireless solution is a fast approach for small cell base station backhaul. Microwave radios are commonly used for backhauling macro base stations. Those point-to-point microwave links use frequencies between 6 and 38 GHz. Such high frequencies usually need line-of-sight (LOS) connection. Small cells preferably utilize near- or non-LOS backhaul solution

Table 6.7   DragonWave Harmony Radio Lite microwave radio features

RF bands	2.3–2.7 GHz
	3.5 GHz
	5 GHz
Duplexing	TDD
Interface	2 × Fast Ethernet/Gigabit Ethernet
Antennas	2 × 2 MIMO
Channel spacing	5/10/20/40 MHz
Throughput	Up to 230 Mbps
Powering	Via Ethernet or via AC/DC
Power consumption	<15 W
Size	19 cm × 19 cm × 5 cm
Antenna	Integrated
Weather protection	Outdoor capable
Synchronization support	Synchronous Ethernet and IEEE 1588v2

since it is difficult to organize LOS from street level small cell to the closest aggregation point. An example small cell backhaul product, DragonWave Harmony Radio Lite, is shown in Figure 6.12 and described in more detail in Table 6.7. The product is a point-to-point microwave solution operating in sub-6 GHz frequency bands and optimized for micro cell backhaul. The product can operate at 5, 3.5 and 2.3–2.7 GHz bands in TDD mode. The low band enables non-LOS operation. The product can use licensed or unlicensed spectrum. The unlicensed spectrum facilitates fast and easy deployment and savings on the license fee. DragonWave Harmony Radio Lite is outdoor capable, has an integrated antenna and has a standard Ethernet interface. It provides synchronous Ethernet and IEEE 1588v2 support for the base station frequency and time synchronization.

6.10 Summary

This chapter introduced small cell product categories including all-in-one micro and pico cells for outdoor and indoor public premises as well as offices, low-power RF heads with high-capacity front haul for outdoor and indoor use and femtocells for homes and small office premises. New small cell solutions will take market share from traditional passive and active DASs. Wi-Fi small cells provide a means of offloading traffic from mobile networks in homes, offices and public premises.

Small cell base station products offer high capacity with an output power ranging from tens of milliwatts to a few watts in very compact packages of a few litres in volume and a few kilograms in weight. Wireless backhaul products with near-LOS or non-LOS capability were also presented offering capacity of more than 200 Mbps in sub-6 GHz bands. 3GPP specifications support different base station categories by offering separate RF requirements for the low-power small base stations.

Reference

[1] H. Holma and A. Toskala, *WCDMA for UMTS*, 5th ed., Chapter 19, John Wiley & Sons (2011).

7

Small Cell Interference Management

Rajeev Agrawal, Anand Bedekar, Harri Holma,
Suresh Kalyanasundaram, Klaus Pedersen and Beatriz Soret

7.1 Introduction

Interference is an old acquaintance with new significance when migrating from carefully planned macro-only deployments to HetNet deployments with denser and denser small cell installations. The deployment of small cells such as pico cells and micro cells poses new challenges. The use of interference mitigation techniques, therefore, plays an important role when deploying small cells to fully maximize the benefits of such investments. Interference mitigation comes in many different forms as summarized in Table 7.1, where the main options offered for LTE are listed. On the network side, neighbouring interfering cells can use different forms of resource partitioning to reduce co-channel interference. Among such options, higher order sectorization and coordinated beamforming implement resource partitioning in the spatial domain and are typically considered feasible for macro eNodeBs that can be equipped with the required number of antennas to enable such solutions. Resource partitioning in the time and/or frequency domain is another option, where especially coordinated time-domain muting

LTE Small Cell Optimization: 3GPP Evolution to Release 13, First Edition.
Edited by Harri Holma, Antti Toskala and Jussi Reunanen.
© 2016 John Wiley & Sons, Ltd. Published 2016 by John Wiley & Sons, Ltd.

Table 7.1   3GPP options for downlink interference mitigation

Network-based resource partitioning	Spatial-domain resource partitioning	Spatial filtering techniques such as higher order sectorization and coordinated beamforming
	Time-domain resource partitioning	Also known as coordinated muting. Examples include 3GPP-defined techniques such as eICIC and (e)CoMP
	Frequency-domain resource partitioning	The frequency-domain resource partitioning can be done on PRB resolution or on carrier resolution if having networks with multiple carriers. Includes options such as using hard or soft frequency reuse between neighbouring cells
Network-based transmit power control	Transmit power control per cell	Adjustment of transmit power per cell to improve the interference conditions. Includes 3GPP-defined femtocell transmit power calibration to reduce interference towards co-channel macro-users
UE-based interference mitigation	Linear interference suppression	Interference suppression by means of linear combining of received signals at the UE antennas, for example, interference rejection combining (IRC)
	Non-linear interference suppression	The UE estimates and reconstructs the interfering signal(s) followed by subtraction before decoding the desired signal. Also includes elements of network assistance for some cases

between macro and/or small cells has gained significant interest and is used for enhanced inter-cell interference coordination (eICIC) and coordinated multipoint (CoMP) or enhanced CoMP (eCoMP) as will be explained in greater details in this chapter. It is furthermore worth noticing that frequency-domain inter-cell interference management on a physical resource block (PRB) resolution can be realized with channel-aware frequency-selective scheduler (FSS) without any explicit inter-eNodeB coordination. Downlink eNodeB transmit power control is another apparatus which is used for closed subscriber group (CSG) femtocells to reduce the generated interference from such cells towards nearby users that are not allowed access, that is, users that are not having a matching CSG cell identity. As will be discussed in greater details in the forth-coming sections, the full potential of the network-based interference coordination options is unleashed by conducting fast inter-cell coordination in coherence with the time-variant traffic fluctuations in the network.

Furthermore, the UE also offers powerful mechanisms for handling interference. UEs with more than one antenna ($M > 1$) can exploit linear interference suppression techniques such as minimum mean square error interference rejection combining (MMSE-IRC) as standardized in Release 11 in the form of UE performance requirements for $M = 2$. In theory, a UE with M antennas have $N = M - 1$ degrees of freedom that can be used for either suppressing N interfering streams or diversity. Another variant of UE-based interference mitigation is to apply non-linear interference cancellation, where the UE estimates and reconstructs the interfering signal(s) followed by subtraction before decoding the desired signal. Such techniques are especially attractive for cancelling interference from semi-static signals such a common reference signal (CRS), broadcast channel and synchronization channels, as supported to a large extent in Release 11 further enhanced ICIC (feICIC). Applying non-linear interference cancellation to data channel transmissions is much more challenging, as the scheduling and link adaptation are highly dynamic, and conducted independently per cell. Hence, getting the most out of non-linear interference cancellation often includes additional network assistance. One such example is network assistance interference cancellation and suppression (NAICS) as included in Release 12 in the form of symbol-level interference cancellation (SLIC). In short, the SLIC implies that UE estimates its dominant interfering cell for each symbol and reconstructs the interfering signal(s) followed by subtraction before decoding the desired signal. The network supplies assistance information such as number of antenna ports and other characteristics of the interfering cells, so the UE can avoid blind estimation of such metrics.

The network-based interference coordination options and the UE-based interference mitigation techniques are naturally not mutually exclusive, but can be exploited jointly. One example of joint network- and UE-based interference mitigation is eICIC as is explained in greater details in Section 7.3. For further information, see also [1].

Although the interference mitigation options in Table 7.1 are outlined for the downlink, similar techniques are to a large extent also applicable for the uplink. However, for the uplink, the user-specific transmit power control mechanism is the most dominant tool for managing co-channel inter-cell interference.

The rest of the chapter is organized as follows: The packet scheduling solutions for interference mitigation are shown in Section 7.2, eICIC in Section 7.3, eCoMP in Section 7.4, CoMP in Section 7.5 and the summary is presented in Section 7.6.

7.2   Packet Scheduling Solutions

3GPP has standardized the interfaces very carefully but not the network-side algorithms. The radio resource management (RRM) algorithms in eNodeB can be developed and optimized by the network vendors together with the mobile operators. Therefore, there may be differences in the algorithms between the different LTE networks globally. The packet scheduler is the key part of LTE RRM algorithms. The packet scheduler has more flexibility in LTE than in 2G or 3G networks since the scheduling optimization can be done in multiple domains: frequency domain, time domain, power domain and spatial antenna domain. The frequency-domain scheduling is a powerful solution for avoiding interfered PRBs and faded PRBs. The frequency-domain scheduling was not possible in CDMA-based systems where the signal is always spread over the whole transmission bandwidth. LTE technology uses frequency-domain transmission both in downlink and in uplink which enables frequency-domain scheduling. The time-domain scheduling is utilized in eICIC and it can also be utilized for quality of service (QoS) differentiation by transmitting high-priority packets first before low-priority packets. The power domain can use power control and can also change the relative power levels of the adjacent cells for the intra-frequency load balancing. The scheduling can also be done in spatial domain with antennas by using CoMP or beamforming. The different scheduling domains are illustrated in Figure 7.1.

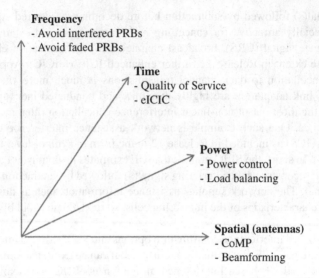

Figure 7.1   LTE packet scheduler can operate in multiple domains

One of the main objectives of LTE scheduler is to avoid inter-cell interference at the cell edge. The cell edge data rate is heavily affected by the interference from the adjacent cells due to frequency reuse. If the scheduler can avoid inter-cell interference, the cell edge data rates can be improved (see Figure 7.2). That is one of the main motivations for developing advanced packet scheduling solution.

The FSS is the main component in the packet scheduler, and we will analyse the FSS solution in more detail. First of all, the scheduler needs to obtain information about the amount of inter-cell interference in the frequency domain. The channel quality indicator (CQI) reporting from UEs can be configured in the frequency domain so that UE reports CQI value, for example, for

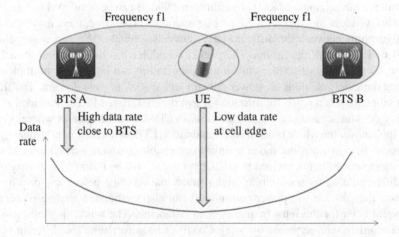

Figure 7.2   Inter-cell interference impacts cell edge throughputs

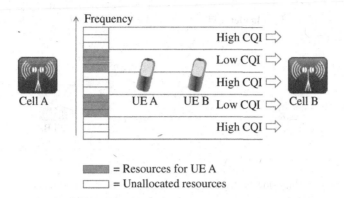

Figure 7.3   Obtaining inter-cell interference information with CQI feedback

every six PRBs. Low CQI value indicates that UE receives high inter-cell interference while high CQI value indicates that UE receives only low inter-cell interference. CQI reporting also provides information about signal fading. The CQI reporting is illustrated in Figure 7.3. Cell A uses part of its PRBs for the data transmission to UE A while part of its PRBs are empty. The adjacent Cell B schedules data for UE B which receives some inter-cell interference from Cell A. UE B provides CQI reports to Cell B and these reports provide information about inter-cell interference in the frequency domain.

The CQI reports can be utilized by Cell B in FSS solution in Figure 7.4. Assume that there are two UEs in Cell B: UE B at the cell edge experiencing inter-cell interference from Cell A and UE C which is located close to Cell B and experiencing no inter-cell interference. Cell B can schedule data to UE B in those PRBs where the inter-cell interference is low. The other PRBs can be utilized for UE C because UE C does not suffer from any inter-cell interference. The FSS solution increases both cell edge data rate for UE B and the total cell capacity for Cell B.

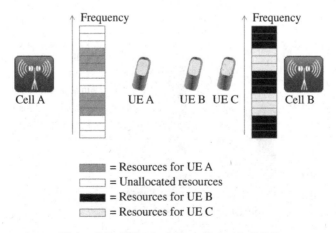

Figure 7.4   FSS scheduling solution in Cell B

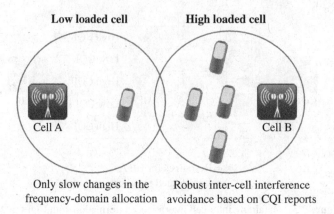

Figure 7.5 Interference shaping can improve the quality of CQI reporting

The quality of CQI reporting is important for the FSS performance. If the quality of CQI reporting can be improved, the FSS benefit would be higher. If Cell A in Figure 7.4 changes the allocations in the frequency domain rapidly, the CQI reporting will be delayed in Cell B and Cell A has already changed the resource allocation before Cell B had time to avoid inter-cell interference from Cell A. The combined CQI reporting and packet scheduling delay is typically 10 ms. One example solution for improving the CQI performance is called interference shaping (Figure 7.5). The idea is that the low loaded cells do not need the gain from FSS. Therefore, it is possible to keep the allocations in the frequency domain constant for tens of milliseconds in low loaded cells and avoid too fast changes. That approach improves the quality of CQI reporting in the adjacent high loaded cells. The principle is shown in Figure 7.5.

The benefit of FSS can be evaluated in the simulations and in the field measurements. Figure 7.6 shows field test results with stationary UEs in different cell edge locations. There

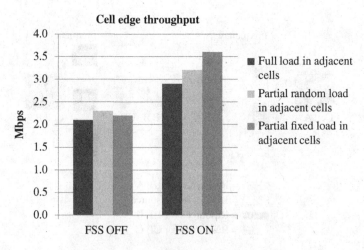

Figure 7.6 Field measurements with FSS

were in total four UEs per cell in the measurements in the multi-cell environment with 10 MHz of bandwidth and proportional fair scheduling. Three different cases were studied in terms of adjacent cell loading: full load in adjacent cells, partial load in adjacent cells with random allocation and partial load in adjacent cells with fixed allocations. FSS OFF case uses wideband CQI reporting which gives no information about the interference or fading in the frequency domain. FSS ON case uses frequency-selective CQI reporting. FSS showed 40–65% improvement in the cell edge throughput depending on the case. The highest gain was obtained with partial load and fixed allocations because CQI reporting can accurately reflect the inter-cell interference. FSS can also provide some gain with the full loading in the adjacent cell because of frequency-dependent fast fading. This FSS gain would be lower for the moving UEs since CQI reporting would not be fast enough to follow the fast fading.

7.3   Enhanced Inter-cell Interference Coordination

7.3.1   Concept Description

The small cell deployment on the same frequency with macro cell requires efficient interference management. The challenge is that the small cell coverage area can be very small in a co-channel deployment due to the difference in the transmission power between the macro cell and the pico cell. If there would be no co-channel macro cell, the pico cell could provide service to larger coverage area. The challenge is illustrated in Figure 7.7.

The target is to increase the pico cell coverage area in the case of co-channel deployment. One option is to coordinate the interference in the time domain, which is defined in 3GPP Release 10 and called enhanced inter-cell interference coordination (eICIC). The solution is shown in Figure 7.8. The macro cell stops its transmission in some subframes to minimize interference to the pico cells. Those muted subframes are called almost blank subframes (ABS). Only cell-specific reference signals and other mandatory signalling channels including physical control format indicator channel (PCFICH), synchronization channels and paging channel are transmitted during ABS. When the interference from the macro cell is reduced during ABS, the pico cell is able to serve UEs which are further away from the pico cell. The ABS muting pattern is periodical with 40 subframes for FDD mode, while the value in

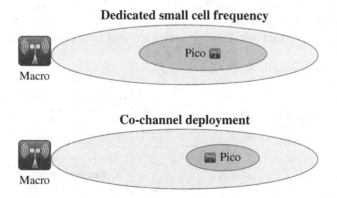

Figure 7.7   Small cell coverage area with dedicated frequency and co-channel deployment

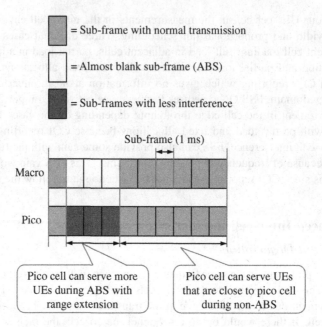

Figure 7.8   eICIC interference coordination in time domain

TDD mode depends on the uplink/downlink configuration. The periodicity of 40 subframes for FDD has been selected to maximize the protection of common channels, including uplink hybrid automatic repeat request (HARQ) performance. For maximum benefit from eICIC, eNodeBs in a given local area are recommended to use overlapping and coordinated ABS muting patterns. The inter-eNodeB X2 application protocol offers support for adjusting the ABS muting pattern [2]. However, the process of adjustment can be too slow as compared to the rapid fluctuations of the traffic. For those cases more dynamic solutions will be discussed in Section 'Fast Dynamic eICIC'.

When the macro cell uses ABS, the pico users are obviously exposed to much less interference, which allows the pico cell to serve users from a larger geographical area as would otherwise have been feasible. This is illustrated in Figure 7.9, where the pico cell uses a large value of the so-called range extension (RE) to increase its service area. Hence, the use of ABS at the macro cell offers enhanced opportunities for load balancing between the macro and pico cell.

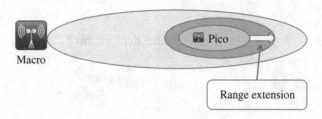

Figure 7.9   Range extension with ABS

In order to get accurate interference measurements to eNodeB, the UE feedback reporting is modified with eICIC. The channel state information (CSI) feedback includes CQI, rank indicator (RI) and precoding matrix indicator (PMI). CQI is used for the link adaptation and the packet scheduling. eICIC causes fast changes in the interference levels by switching on and off the macro cell transmission. Release 10 UE can be configured to perform time-domain-restricted measurements. The principle is that small cell UEs are requested to provide two separate CQI reports: one corresponding to the normal subframes and another one corresponding to ABS. Similarly, the network can also configure time-domain restrictions for RRM measurements such as reference symbol received quality (RSRQ). The latter can, for instance, be used for macro-UEs to make more accurate handover to pico, such that the RSRQ on the pico is measured only during subframes where the macro is using ABS. The latter offers simple, yet efficient, interference-aware load balancing between the macro and pico cells as also illustrated in [3]. Finally, time-domain measurement restrictions can also be configured for radio link monitoring (RLM). As an example, configuration of RLM measurement restrictions is useful for pico users that are typically able to receive service during subframes where the macro uses ABS. Without configuration of RLM measurement restrictions, such pico users would risk triggering undesirable radio link failures. All of the aforementioned measurement restrictions for Release 10 UEs are configured with radio resource control (RRC) messages as specified and are therefore only applicable for connected mode UEs.

We can note that Release 8 and 9 UEs do not support measurement restrictions and the estimated interference is averaged in time domain across non-ABS and ABS subframes. The CQI reported by legacy UEs and by Release 10 UEs is shown in Figure 7.10.

CRS is also transmitted from the macro cell during ABS causing some interference to the pico cell UEs. The CRS transmission power is approximately 9% of the total eNodeB power with 2×2 MIMO configuration. As the CRS transmission is a constant deterministic sequence for each cell, it is possible for the UE to estimate the CRS interference from strongly interfering cells and remove the CRS interference by non-linear interference cancellation (IC). This kind of interference cancellation is included in Release 11 and called further enhanced ICIC (feICIC).

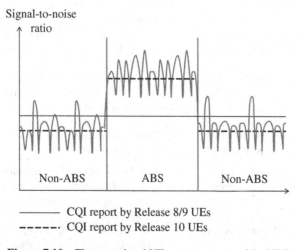

Figure 7.10   Time-restricted UE measurements with eICIC

The interference cancellation capability is typically included together with eICIC in the chip sets. This essentially means that small cell UEs with CRS IC ideally experience zero macro interference when ABS is used. In order to reduce UE implementation complexity, network can provide the UE a list of CRS IC assistance information for the cells where the UE may likely perform CRS IC. The CRS IC assistance information list includes data such as the number of antenna ports, physical cell identity and multicast broadcast single frequency network (MBSFN) subframe configuration of the different cells. Thus, the UE can avoid blind estimation of those parameters. The feICIC CRS IC functionality is a mandatory capability for both FDD and TDD Release 11 UEs. Moreover, UEs could have physical broadcast channel (PBCH) interference cancellation (IC) capability or an implementation with equivalent or better demodulation performance than PBCH IC.

Furthermore, since cells are assumed to be time synchronized for eICIC, collisions of system information block one (SIB1) can cause problems for the pico users in the cell range extended to correctly receive the SIB1. In order to address the problem, the network can send the SIB1 via dedicated signalling to the pico user. This allows sending the SIB1 while the macro is using ABS. Table 7.2 summarizes the benefits, requirements and characteristics for Release 10 eICIC and Release 11 feICIC. Although not discussed here, it is worth noticing from Table 7.2

Table 7.2 Summary of LTE eICIC benefits, requirements and characteristics

Release 10 eICIC	eICIC benefits	Increased offload to pico eNodeB via reduced co-channel macro interference.
	ABS characteristics	ABS are subframes with reduced transmit power (including no transmission) on some physical channels and/or reduced activity. The eNodeB ensures backward compatibility towards UEs by transmitting necessary control channels and physical signals, as well as system information
	Time synchronization	Requires time synchronization between base stations. Several implementation alternatives; GPS and backhaul-based solutions such as IEEE 1588v2
	ABS muting pattern configuration	Distributed dynamic coordination of ABS muting pattern between macro and pico via standardized X2 signalling
	Network RRM	eNodeB RRM functions such as the packet scheduler shall ensure that UEs requiring protection from aggressor nodes are only scheduled when these use ABS
	Release 10 UE requirements	Configuration of time-domain-restricted RLM, RRM and CSI measurements
Release 11 further eICIC (feICIC)	Release 11 UE requirements	Pico–UE interference cancellation (IC) to handle residual interference from macro-ABS transmission. Signalling of IC assistance information from eNodeB to UE
	SIB1 provisioning	SIB1 via dedicated signalling to a connected mode pico-UE in cell range extended region

that eICIC/feICIC also offers benefits for scenarios with co-channel deployment of macro and CSG home base stations (HeNodeB), as also explained in greater details in [2].

7.3.2 Performance and Algorithms

Basic eICIC Performance Considerations

The performance gain offered by (f)eICIC depends on multiple factors, where the primary performance determining factors are summarized in Figure 7.11. First, efficient eICIC operation requires accurate setting of macro cell's ABS muting pattern and load balancing. Second, the gain of eICIC naturally depends on the density of pico cells and their location. For more details, see [4,5]. In general, the more picos that benefit the macro cell using ABS, the higher is the gain from eICIC. Furthermore, the gain from eICIC is generally highest for the cases where both macro and pico cells are outdoor; the gain is reduced for scenarios where the pico cells are indoor. The reason is that the indoor users tend to be served by the indoor pico cells, while the interference from the macro layer is suppressed by the in-building penetration loss. The third factor that impacts the eICIC performance is the UE capabilities, resulting in the highest gain for Release 11 UEs, while worthwhile gains are also achievable for UEs of earlier releases [6]. Finally, eICIC naturally starts to offer benefits when the offered traffic load raises to the point where inter-cell interference between macro and pico cells starts to be a performance limiting factor [2].

The gain eICIC mechanism is further illustrated in Figure 7.12, where it is pictured how multiple picos benefit from each subframe where the macro uses ABS. This represents an attractive net benefit as the aggregated performance gain of the picos exceeds the loss from muting at the macro. Therefore, we can expect that eICIC provides more gains when the number of pico cells increases. Furthermore, as the use of muting at the macro allows offloading more users to the picos, eICIC also offers a net benefit for the remaining macro-users, given that the fraction of offloaded users from applying eICIC exceeds the percentage of subframes

---

ABS adaptation and load balancing

Fast and accurate ABS muting pattern adjustment and dynamic load balancing maximize the benefits of eICIC

Small cell placement and density

Highest eICIC gain observed for cases with outdoor dense small cell deployment. Moderate eICIC gain for cases with indoor cells

---

Terminal support

Moderate gain for pre-Release 10 UEs. Clear gain for Release 10 UEs and even higher gain for Release 11 UEs

---

Offered traffic load

Hardly any gain of eICIC at low offered load. The gain of eICIC increases with offered load. Highest gain achieved at full load condition (interference limited)

Figure 7.11   Summary of eICIC performance determining factors

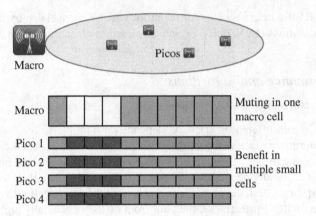

Figure 7.12 eICIC gain in case of multiple small cells

configured as ABS. The former illustrates the importance of coherent load balancing (e.g. via pico cell RE setting) and ABS configuration at the macros to maximize the benefits of applying eICIC.

Figures 7.13 and 7.14 show the performance with/without eICIC in terms of the 5-percentile and 50-percentile end-user-experienced throughput versus the offered load per macro cell area. These performance results are obtained for the 3GPP Release 12 small cell scenario 1 as defined in [7]. The scenario consists of a regular grid of three-sector macro cells with 500 meters inter-site distance and random placement of small cell clusters within each macro cell

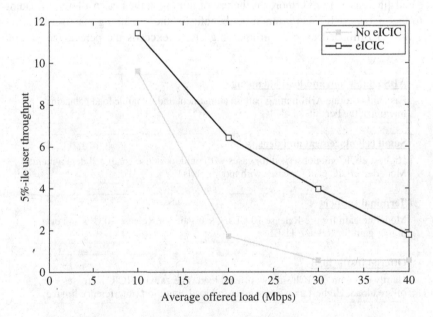

Figure 7.13 5-percentile end-user throughput with and without eICIC

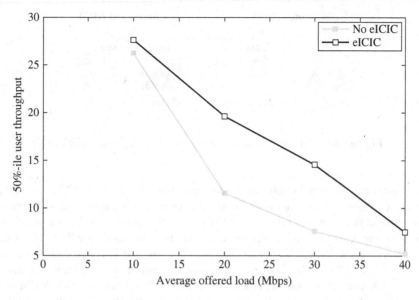

Figure 7.14   50-percentile (median) end-user throughput with and without eICIC

area. One cluster of four pico cells is assumed for each macro cell area. The radio propagation follows the ITU Urban Macro (UMa) and Urban Micro (UMi) models, which include different characteristics for line-of-sight (LOS) and non-LOS (NLOS). The transmit power for the macro cells equals 46 dBm, while the picos transmit with 30 dBm. A dynamic birth–death traffic model is used, where new calls are generated according to a Poisson arrival process. The payload for each user equals 4 Mbit, and once it has been successfully delivered to the user, the call is terminated. The UEs are Release 11 compliant, supporting both CRS IC and configuration of RRM and CSI measurement restrictions. Transmission mode (TM) 4 is assumed with 2×2 SU-MIMO, including dynamic rank adaptation.

From the results in Figures 7.13 and 7.14 it is clearly observed that by enabling eICIC a much higher offered traffic can be tolerated with the same or better end-user throughput. Looking at Figure 7.13, it is observed that the offered traffic can be increased from 20 to 40 Mbps while still having a 5-percentile outage of 2 Mbps, thus corresponding to a 100% capacity gain from enabling eICIC. We need to note that the setting of pico cell RE and macro cell ABS muting is semi-static in these results, meaning that it is only adjusted as a function of the average offered traffic and is the same per layer. At low offered load, the results with and without eICIC are similar indicating that ABS is not used, and only moderate values of the pico cell RE are used. As the offered traffic is increased, it is observed that both the value of the pico cell RE and the usage of ABS increase to cope with the increased inter-layer interference.

Fast Dynamic eICIC

The results above illustrated that there is a need to adjust the number of ABS and the RE parameters dynamically according to the instantaneous UE locations and traffic patterns. We

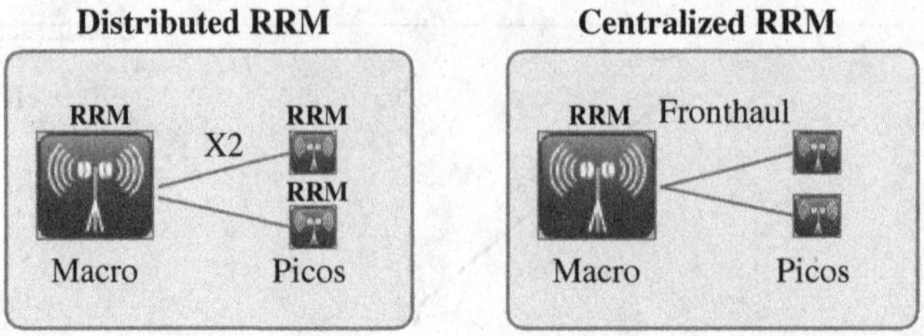

Distributed RRM   **Centralized RRM**

Figure 7.15   Architectures for HetNet radio resource management (RRM)

will next consider network architectures and algorithms for dynamic eICIC configurations. The eICIC configurations are optimized jointly between macro and pico cells: Macro cell has the load information over multiple pico cells while the pico cell has the local load information. The architecture options from HetNet eICIC control is shown in Figure 7.15. The distributed solution has full RRM functionality in each pico cell and the inter-eNodeB coordination is done with X2 signalling. The distributed coordination can be slow (several seconds) or fast (milliseconds). The centralized approach uses baseband hotel where all the RRM functions including packet scheduling are located in the macro cell. The pico cell is just RF head connected with fibre to the baseband hotel.

The majority of existing eICIC studies have focused on slow distributed RRM where a number of small cells in the form of pico cells are deployed in the macro coverage area, that is, using self-organizing network (SON) type of algorithms based on long-term collection of the spatio-temporal load of network. Release 10 provides a mechanism to coordinate and exchange information of the eICIC configuration between eNodeBs, which is typically updated on a time scale of several seconds. It is a semi-static configuration. The classical distributed architecture with slow update frequency is used as the baseline reference in our analysis. We will show that a tighter inter-cell coordination can increase the performance of the network. It is possible to make fast coordination with distributed RRM by enhanced information exchange over X2 [8]. The fast coordination can also be obtained in the centralized RRM solution. In all these cases the RRM algorithms for both the macro and pico cells are implemented in the macro eNodeB.

Slow and fast ABS adaptation are illustrated in Figure 7.16. Slow adaptation uses normal subframes and mandatory ABS subframes where the macro cell is not allowed to transmit data. The number of ABS subframes is changed only with X2 signalling. Fast ABS uses additionally optional ABS subframes, which can be used for the macro cell or the pico cell transmission. The decision can be done on the frame level by the macro cell. The fast RRM gives more flexibility to adapt the resource allocation to the instantaneous capacity requirements.

UE can report CQI only during the normal subframes and during mandatory ABS subframes, but not during optional ABS subframes. All subframes cannot be optional ABS since some normal subframes and some mandatory ABS subframes are needed for UE CQI reporting. The UE CQI measurement configuration with fast ABS adaptation is illustrated in Figure 7.17. It is also implied that the small cell decides which CQI – ABS or normal – shall be used for the link adaptation and packet scheduling decisions for each subframe depending on whether the macro is using ABS or normal transmission.

In case of centralized architecture, all the necessary information is available at the macro cell, including CQI for all UEs in the cluster, instantaneous load information for all cells in

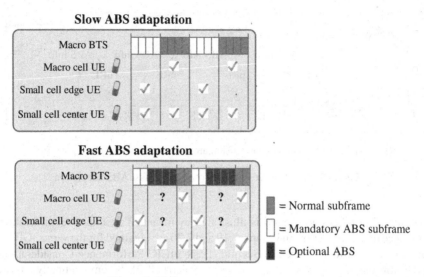

Figure 7.16 Slow and fast ABS adaptation

the cluster and scheduling decision and related metrics. The fast decisions in the centralized architecture are made shortly before each optional ABS on whether to configure as ABS or normal transmission. We will show that the algorithms can also be decomposed to be applicable for the distributed architecture. The macro cell acts as the master for fast ABS decisions based on the information exchange with pico cells over the X2 interface. The macro cell acquires knowledge about the load in the pico cell. The rate of fast ABS adaptation for the distributed architecture is therefore dependent on how frequent the aforementioned information is exchanged between cells as well as the X2 signalling delays. One option is to have the pico cells periodically reporting the required information to the macro cell every N Transmission Time Interval (TTI). Another option is event-triggered reporting where the pico cells inform the macro whenever there is a significant change in the load or in proportional fair scheduling metric. The periodic reporting is shown in Figure 7.18. The pico cell informs the macro cell of the number of RE users in every N TTIs. Macro cell can then decide the ABS configuration and provide the information to the pico cell. Notice that the X2 reporting is independent of the number of optional ABS per period and there is no signalling overhead in the air interface.

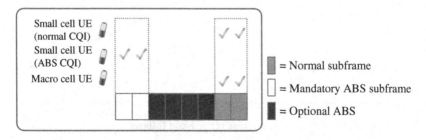

Figure 7.17 UE measurements with ABS adaptation

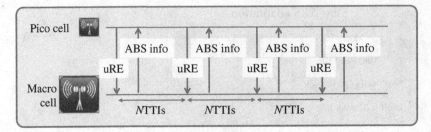

uRE = number of UEs in range extension area

Figure 7.18 Periodic reporting over X2 for fast ABS adaptation

The gain of fast ABS adaptation is illustrated by means of system level simulations. The performance results presented in Figures 7.19 and 7.20 are obtained for the same 3GPP Release 12 small cell scenario 1 as used in Section 'Basic eICIC Performance Considerations'. For both the slow and fast ABS adaptation cases, the pico cell RE is semi-statically adjusted as a function of the average offered traffic. It is clearly observed how the application of fast ABS offers additional benefits in both the 5-percentile and 50-percentile end-user throughput. Thus, using fast ABS adjustment algorithms further boosts the performance of eICIC. The added gain from fast ABS adjustment is on the order of 30–50% as compared to the slow semi-static adaptation of the ABS muting pattern. The gain from fast ABS adjustment is simply a result of more accurately adapting the macro cells muting pattern in coherence with local experienced traffic fluctuations. Additional results and insight on fast ABS adjustment are available in [8], [13] and [14].

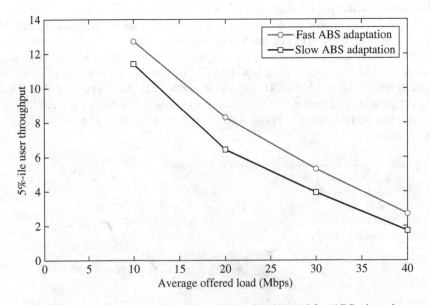

Figure 7.19 5-percentile end-user throughput with slow and fast ABS adaptation

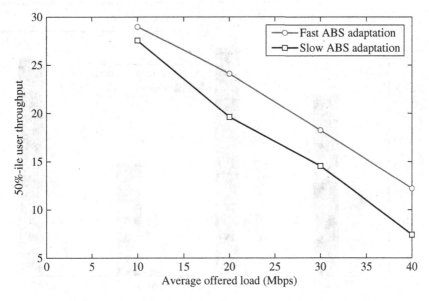

Figure 7.20   50-percentile (median) end-user throughput with slow and fast ABS adaptation

eICIC Performance in a Site-Specific Scenario

In order to shed further light on the performance of eICIC, we next present results from a site-specific use case. The site-specific use case is based on data from a dense metropolitan area with realistic deployment of macro and pico eNodeBs. A three-dimensional digital map of the area is used, which contains information of the building locations (and their dimensions), streets, parks, open, squares and so on. The macro eNodeBs are deployed to provide wide area blanket coverage, while the picos are installed in local areas with high traffic density, where the macros have difficulties carrying all the offered traffic during peak hours. An advanced ray tracing tool is used to compute the radio propagation characteristics between each transmitter and receiver point, taking into account the environment characteristics from the three-dimensional map of the metropolitan area. Similarly as for the 3GPP-defined scenarios, a bursty traffic model is applied, where the arrival of new calls is according to Poisson process, assuming a finite payload per user. The spatial traffic distribution is, however, highly non-uniform in line with observations from the considered area. In essence, the considered dense urban metropolitan area is rather irregular in coherence with real-life operator installations, meaning that propagation characteristics vary significantly depending on localized environment, the inter-site distance between macro cells varies from approximately 100 meters up to several hundred meters. Similarly, the placement of picos is also irregular, so the equivalent number of picos per macro cell area varies from zero up to seven for the considered use case. The performance gain of eICIC therefore varies significantly from one localized area to another in the metropole. In order to illustrate the former, we have analysed the eICIC in four different local areas, denoted Area 1–4. Area 1 corresponds to the coverage of one macro cell with four distantly placed picos; Area 2 is illuminated by 4 macro cells and containing 11 picos; Area 3 has similar characteristics as Area 1; while Area 4 has six picos deployed in the coverage area

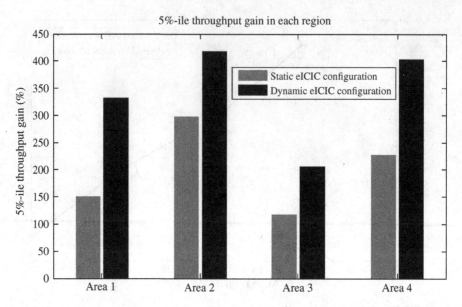

Figure 7.21   Gain in 5-percentile (cell edge) user throughput from using eICIC in the four local areas

of a single macro cell. The highly irregular nature of the considered environment naturally calls for different eICIC parameter settings per area to unleash the full performance potential. We are therefore focusing on the cases where the ABS muting pattern is set either semi-statically per localized area or fast dynamically adjusted according to the algorithms outlined in Section 'Fast Dynamic eICIC'. Furthermore, dynamic load balancing at each connection setup is assumed, facilitating both intra- and inter-layer load balancing.

Figure 7.21 shows the gain in the 5-percentile user throughput from using eICIC in the different areas. As expected, also for this case there is an attractive additional benefit from using fast dynamic adjustment of ABS muting patterns in coherence with local conditions. The results show that the proposed algorithm for fast dynamic ABS muting control is able to adapt efficiently to the experienced conditions. Area 2 and Area 4 naturally experience the highest eICIC benefits due to the higher concentration of picos in those locations.

Figure 7.22 shows the capacity gain for each of the four areas from using eICIC. In this context, the capacity is defined as the maximum offered capacity that can be tolerated, while still being able to serve at least 95% of the users with a minimum data rate of 2 Mbps. Hence, the capacity gain results express the additional traffic growth that can be accommodated per area by enabling dynamic eICIC with fast ABS adaptation. Also from these results it is visible how the benefit of eICIC varies from one area to another. Most prominent, the eICIC capacity is in excess of 100% for Area 4, meaning that a traffic growth of a factor of ×2 can be tolerated in this area by enabling eICIC.

eICIC Performance Measurements

The following example illustrates eICIC benefit in action based on simple measurements. The measurement setup is illustrated in Figure 7.23 and the corresponding results in Figure 7.24.

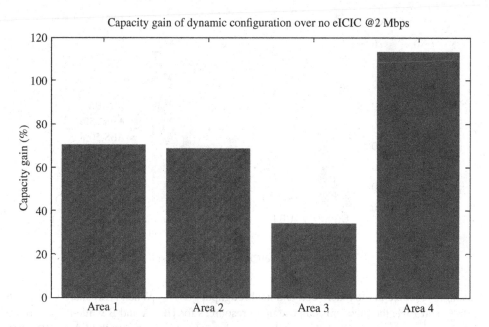

Figure 7.22 Capacity gain offered by fast dynamic eICIC in the four local areas

The measurement has single macro cell, two pico cells and three UEs. UE C is always connected to macro cell. Also UEs A and B are connected to the macro cell without RE. When eICIC and RE are activated, UE A gets connected to pico cell A and UE B to pico cell B. Four different cases are illustrated: one case without eICIC and three cases with different amounts of ABS 25%, 50% and 75%. All three UEs share macro cell capacity without eICIC and UE C gets highest data rate since it is closest to the macro cell. When eICIC is activated with ABS 25%, UEs A and B get connected to the pico cells but can use only 25% of TTIs for the data reception. Therefore, the throughput of UEs A and B goes down. On the other hand, UE C

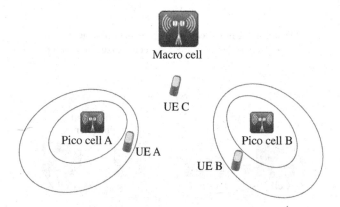

Figure 7.23 eICIC measurement setup

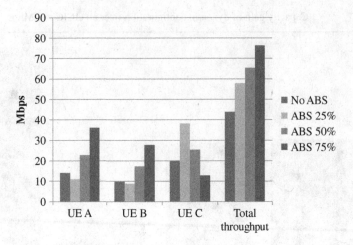

Figure 7.24   eICIC measurement results

has more resource in the macro cell and its throughput goes up. When ABS share is increased to 50% and 75%, the pico cells can use more resources for UEs A and B while the macro cell has fewer resources for UE C. The total system throughput increases with higher ABS share, since two UEs can take the benefit of ABS by connecting to pico cells.

7.4   Enhanced Coordinated Multipoint (eCoMP)

FSS can avoid inter-cell interference with UE CQI reporting. Another approach would be to coordinate the multi-cell transmission between the base stations instead of relying on UE feedback. The coordination can use distributed architecture where the decisions are carried by eNodeBs and the information is exchanged over X2 interface. Another architecture option can use a new network element for centralized scheduler. The eNodeBs would provide load and interference information to the centralized element which would then coordinate the scheduling of the individual eNodeBs. These two architecture options are shown in Figure 7.25. This kind

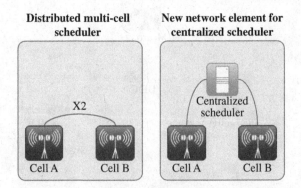

Figure 7.25   Multi-cell coordination architectures

of coordination is called enhanced coordinated multipoint (eCoMP). The 3GPP term for the feature is CoMP with non-ideal backhaul. The multi-cell scheduler has become more attractive since it can utilize the 3GPP Release 11 UE features that allow more accurate reporting of the neighbour cell channel conditions.

We will analyse the differences of these two implementations. We assume that there is some latency in X2 interface between eNodeBs and between centralized scheduler and eNodeBs. The algorithms need to be designed in such a way that latencies at least up to 10 ms can be tolerated. The distributed multi-cell scheduler can access the local scheduler information without any delay while the information from adjacent cells includes some delay. The centralized scheduler can access local scheduler information only with some delays. Therefore, the distributed multi-cell scheduler can have a benefit compared to the centralized solution because the distributed solution can take the local instantaneous scheduler information into account when optimizing the multi-cell scheduling. The distributed solution is also compatible with earlier 3GPP features like intra-frequency load balancing, also known as mobility load balancing (MLB), which uses distributed solution for balancing the loading between two adjacent cells.

We also need to consider the amount of signalling needed in the information exchange. The distributed solution can be designed so that exchanging raw information, such as CSI of all users, is not required. Each eNodeB can access the CSIs of its own UEs and exchange only cell-level metrics between adjacent eNodeBs. Each eNodeB makes a local decision based on the CSIs of its own UEs, metrics maintained by its own scheduler and the metrics received from adjacent eNodeBs. The amount of X2 signalling can be minimized by exchanging only cell-level metrics instead of UE-specific information as shown in Figure 7.26. The centralized scheduler benefits if it can obtain UE-specific CSI information since it does not have direct access to any local scheduler information. UE-specific information exchange may increase the amount of signalling.

The scalability of the multi-cell coordination in the large networks needs to be considered. Since the interference caused by a cell typically affects closest neighbouring cells, the coordination to overcome interference typically requires only local coordination instead of coordination over large areas. The most relevant interfering neighbouring cells can change depending on the UE locations, antenna tilts and network expansions. The preferred solution is a flexible cluster configuration instead of predefined coordination clusters. The flexible cluster means that each cell can have dynamically defined neighbour cells in the multi-cell coordination. This concept is called liquid clusters and is shown in Figure 7.27. The coordination capability with fixed and liquid clusters in system simulations is illustrated in Figure 7.28. If we have three-cell fixed cluster, only 60% of UEs have its strongest interferer in the coordination cluster. If the fixed cluster size is increased to nine cells, the probability grows to 72%. If

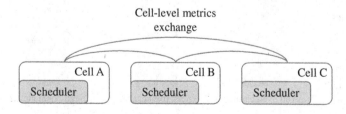

Figure 7.26 Scheduling metrics exchange in distributed multi-cell scheduling

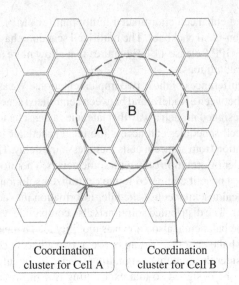

Figure 7.27   Liquid cluster concept in multi-cell scheduling

seven-cell cluster is used with liquid cluster selection, the probability is increased even above 99% so that the strongest interferer is in the coordination cluster. That gives a good starting point to consider the algorithms and performance of the multi-cell coordination.

Table 7.3 summarizes the differences between distributed and centralized multi-cell coordination solutions. Both solutions can be utilized for providing multi-cell coordination but there is no motivation to bring a new network element to the architecture because the distributed solution can provide the same functionalities, or maybe even better capabilities than the centralized solution.

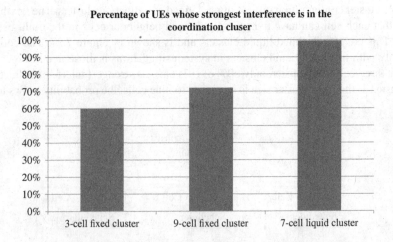

Figure 7.28   Coordination capability with fixed and liquid clusters

Table 7.3 Comparison of distributed and centralized multi-cell coordination

	Distributed multi-cell coordination	Centralized multi-cell coordination
Interworking with local scheduler	Distributed scheduler can access the local scheduler information	Centralized scheduler does not have fast access to the local scheduler
Interworking with other 3GPP features	Compatible with other 3GPP features like intra-frequency load balancing which is also integrated in eNodeB	Interworking with local schedulers by X2 signalling messages
Signalling volumes	Low signalling volumes since only cell-level metrics exchanged	Potentially higher signalling volumes since UE-specific information is carried
Coordination area and scalability	Liquid cluster enables flexible coordination area	Fixed cluster size typically assumed
Robustness	No single point of failure	Higher reliability required for a centralized element

The eCoMP gains with ideal backhaul are studied in the system simulations and shown in Figure 7.29 for a 2 × 2 single-user MIMO system. Both full buffer and bursty traffic cases are shown. Three-sector site configuration was assumed in the studies. Release 11 UEs are used for providing multiple CSI feedback, which is assumed ideal in these simulations, corresponding to different hypotheses of CoMP set cells being muted versus transmitting and a MMSE – maximal ratio combining (MRC) receiver is assumed at the UE. The reference case is single-cell scheduling without any multi-cell coordination, and no additional overhead for TM 10 is assumed. The first step is to consider multi-cell scheduling within each eNodeB. The

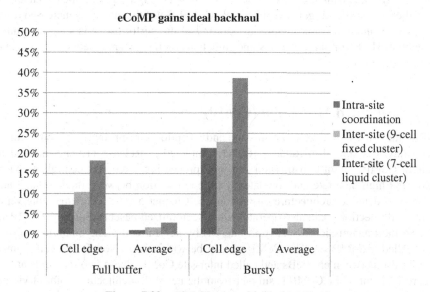

Figure 7.29 eCoMP gains with ideal backhaul

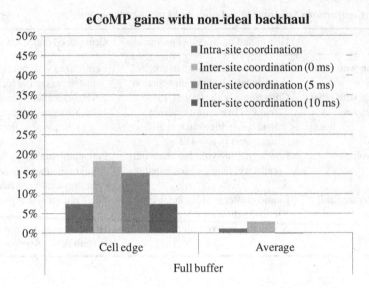

Figure 7.30   eCoMP gains with non-ideal backhaul

intra-site coordination gives 7% cell edge improvement with full buffer and 21% with burst traffic. The gains with bursty traffic are higher than with full buffer because there are typically unused resources available in the adjacent cells in case of bursty traffic. If we also consider inter-eNodeB coordination with nine-cell fixed cluster, the gains are increased to 10% and 23% with full buffer and with bursty traffic. The liquid cluster solution with seven cells gives 18% to 39% gains. The gains from eCoMP in the average data rates are very low.

The eCoMP gains with non-ideal backhaul are shown in Figure 7.30. If the backhaul latency is 5 ms, there are still clear gains visible from eCoMP. But if the latency is increased to 10 ms, the gains from inter-site coordination disappear and the performance is similar to intra-site coordination. These simulations give some guidelines for the transport network design in terms of delay requirements.

7.5   Coordinated Multipoint (CoMP)

CoMP solution uses multi-cell transmission and reception for improving the radio efficiency. CoMP study was started in Release 10 and completed in Release 11. The most advanced version of CoMP – joint transmission and joint processing – requires centralized baseband solution and high data rate and low latency fibre connection between the baseband and the RF. For more detailed architecture discussion, see Chapter 5. The less demanding version – dynamic cell selection – can also operate without centralized baseband. Uplink CoMP allows to receive the transmission signal from one UE by several cells. The combination of these cells is called a CoMP set. The CoMP set can be within one eNodeB and called intra-site CoMP, or also between eNodeBs and called inter-site CoMP. These two scenarios are shown in Figure 7.31. Intra-site CoMP is simpler from the network architecture point of view since the signal combination takes place within one eNodeB.

Intra site CoMP

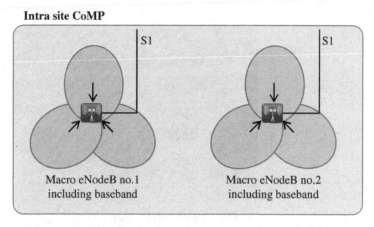

Inter-site CoMP

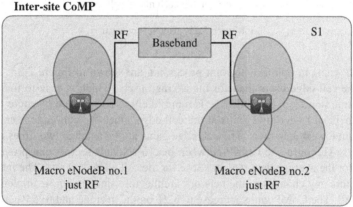

Figure 7.31   Intra-site and inter-site CoMP

The benefit of uplink CoMP is studied in the system simulations both in macro cellular network and in heterogeneous network with co-channel macro and pico cells. The CoMP gain is expected to be higher in HetNet scenario than in macro network. The macro cell transmission power is considerably higher than the transmission power of pico cell. UE selects the cell based on the downlink power which means that UE is much closer to pico cell before making cell reselection. Therefore, the uplink reception may be better via pico cell while the downlink reception works better from the macro cell. The studied HetNet scenario is shown in Figure 7.32 with four pico cells under each macro cell.

The CoMP performance depends on the accuracy of the channel estimation. Demodulation reference signals are utilized for the channel estimation. eNodeB in the same CoMP set needs to estimate the channel of more than one UE transmitted in the same subband. In order to distinguish channels from different UEs, it is important to provide orthogonality of reference signals. Release 10 improves the channel estimation by providing inter-cell orthogonality of reference signals but the uplink CoMP benefits can be obtained also for legacy Release 8 and 9 UEs.

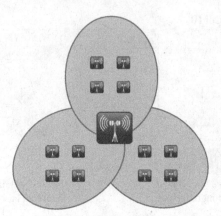

HetNet scenario with four pico
cells under each macro cell

Figure 7.32   HetNet scenario

The CoMP gains in uplink with joint processing are shown in Figure 7.33. The gains are higher for the cell edge users than for the average users, which is easy to understand since the UE transmission from the cell edge is more likely to be received by more than one cell. It is also beneficial to get high gains at the cell edge, since the cell edge users are the ones suffering from lowest data rates. The cell edge gains are on average two times more than the average gains. The intra-site CoMP between macro eNodeB sectors can provide quite nice gains: 19% for the average data rate and 36% for the cell edge data rate. The intra-site CoMP does not require any changes to the network architecture since it can be implemented within single eNodeB. The CoMP gains increase to 32% on the average and to 54% at the cell edge when up to nine cells are included into CoMP set from different eNodeBs. We can note that

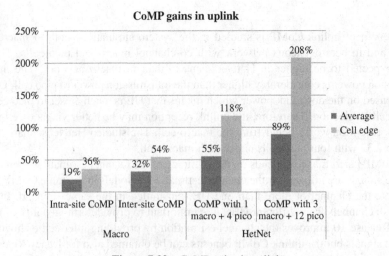

Figure 7.33   CoMP gains in uplink

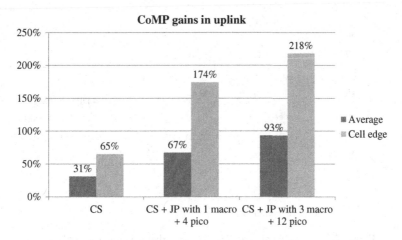

Figure 7.34   CoMP gains in uplink in HetNet scenario with cell selection (CS) and joint processing (JP)

the intra-site CoMP gives 60% of the benefit of the inter-site CoMP. The highest gains are obtained in HetNet scenario where CoMP set extends over multiple macro cells and pico cells. The average gains are 89% and cell edge gains 208% [9].

Figure 7.34 shows CoMP gains in HetNet scenario with cell selection (CS) and joint processing (JP). The first option uses just cell selection and can provide some gains: 31% for the average data rates and 65% for the cell edge data rates. The gains more than double when joint processing reception is applied between macro site and four pico cells under the macro coverage area. The gains further increase when CoMP set is extended over multiple macro sites. The uplink CoMP measurements are illustrated in Chapter 13.

The downlink CoMP gains are expected to be lower than the uplink gains. The uplink CoMP uses multi-cell reception which can provide clear gains while the downlink CoMP gains rely on the selection diversity or joint transmission, which also adds interference. One challenge in downlink is the availability of the channel estimates at the eNodeB transmitter. UE first needs to estimate the multipath channel from all cells in CoMP set, and then UE can provide the CSI to the network in the uplink feedback channel. The feedback adds delay and quantization errors. A UE-specific measurement set is defined as a set of cells that a given UE performs measurement on to determine CSI feedback. The multi-cell CSI measurement requirements are defined in Release 11. Therefore, the downlink CoMP requires Release 11 UEs while the uplink CoMP benefits can also be obtained for the legacy UEs.

The downlink CoMP can use joint transmission (JT) or dynamic point selection (DPS) which is also known as dynamic cell selection (DCS). DPS is a simple but effective downlink CoMP scheme that switches the serving transmission point to the UE based on the UE's channel estimate feedback and the cell loading conditions. The switching can be done very fast even on subframe-to-subframe basis without any handover signalling. It provides both macro diversity and fast load balancing gains. The macro diversity benefit is obtained by choosing the best serving transmission point according to the UE's current channel conditions. Fast load balancing gains are realized by transmitting to the UE from the less-loaded transmission point. The UE estimates the CSI from up to three cells.

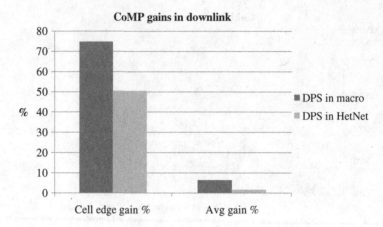

Figure 7.35 CoMP gains in downlink with dynamic point selection (DPS)

Figure 7.35 shows the gains with DPS both in three-sector macro cell environment and in HetNet scenario, with ideal backhaul assumed between the baseband units. The cooperation cluster is assumed to include the immediate neighbours of each cell. The selected algorithm is described in more detail in [10]. The cell edge gain in the macro cell is 75% and in HetNet case about 50% which illustrates the power of DPS solution. There is even a small gain in the average data rate. These results assume a realistic handover margin of 2 dB, and DPS is able to fully overcome the negative effects of handover margin. The gains for the HetNet case are slightly smaller because the baseline with RE and eICIC mechanisms provides much of the load balancing gains. The CoMP performance factors are discussed in [11]. 3GPP CoMP simulations can be found from [12].

DPS solution can be implemented with reasonably low backhaul requirements – no need for centralized baseband. UE's RRC connection and the S1 connection stay at its anchor eNodeB while its serving transmission point can be switched rapidly within CoMP set. If the current serving transmission point is different from the anchor transmission point, then the data need to be forwarded from the anchor transmission point to the serving transmission point. Figure 7.36 shows the transmission requirements for an offered bursty load of 10 Mbps

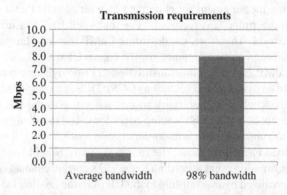

Figure 7.36 Transmission requirements with DPS for an offered load of 10 Mbps per cell in a 3GPP macro case-1 deployment

in a hexagonal-sectorized 3GPP macro case-1 deployment with an inter-site distance of 500 m. The average requirement is below 1 Mbps but it is more important to look at the worst-case requirements. The 98% requirement is 7.9 Mbps which is still very low compared to the cell peak data rate. The transport requirements tend to be lower for full buffer traffic because the load is more balanced compared to bursty traffic.

7.6 Summary

HetNet deployments bring a clear need for an efficient interference management since the small cell deployment will happen with less planning and with less optimization. This chapter illustrated the main features and their benefits. 3GPP has added a number of features in the latest releases for more efficient management of inter-cell interference. The features are summarized in Table 7.4. Release 8 allows avoiding inter-cell interference with frequency-domain CQI reporting and packet scheduling. That approach is called inter-cell interference coordination (ICIC). It is also possible to utilize signalling over X2 interface. The measurements show that the packet scheduler solutions can provide even 50% improvements in the cell edge data rates. Release 10 adds eICIC capability where the interference is managed in the time domain. eICIC is designed for HetNet deployments and requires base station synchronization, but the backhaul requirements are still low. eICIC can enhance median data rates by more than 50% and cell edge data rates by more than 100%. eICIC requires dynamic adaptation of the subframes and parameters depending on the user locations. Release 11 improved eICIC with UE interference cancellation. Release 11 defined CoMP which is the most advanced multi-cell transmission solution with joint processing. Joint processing sets highest requirements for the transport: in practice, direct fibre connection is required between the baseband unit and the RF head. Release 12 added eCoMP that also allows to use non-ideal backhaul without direct fibre.

Table 7.4 Comparison of different interference mitigation solutions

	Inter-cell interference coordination (ICIC)	Enhanced inter-cell interference coordination (eICIC)	Coordinated multipoint (CoMP)	Enhanced coordinated multipoint (eCoMP)
Operating domain	Frequency domain	Time domain	Additionally spatial domain (antennas)	Additionally spatial domain (antennas)
Operating principle	CQI feedback in frequency domain	Time-domain resource sharing in HetNet	Multi-cell transmission and reception	Fast multi-cell coordination over non-ideal backhaul
Base station time synchronization	Not needed (only frequency synchronization)	Yes	Yes	Yes
Transport requirements	No requirements	Low requirements – only control plane	High transport requirements for joint processing	Low requirements
3GPP Release	Release 8	eICIC in Release 10 feICIC in Release 11	Release 11	Release 12

eCoMP uses resource coordination between eNodeBs via X2 interface. eCoMP can improve cell edge data rates by 20%.

References

[1] B. Soret, K. I. Pedersen, N. Jørgen, and V. Fernandez-Lopez, 'Interference Coordination for Dense Wireless Networks', *IEEE Communications Magazine*, 53, 102–109 (2015).

[2] K. I. Pedersen, Y. Wang, S. Strzyz, and F. Frederiksen, 'Enhanced Inter-cell Interference Coordination in Co-channel Multi-layer LTE-Advanced Networks', *IEEE Wireless Communications Magazine*, 20, 120–127 (2013).

[3] B. Soret, H. Wang, K. I. Pedersen, and C. Rosa, 'Multicell Cooperation for LTE-Advanced Heterogeneous Network Scenarios', *IEEE Wireless Communications Magazine*, 20(1), 27–34, (2013).

[4] Y. Wang, B. Soret, and K. I. Pedersen, 'Sensitivity Study of Optimal eICIC Configurations in Different Heterogeneous Network Scenarios', IEEE Vehicular Technology Conference (VTC), September 2012.

[5] B. Soret and K. I. Pedersen, 'Macro Cell Muting Coordination for Non-uniform Topologies in LTE-A HetNets', IEEE Vehicular Technology Conference (VTC) Fall 2013, September 2013.

[6] B. Soret, Y. Wang, and K. I. Pedersen, 'CRS Interference Cancellation in Heterogeneous Networks for LTE-Advanced Downlink', IEEE International Conference on Communications ICC 2012 (International Workshop on Small Cell Wireless Networks), pp. 6797–6801, June 2012.

[7] 3GPP 36.872. Technical Report 'Small Cell Enhancements for E-UTRA and E-UTRAN – Physical Layer Aspects', v. 12.1, 2013.

[8] B. Soret and K. I. Pedersen, 'Centralized and Distributed Solutions for Fast Muting Adaptation in LTE-Advanced HetNets', *IEEE Transactions on Vehicular Technology*, 64, 147–158 (2014).

[9] Y. Huiyu, Z. Naizheng, Y. Yuyu, and P. Skov, 'Performance Evaluation of Coordinated Multipoint Reception in CRAN Under LTE-Advanced Uplink', 7th International ICST Conference on Communications and Networking in China (CHINACOM), 2012.

[10] R. Agrawal, A. Bedekar, R. Gupta, S. Kalyanasundaram, H. Kroener, and B. Natarajan, 'Dynamic Point Selection for LTE-Advanced: Algorithms and Performance', IEEE Wireless Communications and Networking Conference (WCNC), 2014.

[11] B. Mondal, E. Visotsky, T. A. Thomas, X. Wang, and A. Ghosh, 'Performance of Downlink CoMP in LTE under Practical Constraints', PIMRC, 2012.

[12] 3GPP 36.819. Technical Report 'Coordinated Multi-point Operation for LTE Physical Layer Aspects', v. 11.2, 2013.

[13] A. Bedekar and R. Agrawal, 'Optimal Muting and Load Balancing for eICIC', International Symposium and Workshops on Modeling and Optimization in Mobile, Ad Hoc and Wireless Networks (WiOpt), pp. 280–287, May 2013.

[14] B. Soret, K. I. Pedersen, T. E. Kolding, H. Kroener, and I. Maniatis, 'Fast Muting Resource Allocation for LTE-A HetNets with Remote Radio Heads', Proceedings IEEE Global Communications Conference (GLOBECOM), December 2013.

8

Small Cell Optimization

Harri Holma, Klaus Pedersen, Claudio Rosa, Anand Bedekar and
Hua Wang

8.1 Introduction

This chapter presents the optimization of small cells focusing especially on the radio resource management. The optimization considers the following aspects:

- Mobility: improvement of the reliability of high-speed mobility in the dense small cell environment.
- Data rates: utilization of the macro and small cell resources simultaneously for the maximization of user data rates.
- Interference management and capacity: minimization of interference between small cell layers in order to maximize the network capacity.
- Power savings: minimization of the total network power consumption.
- Multivendor case: operation and optimization of macro and small cell layers from different vendors.

LTE Small Cell Optimization: 3GPP Evolution to Release 13, First Edition.
Edited by Harri Holma, Antti Toskala and Jussi Reunanen.
© 2016 John Wiley & Sons, Ltd. Published 2016 by John Wiley & Sons, Ltd.

Section 8.2 presents the mobility optimization and Section 8.3 the performance of inter-site carrier aggregation where user equipment (UE) has simultaneous connection to the macro and small cells and can receive data simultaneously from both layers. Section 8.4 illustrates the optimization of the very dense small cell networks, also called ultra dense networks (UDN). The power savings are discussed in Section 8.5 and multivendor cases in Section 8.6. The chapter is summarized in Section 8.7.

8.2   HetNet Mobility Optimization

The long-term evolution (LTE) mobility is network controlled and UE assisted for radio resource control (RRC) connected mode UEs, while RRC Idle mode behaviour relies on UE autonomous cell selection based on UE measurements. This implies that all handover decisions for RRC connected mode are taken by the network, often triggered by radio resource management (RRM) reporting events from the UE. The most commonly used UE RRM reporting for triggering a handover is event A3, which is defined as the target cell being offset decibels stronger than the serving source cell. For RRC Idle mode, the network can provide additional guidance (a.k.a. Idle mode traffic steering) by broadcasting cell reselection parameters and setting different priorities per frequency layer.

Figure 8.1 pictures a typical HetNet environment with different cell types, as well as a mixture of terminals that are either semi-stationary or moving along certain trajectories at different velocities. Typically, the path loss slope and the spatial correlation distance of shadow fading are different from macro and small cells to UEs, generally resulting in a steeper

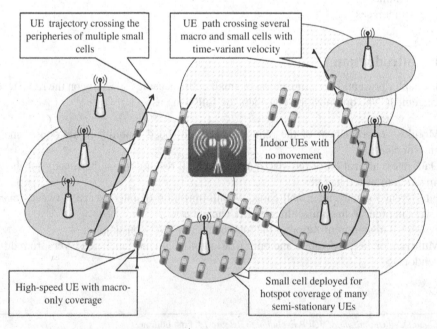

Figure 8.1   High-level sketch of LTE HetNet scenario with a 3-sector macro base station and multiple small cells and UEs

gradient of the received signal strength from small cells as compared to macro cells. Timely and accurate handover between co-channel-deployed macro and small cells is therefore essential for ensuring acceptable mobility performance. Especially small cell outbound handover is challenging for high-velocity users, while good mobility performance for users at lower speed can be achieved with basic LTE handover methods [1]. In this context, the following aggregates are worth highlighting for optimizing the co-channel HetNet mobility:

- *Mobility state estimation and time-to-trigger (TTT) scaling [Release 8]*: Based on the past number of experienced handovers, the RRC Connected mode UEs estimate their mobility state, enumerated as either normal, medium or high. UEs in the high mobility state are those experiencing the highest handover rate. Depending on the UEs' mobility state, handover parameters such as TTT are scaled to make the RRC Connected mode mobility performance more robust over the range of possible UE speeds. The TTT parameter influences the UE measurement reporting process and has an impact on the overall handover time. A lower TTT value is used for UEs operating in the high mobility state, thus essentially facilitating faster handovers for high-velocity users.
- *Inter-eNodeB mobility history signalling [Release 8]*: At the network, UE mobility history information is signalled on the X2 interface between eNodeBs. The UE mobility history information contains knowledge of previously serving cells for the terminal while in RRC Connected mode. The former also includes time-of-stay per cell and cell type; enumerated as very small, small, medium, large, and so on. The network can utilize the UE mobility history information to estimate handover rate of the terminal and, based on that, adjust mobility parameters as well as decide if the user should be allowed handover to small cells, or mainly be kept at the macro layer.
- *Signalling of UE mobility information [Release 12]*: Mobility history reporting by the UE on RRC Idle to RRC Connected transition allows the network to more accurately estimate the UE mobility by using mobility history information stored and reported by UE. The UE indicates the availability of the stored history to the network. The UE stores its mobility state and global cell identity (GCID), or physical cell identity (PCI) if GCID is not available, as well as the duration of stay in the 16 most recently visited LTE cells, regardless of UE RRC state or PLMN (i.e. can also include emergency camped cells). Time spent outside LTE (in other radio access technology (RAT) or in out-of-service state) and the duration of such 'stay' is also recorded (without cell identity information).
- *Target cell-dependent TTT [Release 12]*: The network can configure UEs to use different TTTs depending on the target cell. This allows to use large TTT for small cell inbound handover to reduce the probability of fast moving users to be handed over to small cells. Still, users served by small cells can have low values of TTT to ensure safe outbound handover back to the macro layer. This is supported by the specifications as follows: The network can signal to the UE a list of PCIs for each measurement object (i.e. carrier frequency) for which an alternative TTT can be used. Default TTT is used for the target cells not in the list.

With the above aggregates, it is also possible to ensure efficient and robust mobility for HetNet co-channel cases. As an example, the UE in Figure 8.1 with peripheral small cell crossings should only hand off to those small cells if it is moving at low to moderate speed, while otherwise be kept at the macro layer if it is travelling at high speed.

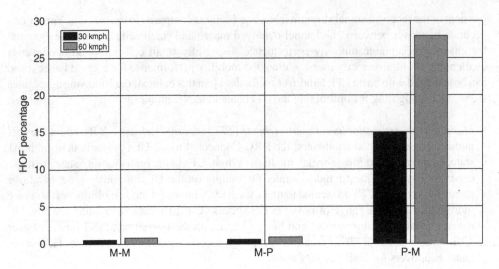

Figure 8.2 Handover failure (HOF) percentage for macro–macro (M–M), macro–pico (M–P) and pico–macro (P–M) co-channel handovers

In order to further illustrate the differences in handover performance for co-channel HetNet cases, Figure 8.2 shows the handover failure (HOF) percentage separately for macro–macro (M-M), macro–pico (M-P) and pico–macro (P-M) handovers. These results are obtained for the case with two picos per macro cell area. HOF event is declared if radio link failure (RLF) occurs after TTT expires, during the handover execution time as defined in [8]. The results in Figure 8.2 are obtained under the assumption of using the event A3 for triggering handovers, using a conservative setting with 3 dB handover offset and a TTT of 480 ms. As can be seen from these results, the performance of P-M handovers is significantly worse than the M-M and M-P handover cases – especially for the higher UE speeds. The worse handover performance for the P-M case is primarily a result of using too high TTT and handover offset, which essentially results in too late outbound handover from the pico. Furthermore, allowing the users moving at 60 kmph to be handed over to the picos is often not attractive, as they quickly leave the pico coverage again, resulting in an unnecessary high handover rate with an undesirable short time-of-stay in the pico. For additional results see References 2 and 6.

However, by using differentiated TTT settings for the different handover settings as supported for Release 12, the desired mobility robustness can also be maintained for the co-channel HetNet cases. This is illustrated in Figure 8.3, where the average experienced HOF percentages are reported for different settings of the TTT, assuming a handover offset of 2 dB for the A3 event. In these results the TTT for P–M is set to 160 ms, while the TTT for M–M is 256 ms. As can be seen, good performance is achieved by using a relatively long TTT for pico inbound handovers (M–P), as this significantly lowers the probability of high-speed users from being handed over to picos (i.e. will also not experience P–M handover). For users that do experience P-M handovers, the TTT should take a low value to ensure quick outbound handovers, while the TTT should take medium values for M–M to low HOF and ping–pong probabilities. Finally, the HOF probability can be further reduced by enabling self-organizing network (SON)-based techniques such as mobility robustness optimization (MRO) for whether

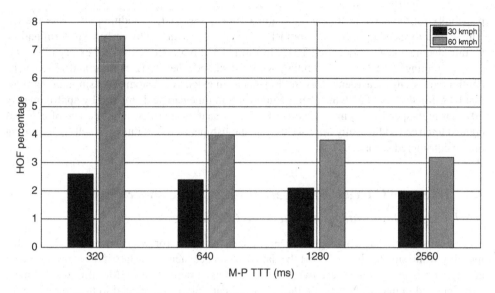

Figure 8.3 Handover failure (HOF) percentage for different TTT settings

the cell individual offset is further adjusted for each cell boarder to avoid too late or two earlier handovers; see more detail in Reference 3.

For HetNet deployments with macros and small cells on different carrier frequencies, the mobility performance does not naturally suffer from interference between macro and small cells. However, for such cases the main challenge is for macro-UEs to discover small cells on other carriers in due time without performing unnecessary inter-frequency measurements. Here the dilemma is that while frequent inter-frequency measurement by macro-UEs would enable timely small cell discovery on neighbour carriers, it comes with a cost in terms of both UE power consumption and measurement gaps. Current LTE specifications include options of enabling periodical inter-frequency measurements every 40 or 80 ms, using measurement gaps of 6 ms. Typically, the network first enables inter-frequency measurements for the UEs when the serving cell signal strength (or quality) drops below a certain threshold (i.e. corresponding to UE reporting event A2) as these are the conditions where inter-frequency handovers are most likely to offer benefits. Alternatively, location-aware methods for automatic suspend and resume of macro-UE inter-frequency RRM measurements for small cell detection could be controlled depending on whether the UE is likely to be in close vicinity of deployed small cells [4]. Examples of the former include using RF finger printing techniques (e.g. based on collected UE RRM measurements). Inter-frequency small cell discovery is less challenging for UEs supporting carrier aggregation, as those may be able to perform concurrent reception on carrier A and inter-frequency measurements on carrier B (depending on the UE category and implementation).

In the rare event of RLF due to, for example, too early or too late handover, mechanisms for efficient recovery are available. Among those include re-establishment attempts initiated by the UE. If the eNodeB receiving the re-establishment request is not prepared for the UE, the eNodeB can fetch the context from the eNodeB previously serving the UE. The context fetch functionality is standardized as part of Release 12. Furthermore, Release 12 also offers the possibility for the network to configure a new timer for the UE to additionally lower the UE outage time to speed up recovery from RLF. This goal is achieved by early T310 termination, by introducing a new timer called T312. In short, T312 is started upon TTT expiry if T310 is already running and T312 is not running. RLF is declared when T312 expires, and T312 can be configured for any measurement event. Results from extensive system-level simulations show that by using the new T312, the average interruption time can be decreased by approximately 30% for high-speed users in co-channel HetNet scenarios. Furthermore, the use of eNodeB context fetch can additionally improve the re-establishment success rate by 20–40% depending on the considered scenario.

8.3 Inter-site Carrier Aggregation with Dual Connectivity

8.3.1 User Data Rates with Inter-site Carrier Aggregation

Macro cells with wide coverage area are a great solution for wide coverage and reliable mobility while small cells can boost the data rates and capacity. The target of inter-site carrier aggregation is to combine these two benefits in HetNet deployments. Mobility management is maintained in macro layer while the small cell capacity is aggregated to the user plane for higher throughput. The solution uses dual connectivity (DC) between macro cells and small cells. The concept is illustrated in Figure 8.4 and described in more detail in Chapter 4. This chapter focuses on the performance and usage of inter-site carrier aggregation.

The user throughput performance of inter-site carrier aggregation is first studied assuming that the small cells are realized as remote radio heads (RRHs) with centralized baseband processing at the macro, and virtually zero latency fibre-based fronthaul connections between macro and RRHs. This architecture option is already feasible with Release 10 carrier aggregation. The best HetNet performance can be obtained by using direct fibre to the small cells but in practice many small cells use non-ideal backhaul, which add the requirement of independent radio functionalities in the macro and small cell layers. Release 12 inter-site carrier aggregation is designed to work with non-ideal backhaul which means that no direct fibre connection is required to the small cell. The architecture options are shown in Figure 8.5. Release 12 solution is based on DC feature and bearer split where the UE is simultaneously connected to macro cell and small cell and it can receive data of the same bearer from each of the two cells.

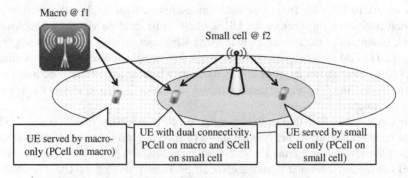

Figure 8.4   Inter-site carrier aggregation between macro and small cells

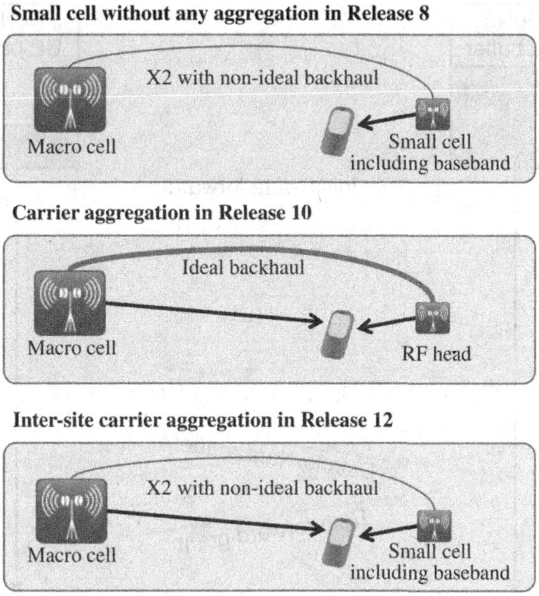

Small cell without any aggregation in Release 8

X2 with non-ideal backhaul

Macro cell

Small cell
including baseband

Carrier aggregation in Release 10

Ideal backhaul

Macro cell

RF head

Inter-site carrier aggregation in Release 12

X2 with non-ideal backhaul

Macro cell

Small cell
including baseband

Figure 8.5   HetNet architecture options

Figure 8.6 generally illustrates the network and UE functioning with DC and bearer split between a macro and a small cell. Data from the core network are first transferred to the macro cell, which operates as the master eNodeB (MeNB). In the macro the data flow is split, so some data are transmitted via the macro to the UE, while other data are transferred over X2 to the small cell, which operates as the secondary eNodeB (SeNB). The MeNB and SeNB have independent medium access control (MAC) entities and physical layer processing, including independent hybrid automatic repeat request (HARQ) and link adaptation. In order to support Release 12 inter-site carrier aggregation the UE is required to have multi-carrier transmission capabilities in uplink so that it can feedback separate channel state information (CSI) and HARQ acknowledgements (ACK) to the macro and small cells.

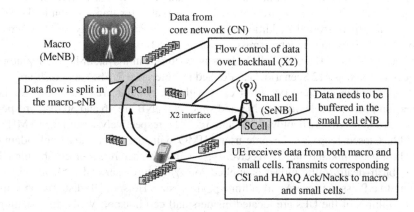

Macro
(MeNB)

Data from
core network (CN)

Flow control of data
over backhaul (X2)

Data flow is split in
the macro-eNB

PCell

Small cell
(SeNB)

Data needs to be
buffered in the
small cell eNB

X2 interface

SCell

UE receives data from both macro and
small cells. Transmits corresponding
CSI and HARQ Ack/Nacks to macro
and small cells.

Figure 8.6   High-level sketch of assumptions for a user in DC between a macro and a small cell

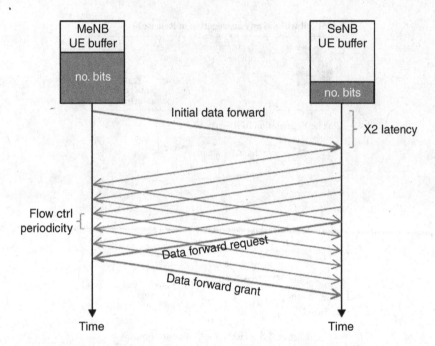

Figure 8.7   Schematic illustration of the X2 flow control mechanism

The performance of DC with split bearers depends on multiple factors, of which flow control between the MeNB and SeNB over the X2 interface is of great importance. The flow control mechanism is presented schematically in Figure 8.7. It is a simple request-and-forward-based scheme, where the SeNB (in accordance with 3GPP specifications) is in charge of requesting data from the MeNB. Received data from the MeNB are buffered in the SeNB until they have been successfully transmitted over the air interface to the UE via the SCell. The data requests from SeNB to MeNB are sent periodically on a per-user basis. As by definition, the design target of the flow control algorithm is to avoid data overflow and underflow in the SeNB. More specifically, the goal is to always have data available for transmission in the SeNB (so that users can benefit from the additional resources in the small cell layer) while limiting the additional delay introduced by transmission via the SeNB.

The performance of DC with split bearers is evaluated in a system-level simulator in line with the 3GPP Release 12 Scenario 2a as defined in Reference 5. The small cells are randomly deployed in condensed clusters with 4 cells within a circular area with 50 m radius. The transmission powers for the macro eNB and small cell eNB are 46 and 30 dBm, respectively. Macros and small cells are deployed at 2 and 3.5 GHz, respectively, assuming 10 MHz carrier bandwidth. Closed loop 2 × 2 single-user MIMO with pre-coding and rank adaptation is assumed for each link and the UE receiver type is interference rejection combining (IRC). A dynamic birth–death traffic model is applied for generating user calls, where call arrival is according to a Poisson process with a finite payload size. Hotspot UE distribution is assumed where two-thirds of the UEs are located inside small cell hotspots while the remaining UEs are uniformly distributed within the macro cell area. For cases with non-ideal backhaul, an

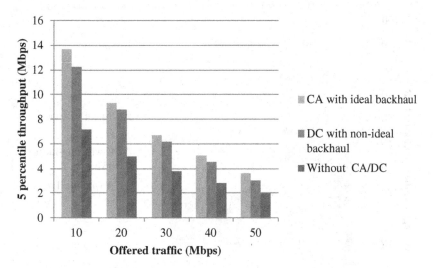

Figure 8.8   5 percentile user throughput with/without DC under different backhaul configurations

X2 latency of 5 ms is assumed. The flow control between macro and small cells is performed periodically every 5 ms. The schedulers in the macro and small cells exchange information on the past average scheduled throughput for UEs that are configured with DC every 50 ms.

As can be observed from Figures 8.8 and 8.9, both the 5 percentile and median user through-put performance with DC are significantly higher than without carrier aggregation or DC. Also, with non-ideal backhaul the 5 percentile and median user throughput are relatively close to the performance assuming ideal fibre-based fronthaul connection. To better quantify the gains of inter-site carrier aggregation, the system performance is compared for a target 5 percentile

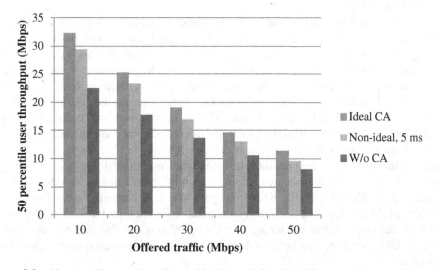

Figure 8.9   50 percentile user throughput with/without DC under different backhaul configurations

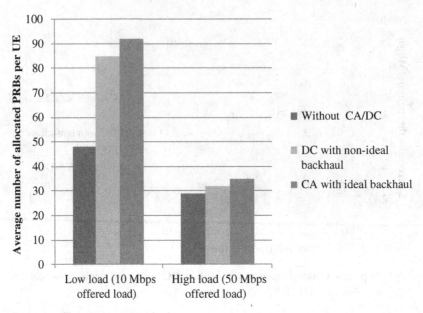

Figure 8.10 Average number of allocated PRBs per UE with and without DC under different load conditions

user throughput performance of 4 Mbps. Under such requirements, the maximum tolerable offered load increases from 30 Mbps (without carrier aggregation or DC) to approximately 45 Mbps for cases with either DC or carrier aggregation, corresponding to a capacity gain of about 50%. With efficient flow control, DC with split bearer over X2-type non-ideal backhaul connections can achieve approximately 80% of the gain available with ideal fibre-based fronthaul connections.

The gain mechanism with inter-site carrier aggregation and DC is multifold. First, terminals configured with carrier aggregation/DC benefit from higher transmission bandwidth by being able to utilize the radio resources in both macro and small cell layers. This is especially visible in low load conditions in Figure 8.10, in which case the average number of allocated physical resource blocks (PRBs) per UE with inter-site carrier aggregation almost doubles as compared to the case with neither carrier aggregation nor DC. This is because the probability of having a single user accessing all the available radio resources in both the macro and the small cells is higher at low load. At higher load the gain from larger allocated bandwidth becomes marginal as the average number of users in a system starts to increase.

Second, the system can benefit from increased multi-user diversity order and faster inter-layer load balancing if terminals are configured with DC/carrier aggregation between macro and small cells. Contrary to the bandwidth gain, the load balancing gains become more noticeable as the load increases. As clearly illustrated in Figure 8.11, the load is more balanced among the cells when either carrier aggregation or DC is used. This is especially evident in the macro layer, resulting in a better utilization of the available radio resources, as shown in Figure 8.12.

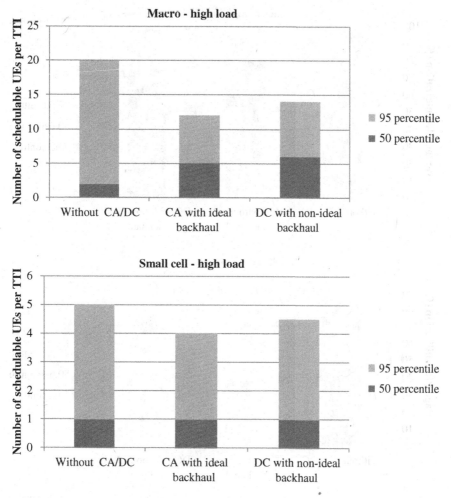

Figure 8.11 Number of UEs served in each cell with and without DC at high offered load of 50 Mbps

8.3.2 Mobility with Dual Connectivity

DC can be utilized for improving mobility robustness. We will next analyse the impact of DC on the number of primary cell handovers and on the HOF. We use two different HetNet scenarios. The first scenario is 3GPP simulation model using static and random base station deployment models. The 3GPP HetNet model is a so-called generic model, which aims at capturing general effects that would typically occur if observing a large number of dense urban environments. The second model is based on site-specific use cases where a particular urban area is explicitly modelled using three-dimensional topography maps and ray-tracing propagation characteristics. We consider two dense metropolitan urban areas, based on data from Europe and Tokyo.

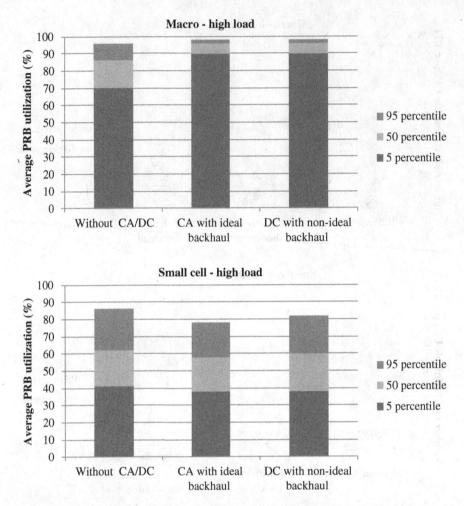

Figure 8.12   Average PRB utilization in each cell with/without DC at high load (50 Mbps offered load)

For the considered European city area, the macro sites are deployed at different heights, taking the local environment into account in order to have good wide area coverage. A total of 48 small cells are deployed at 5 m height in street canyons or at open squares. The small cells are deployed to further improve 5 percentile outage throughput of the network [6]. The former brings to a deployment where the small cells are clustered in certain areas, while other areas contain only few or no small cells. For the areas with small cells, there are on average two small cells per equivalent macro cell coverage area. The small cells are equipped with omni-directional antennas. The Tokyo area has more regular streets and buildings layout. Buildings have an average height of 24 m and a maximum of 150 m, and the streets are wider. A total of 20 macro sites (all with 3 sectors) are deployed per square kilometre. The macro antennas are placed 5 m above buildings in the local area. A total of 70 small cells are deployed at 5-m height. The small cells are mainly placed near the tallest buildings, in order to offload

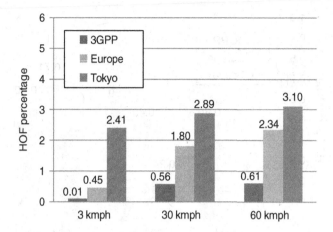

Figure 8.13 Handover failure (HOF) percentage for different scenarios without DC

and carry more traffic where the radio signal from macro cells is typically weaker and there is higher traffic. For the site-specific scenarios, the users move along the streets, while users are free-moving for the 3GPP scenario as defined in Reference 7.

We focus on the RRC connected mode downlink performance. The mobility management is based on UE measurements and controlled by eNodeBs. The UE measurement quantities are reference signal received power (RSRP) or reference signal received quality (RSRQ). The RSRP is essentially a measure of the received power of the transmitted reference signal from a cell, while the RSRQ equals the RSRP/RSSI ratio, where RSSI is the received signal strength indicator, or equivalently the total wideband received power. The UE can be configured to perform measurements of RSRP and RSRQ from its serving and surrounding cells. The RSRP and RSRQ are measured at the physical layer, followed by additional time-domain filtering at Layer-3 using a first-order auto regressive filter. For additional details and background on the assumed RRM measurement events for PCell and SCell mobility/cell-management events, see References 8 and 9.

Figure 8.13 shows the HOF rate for the case without DC, while Figure 8.14 shows the corresponding performance results when DC is enabled. In this context, a HOF event is declared if RLF occurs after TTT expires, during the handover execution time as defined in [8]. Note that RLF is only declared based on radio link monitoring of PCell connection quality by the UE. As observed from Figure 8.13 (results without DC), the probability of experiencing HOFs is relatively low at the different speeds. The lowest HOF probability is observed for the generic 3GPP simulation scenario, while higher HOFs are found for the site-specific cases. A closer inspection of the HOF statistics from the site-specific scenarios reveals that those errors primarily occur at street intersections, especially when the UE is turning a corner. The former is a result of the more detailed modelling for the site-specific scenarios with explicit representation of buildings, and the related radio propagation characteristics in street canyons, where a user passing through street intersection is more likely to experience fast changes of both the desired and interfering signalling strength. Furthermore, the HOF events are primarily observed when attempting intra-frequency handover between two small cells or when conducting inter-frequency handovers between the two layers. SON-based techniques

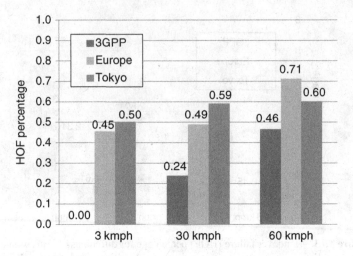

Figure 8.14 Handover failure (HOF) percentage for different scenarios with DC

such as MRO can help further reduce the HOF probability as a result of tuning the handover-related parameters [3].

It is observed from the simulations that the HOF percentage is significantly lower with DC enabled when comparing the results in Figures 8.14 and 8.13, thus demonstrating that the use of DC offers benefits in terms of mobility robustness also. The improved HOF performance from using DC comes from always having the PCell at the macro layer, while utilizing the small cells by configuring those as SCell whenever possible for the user. Keeping the PCell at the macro layer essentially means that the HOF probability is corresponding to the macro-only scenario, and therefore not affected by the small cells.

Figure 8.15 shows the empirical cumulative distribution function (cdf) of the small cell time-of-stay, assuming the use of DC. Here it is worth noticing that the small cell

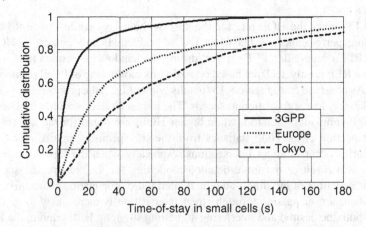

Figure 8.15 Small cell time-of-stay for different scenarios, assuming 3 kmph user speed and use of DC

time-of-stay is orders of magnitudes higher for the site-specific scenarios as compared to the generic 3GPP simulation case. The result is essentially showing a better utilization of the small cells in the site-specific cases. The former is again a result of the more accurate environment and propagation modelling for the site-specific cases, where especially the small cell signal coverage is more widespread in the explicitly represented street canyons. Furthermore, as reported in [9, 10] the use of DC increases both the macro and small cell time-of-stay as compared to the cases without DC, where the UE is subject to more frequent PCell handovers (both intra- and inter-frequency). However, one of the costs of operating with DC is larger RRC signalling overhead from having to manage both PCell and SCell for the users.

8.4   Ultra Dense Network Interference Management

8.4.1   Ultra Dense Network Characteristics

Very dense deployment of small cells on a set of dedicated carriers (or on a single carrier) can bring noticeable benefits by providing more transmission resources. The former is also known as ultra dense networks, which can be deployed in either outdoor or indoor traffic hotspots where additional capacity is needed. However, UDN deployments also come with a number of challenges, where co-tier interference is high on the list. Assuming that the cells are deployed with equal maximum transmit power, the interference footprint tends to be rather spread, where many users are subject to a larger number of interfering signals without a clear dominant interference contribution, and only some users are subject to a dominant interfering source (a.k.a. aggressors). Thus, the aggressor–victim relation is much more diffuse for UDN and hence calls for different inter-cell interference coordination (ICIC) solutions as compared to, for example, co-channel macro–pico scenarios, where the macro is a well-defined aggressor for the pico victim users (see detail in Chapter 7).

UDN are furthermore characterized by typically only having each cell simultaneously serving a single or few users, while several cells may have no users to serve at certain time instances. As an example [10], studies of the 3GPP Release 12 dense outdoor small cell scenarios with 10 small cells within a localized circular area with radius of 50 m show that even at high offered traffic load, only approximately half of the cells are scheduling users in each transmission time interval. This finding is for the case where the traffic arrival is according to a homogeneous Poisson process, assuming a finite payload for each user. It is furthermore found that only approximately 30% of the users have a dominant interference ratio (DIR) above 3 dB, as typically required to achieve worthwhile gains from mitigating the dominant interferer. The dominant interferer for a user is furthermore found to be highly time-variant due to the small cells downlink transmit power variations depending on whether they have schedulable users. Hence, the findings from [11] essentially tell us that only a subset of users in an UDN can gain from ICIC, and for those clear victim users with a high DIR, the ICIC mechanism needs to be rather dynamic as the cell playing the role of aggressor is likely to vary during the call duration for the user. Those are properties that are different from what is typically observed for other network deployments such as macro-only carriers or co-channel macro–pico scenarios.

In the following, two examples of efficient network-based ICIC schemes for UDN cases are presented. First, a proactive ICIC scheme is presented, which relies on coordinated time-domain muting among the small cells. The basic principle of the proactive ICIC scheme is to constantly optimize the system performance. Second, a reactive carrier-based ICIC scheme is presented, which is first activated if interference problems are detected that prevent the users from having their minimum data rate requirement fulfilled. Performance results for those two solution candidates are presented by dense outdoor and indoor small cell deployments, respectively.

8.4.2   Proactive Time-Domain Inter-cell Interference Coordination

The principle of coordinated muting by using almost blank subframes (ABS), as exploited for the eICIC concept as explained in Chapter 7, can in principle also be used for UDN deployments. However, only small cells that act as a dominant interferer for a victim user should start using ABS. Furthermore, the nearby small cells that also potentially act as aggressor node should ideally coordinate which subframes to mute (i.e. configure as ABS). The combinational problem of coordinated muting among small cells is specially challenging when one small cell is identified as being simultaneously aggressor and serving cell for a victim UE that experiences a dominant interferer from another small cell. Such coordination can be attained by pre-assigning some 'good' subframes and 'bad' subframes for the different small cells, where bad subframes are those that may be muted. The pattern of such pre-assigned subframes can either be set a priori or by using slowly adapting SON coordination algorithms. Furthermore, as for the eICIC concept, the users that are subject to significant time-variant interference fluctuations shall be configured with time-domain measurement restricts in coherence with their dominant interfering cells usage of ABS and normal transmissions to ensure good link adaptation and scheduling based on trustworthy CSI feedback.

Following the concept originally presented in [10], the proactive time-domain ICIC scheme can be realized as follows. The muting actions are only taken if a small cell is identified as an aggressor to a victim user. Otherwise, normal transmission is used on all subframes. If the difference between the received signal from the serving cell and the user's dominant interfering cell is below a threshold (set to 10 dB), while the DIR is above 3 dB, the user is marked as a victim user. The cell that is the dominant interferer for the victim user is denoted as the aggressor cell, and is requested to mute. The muting action is reverted when the victim user that triggered the muting leaves the system. The network can identify potential victim users via their reported RRM measurements such as RSRP. Note that given the suggested criteria for classifying a user as a victim, its signal-to-interference noise ratio (SINR) will be improved by at least 3 dB by muting its aggressor. The basic principle of the algorithm is pictured with an example in Figure 8.16. Here the small cell no. 1 detects that UE no. 1 fulfills the criteria for a victim user. Small cell no. 1 therefore takes the action of requesting the victim user's dominant interferer (in this example, small cell no. 3) to increase its muting, followed by small cell no. 3 muting more of its preconfigured 'bad' subframes. Furthermore, UE no. 1 is configured with time-domain restricted CSI measurements in coherence with the pattern of 'bad' and 'good' preconfigured subframes of small cell no. 3. Similarly, when UE no. 1 ends its session, small cell no. 1 would inform small cell no. 3 via the backhaul that it can change the earlier muted subframe(s) for protecting UE no. 1 back to normal transmission. Assuming that UEs nos. 3 and 4 in Figure 8.16 are not detected at victim users, no muting is triggered at their neighbouring cells.

Figure 8.17 shows the performance gain from using the proposed proactive time-domain ICIC mechanism. The gains in 5 percentile and 50 percentile are reported versus the average

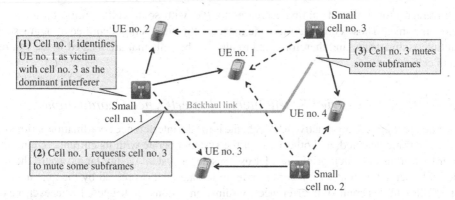

Figure 8.16   Sketch of the basic principle of the proactive time-domain ICIC scheme

offered load per dense cluster of 12 small cells. These results are obtained by the Release 12 small cell simulation assumptions for dense outdoor small cell clusters at 3.5 Hz, using 10 MHz bandwidth and without macro cell interference (i.e. macros are assumed operating at other carrier frequencies). The traffic model is dynamic arrival according to a homogeneous Poisson process with finite payload per call. In line with expectations, the benefits from the considered ICIC scheme are modest at low offered traffic as the inter-cell interference is less dominant. This also causes the ICIC scheme to mute (i.e. use ABS) with a probability of less than 5% for the offered load of 50 Mbps. When the offered load is increased to 100 Mbps (i.e. close to the clusters' capacity limit) the gain from the ICIC scheme becomes attractive as there starts to be much more inter-cell interference to combat. At 100 Mbps offered load,

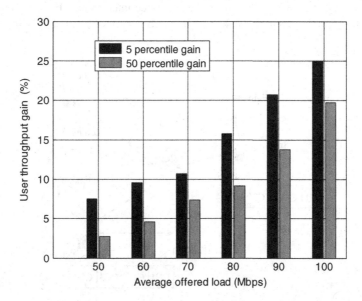

Figure 8.17   User throughput gain versus offered load per small cell cluster from using proactive time-domain muting

the average probability of muting increases to 7%, with some cells muting up to 45% of their subframes. Those cells that mute the most are the ones being identified as aggressor for multiple victim users, and therefore tend to also be the cells that are located in the centre of small cell cluster.

8.4.3 Reactive Carrier-Based Inter-cell Interference Coordination

The basic principle of any reactive ICIC scheme is to take interference coordination actions only if problems are detected, and otherwise let the system operate with its current configuration. In order to illustrate such principles, let us consider a system design target where the users should be served with at least a certain minimum data rate, expressed by the guaranteed bit rate (GBR). In this context, interference coordination actions are triggered whenever the data rate for a user is found to fall below its minimum target. Assuming that each small cell is able to transmit on multiple component carriers (CCs), the ICIC can be conducted in the frequency domain by switching CCs on/off at the small cells in a coordinated manner – also known as carrier-based ICIC. Carrier-based ICIC is especially applicable for small cells in the 3.5 GHz band where there is more spectrum available. By default, all the small cells utilize all the available CCs (i.e. reuse one strategy). The small cells monitor the throughput offered for each user, and only if the throughput is found to fall below the promised GBR is the ICIC framework triggered. The example pictured in Figure 8.18 shows a situation where the UE no. 1 served by small cell no. 1 is detected as a victim user, as the network is unable to serve it the promised GBR. In this example, each small cell can use up to four CCs, where used CCs are indicated with grey background colour.

As only small cell no. 1 is using three of the available CCs, it can choose to enable CC3 to increase the available bandwidth. It can also decide to request some of the interfering small cells to mute certain CCs to reduce the generated interference. For each of the possible hypothesis to improve the performance of the victim user (UE no. 1) above its GBR, the

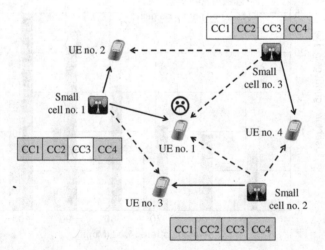

Figure 8.18 Basic principle of carrier-based ICIC. Solid lines indicate serving cell links, while the dashed lines are interfering links

corresponding value (benefit minus cost) is estimated, followed by implementing the action that results in the highest value. The hypothesis corresponding to taking more CCs into use for small cell no. 1 will result in benefit for that cell, but also a potential cost in the neighbouring cells that will experience interference from the CC. Similarly, if a CC is switched off in small cell no. 2 it will result in a performance loss (cost) for UE no. 3, while UE no. 1 and UE no. 4 will experience less interference (benefit). In the interest of complexity, not all the possible hypotheses need to be evaluated, but only those that involve neighbouring cells that act as dominant interferer for the identified victim user. Only hypotheses that result in a positive value without causing other users to fall below their GBR are considered valid options; see additional details in Reference 11.

When UE that previously triggered the proactive ICIC framework leaves the system, the prior actions taken to improve the performance for that user are reverted if, and only if, the value of such a change is positive. The benefit and cost calculations require information to be shared between the small cells over the backhaul. However, the exchange of cost–benefit information is only triggered when a victim user is detected, and signalling delays of several tens of milliseconds are not considered critical, as the scheme does not aim at fast on/off switching of CCs.

The performance of the reactive carrier-based ICIC scheme is evaluated for dense small cell indoor environment. The considered scenario is Release 12 small cell scenario 3 as defined in Reference 12, where small cells are deployed inside a so-called dual stripe building structure with apartments of size 10 m × 10 m. Each of the two parallel building blocks contains a total of $2 \times 5 = 10$ apartments per floor. One small cell is placed randomly within each apartment. The small cells are assumed to operate at 3.5 GHz, each being equipped with 4 CCs, forming a total bandwidth of 20 MHz. UEs are supporting carrier aggregation, such that a single UE can be simultaneously scheduled on multiple CCs from the same small cell. A dynamic traffic model with finite payload per user is assumed, where the minimum data rate target equals 3 Mbps (GBR). Figure 8.19 shows the percentage of users with served bit rate below the GBR (denoted outage probability), as a function of the average number of users per small cell. Considering a target outage probability of maximum 5%; the presented results show that the average offered load can be increased from 2.7 to 4.0 users per small cell if using the proactive carrier-based ICIC scheme. This is equivalent to a capacity gain of 50%.

8.5 Power Saving with Small Cell On/Off

The network power consumption is a concern when a large number of small cells are added to the existing macro layer. Even if the power efficiency of a single base station improves, it does not compensate the high density of the small cells. Therefore, there is a clear need to have new system-level solutions for the minimization of the power consumption. Since the small cells cover only very limited area, it is quite likely that there are no users in some of the small cells at certain times outside the busy hours. The reference signals and other common channels need also to be transmitted even in the case when there are no users connected to the cell. One option is to switch off those small cells that are not needed. It is still important to maintain the network coverage area and be able to switch on the cell again when needed. An example case is shown in Figure 8.20 where the macro cell provides the full coverage and the small cells can be switched off when there are no users or low number of connected users.

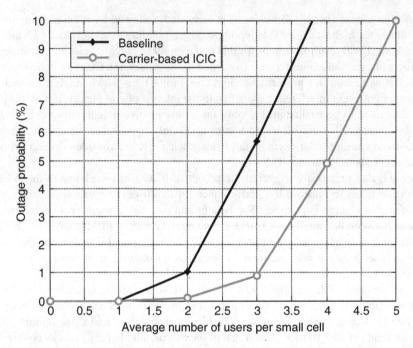

Figure 8.19 Outage probability as a function of average number of users per small cell. An outage event is a user that is served with less than 3 Mbps GBR

Switching off small cells brings an additional benefit in terms of minimization of inter-cell interference.

While switching off the small cells during low load is simple, more intelligence is required to identify when to switch on the small cells again. When the load in the macro cell increases, some of the small cells should be switched on again. A number of different proposals can be considered:

- *Discovery signal in 3GPP Release 12*: The small cell in dormant state can transmit the new discovery signal with low density allocation in time and in frequency domain. The network informs UE about the discovery signal and the timing information, which allows UE to detect the discovery signal of multiple small cells at the same time.

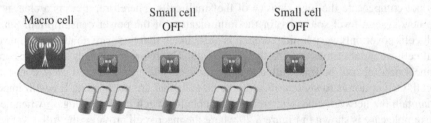

Figure 8.20   Switching off small cells when the load is low

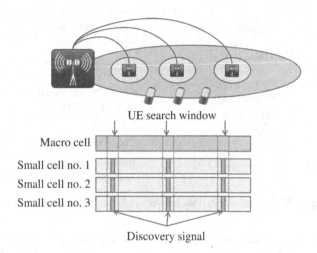

Figure 8.21   Release 12 discovery signal

- *Pre-configuration*: The macro cell includes a predefined list of small cells that should be switched on first. That information is based on the earlier statistical learnings.
- *Small cell uplink measurements*: The small cells can measure the uplink interference levels even if they do not transmit any data. If there is high uplink interference, it implies that there must be UEs close to that cell.
- *UE measurements*: Activation of reference signal transmission in small cells and requesting UE measurements to find out of those UEs can receive the small cell signal.
- *Location measurements*: UE positioning information relative to the small cell locations can be used to define which small cells would best serve the UEs.

The discovery signal solution enables UEs to find small cells that are in dormant state. Figure 8.21 illustrates the concept. The macro cell can provide assistance information to UE to indicate the measurement windows where the UE can find the small cell discovery signals. The benefit of the solution is that UE does not need to search for the small cells all the time but only on those predefined windows. The small cell transmission activity can be kept low to minimize the power consumption. The transmission activity of the discovery signal can be very low, for example, with 100 ms period or more.

8.6   Multivendor Macro Cell and Small Cells

The mobile networks 10 years ago commonly had different radio vendors in 2G and in 3G networks. Another typical multivendor case is home femto cells that use different vendor than macro cells. But things are changing and there are multiple reasons why the multivendor networks are becoming more difficult lately; see Figure 8.22:

- LTE carrier aggregation combines all LTE bands together. The carrier aggregation from Release 10 works only within one vendor in practice.

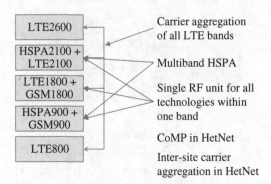

Figure 8.22   Interconnections between radio technologies and cell layers

- Multiradio RF implementation allows to use single RF unit for all technologies within one band: HSPA900 and GSM900 can share the RF unit; LTE1800 and GSM1800 can share the RF unit; and LTE2100 and HSPA2100 can share the RF unit.
- Also dual-band RF units are available combining for example 800 and 900 MHz bands.
- Tighter interworking between macro cells and small cells with CoMP and inter-site carrier aggregation. CoMP works only within one vendor while the inter-site carrier aggregation can also be implemented between vendors over open X2 interface.

The target is to optimize the complex implementation and improve the system performance, which leads to higher interworking between frequencies, technologies and cell layers. Such tighter interworking can make the multivendor case challenging even if the standardized and open interfaces enable multivendor rollouts. S1 interface between radio and core open and different radio and core vendors are used typically. Also X2 interface between macro eNodeBs is open and allows to use different vendors in different geographical areas. The X2 interface between macro and small cell layer is standardized as well and makes it possible to have different cell layers from different vendors. The challenge still is that the eNodeB algorithms are not standardized and may require coordination between the different layers. Figure 8.23 illustrates the main interfaces.

The interface between baseband and RF can use Open Base Station Architecture Initiative (OBSAI) or Common Public Radio Interface (CPRI) but those interfaces are generally not open and only work within one vendor.

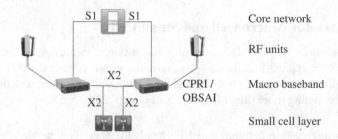

Figure 8.23   Interfaces in HetNet deployments

Femto cells commonly use different vendor than the macro cells. Femto cell case is simpler for multivendor because the interference management is easier in the indoor deployments; femto cell transmission power is low; and the femto gateway hides the large number of small cells from the macro layer.

Efficient operation of heterogeneous networks requires automation to control opex and achieve optimal network performances. Automation is provided by means of SON-based mechanisms. Although SON is largely specified for LTE the SON logic (algorithms) is typically vendor specific. SON is not a single feature but rather a family of several use cases; for those use cases whose logic is fully in the eNodeB, an open X2 assures multivendor capability. For those use cases whose logic is centralized, the multivendor tools at the network management level weds SON to other vendors' systems. The network management interfaces are not fully standardized and some part of integration and adaptation is required.

8.7 Summary

This chapter showed that inter-site carrier aggregation and dual connectivity are beneficial for increasing user data rates in HetNet environment and improving the reliability of mobility. The results show that inter-site carrier aggregation provides over 50% gains at low to medium loads for cell edge performance and over 30% for the median throughput. It is observed that the handover failures are significantly lower with dual connectivity. The improved mobility comes from always having the primary cell at the macro layer, while utilizing the small cells by configuring those as secondary cells.

The chapter also illustrated that advanced inter-cell interference coordination solutions can increase the throughputs in ultra dense small cell networks. The proactive time-domain solution gives up to 20–25% gain and the reactive carrier-based solution even more.

The chapter discussed the base station power-saving options in HetNet scenarios as well as multivendor deployment aspects.

References

[1] S. Barbera, P. H. Michaelsen, M. Saily, and K. Pedersen, 'Mobility Performance of LTE Co-channel Deployment of Macro and Pico Cells', IEEE Proceedings WCNC, Paris, France, 1–4 April 2012, pp. 2890–2895.

[2] S. Barbera, P. H. Michaelsen, M. Saily, and K. Pedersen, 'Improved Mobility Performance in LTE Co-channel HetNets Through Speed Differentiated Enhancements', IEEE Proceedings Globecom, Workshop on Heterogeneous, Multi-hop, Wireless, and Mobile Networks, Anaheim, CA, December 2012.

[3] S. Hämäläinen, H. Sanneck, and C. Sartori, editors. LTE Self-Organising Networks (SON), 1st ed., John Wiley & Sons, Ltd (2012).

[4] A. Prasad, O. Tirkkonen, P. Lunden, O. N. C. Yilmaz, L. Dalsgaard, and C. Wijting, 'Energy Efficient Small Cell Discovery Techniques for LTE-Advanced Heterogeneous Network Deployments', IEEE Communications Magazine, 51, 72–81 (May 2013).

[5] 3GPP TR 36.872, 'Small Cell Enhancements for E-UTRA and E-UTRAN – Physical Layer Aspects', v.12.1.0, 2013.

[6] C. Coletti, L. Hu, H. Nguyen, I. Z. Kovacs, B. Vejlgaard, R. Irmer, and N. Scully, 'Heterogeneous Deployment to Meet Traffic Demand in a Realistic LTE Urban Scenario', IEEE Proceedings on Vehicular Technology Conference (VTC Fall), September 2012.

[7] 3GPP TR 36.839, 'Evolved Universal Terrestrial Radio Access (E-UTRA); Mobility Enhancements in Heterogeneous Networks', v.11.1, 2013.

[8] K. I. Pedersen, S. Barbera, P.-H. Michaelsen, and C. Rosa, 'Mobility Enhancements for LTE-Advanced Multilayer Networks with Inter-site Carrier Aggregation', IEEE Communications Magazine, 51, 64–71 (May 2013).

[9] S. Barbera, K. I. Pedersen, P.-H. Michaelsen, and C. Rosa, 'Mobility Analysis for Inter-site Carrier Aggregation in LTE Heterogeneous Networks', IEEE Proceedings on VTC, September 2013.

[10] V. Fernandez-Lopez, B. Soret, and K. I. Pedersen, 'Effects of Interference Mitigation and Scheduling on Dense Small Cell Networks', IEEE Proceedings on Vehicular Technology Conference (VTC-2014 Fall), September 2014.

[11] B. Soret, K. I. Pedersen, N. Jørgensen, and V. Fernandez-Lopez, 'Interference Coordination for Dense Wireless Networks', *IEEE Communications Magazine*, 53, 102–109 (January 2015).

[12] 3GPP TR 36.872, 'Evolved Universal Terrestrial Radio Access (E-UTRA); Mobility Enhancements in Heterogeneous Networks', v.12.1, 2013.

9

Learnings from Small Cell Deployments

Brian Olsen and Harri Holma

9.1 Introduction

This chapter gives an overview of the main learnings from the small cell deployments in live networks. The results are collected from a large number of cases from multiple networks in Asia Pacific, United States and Europe. Section 9.2 discusses the small cell motivations from the operator point of view and Section 9.3 discusses the challenges. The main learnings are summarized in Section 9.4. The installation considerations are presented in Section 9.5. A small cell case study from the United States is illustrated in Section 9.6 and the chapter is summarized in Section 9.7.

9.2 Small Cell Motivations by Mobile Operators

The traffic volumes have grown rapidly in the mobile networks – more than 100% annual growth pushes the capacity limits of the macro cell network. It is not possible to add more frequencies to the macro sites in busy areas simply because all the spectrum is already fully

LTE Small Cell Optimization: 3GPP Evolution to Release 13, First Edition.
Edited by Harri Holma, Antti Toskala and Jussi Reunanen.
© 2016 John Wiley & Sons, Ltd. Published 2016 by John Wiley & Sons, Ltd.

utilized. Acquiring more spectrum through auctions takes time and can be very expensive, and may also require new frequency variants in the devices resulting in an additional delay in providing capacity offload. Overtime as the capacity demand of the mobile networks grows, additional capacity sites are sometimes needed to augment the capacity and adjust the coverage grid so that there is a higher density of macro sites in high traffic areas. The next evolution to this concept is to build small cells in hotspot traffic areas as an underlay to the existing macro cell network. The primary motivation behind using small cells is to use a non-uniform network topology that more closely mirrors the non-uniform traffic distribution by placing the network closer to the end user to improve the overall performance and wireless capacity. There are a number of reasons why there is significant interest from operators to utilize this approach:

- Customer high traffic areas are non-uniformly distributed so small cells can often be the more practical and efficient solution to address specific hotspot congestion areas.
- The smaller physical form factor of small cells will help mitigate zoning and installation concerns that often impede rapid deployment.
- The current and expected wireless data usage growth is outpacing the expected pipeline of usable wireless spectrum assets that could be deployed to support the hotspot traffic.
- Improving the signal level and signal quality is important especially in the indoor locations to guarantee reliable data connections.
- Existing wireless access technologies such as LTE are already very spectrum efficient, so further site densification of the network is needed.

Finding a site location for the macro base station can be difficult, expensive and take a lot of time to get the site permissions. The site lease for the full macro site can be expensive which contributes to the reoccurring radio network operating expenses (OPEX). The compact small cell product can be placed close to the area where higher signal level and more capacity are required. The implementation cost and the reoccurring costs of the small cell can be much lower than that of the macro site. In short, the small cell motivations can be summarized as follows:

- More capacity specifically in hotspot areas of the network
- Better signal level and signal quality
- Lower OPEX in terms of site costs
- Faster rollout of new sites

An example evolution of network topology is shown in Figure 9.1. The starting point is macro-only network with a few traffic hotspots and in-building coverage challenges. Small cells are added to provide more capacity in the hotspots and better indoor coverage.

9.3   Small Cell Challenges and Solutions

From an operator's perspective there are several challenges to building, optimizing and operating small cells. Due to their low transmit power, small cells will have a smaller coverage footprint, so it is imperative that traffic hotspots are identified accurately so that a small cell capacity solution will yield the intended capacity relief. Further development regarding the accuracy of geolocation of customers will be required to be able to surgically place small cells

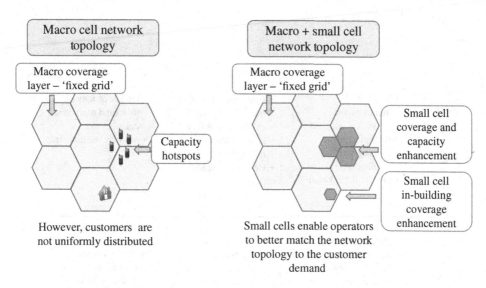

Figure 9.1   Network topology evolution to include small cells

in the proper locations. For example, if a small cell covers only a 100 m footprint the geolocation techniques used to estimate customer traffic must be within this criterion; otherwise, there is a risk that a small cell is placed too far away from the hotspots and the small cell may actually cause unintended interference into the macro network and decrease system capacity. The location of small cells is so critical that this may present additional challenges when the time comes for zoning and site acquisition of suitable site candidate locations.

Another challenge is providing suitable cost-effective backhaul solutions to the small cells that scale down proportionally with the expected decrease in radio access network (RAN) equipment and leasing costs.

As the number and density of small cells increases over time the next challenge will be to optimize the coordination with the existing macro layer to further enhance the capacity. There are numerous features being standardized as discussed in Chapter 7 that will assist with creating a seamless experience and maximizing the offload to the small cell network while mitigating system interference issues. Optimization features as described in Chapters 7 and 8 will also play a key role in the operators' ability to manage the integration and optimization of potentially tens of thousands of small cells that may be needed.

The challenges and enablers of the small cell deployments are shown Figure 9.2.

9.4   Summary of Learnings from Small Cell Deployments

The small cell products have been deployed in a number of commercial networks which allows engineers to get useful hands-on learnings from the field. The 10 main learnings from the small cell measurements are summarized in Figure 9.3:

1. The site location is very critical for maximizing the benefit from the small cell. The location is important relative to the users and relative to the macro cell signal level. If

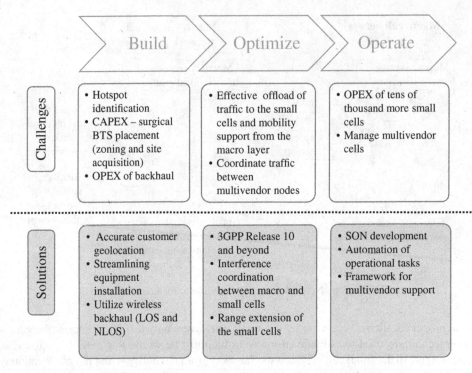

Figure 9.2 Challenges and solutions of large-scale small cell deployments

the small is located very close to the users, it can collect a lot of traffic and offload users from the macro cell. If the small cell is located in areas with weak macro signal, the coverage is larger compared to placing the small cell in an area with strong macro signal. The coverage-driven small cells tend to have larger coverage areas than capacity-driven small cells. The small cell installation should avoid direct line of sight to the macro cell because the small cell dominance area would be very small. One option is to use higher gain antenna in the small cell to improve the dominance area.

2. The measurements show that the radio quality in the small cells is generally good: small cells improve signal levels and signal-to-noise ratio (SNR) which make the overall network quality and application performance better.

3. The UE data rates improve with the small cells driven by higher signal quality and less simultaneous users.

4. The co-channel deployment between small cell and macro cell works fine in practise. Even multivendor co-channel case has been tested and shown good performance. UEs did not support enhanced inter-cell interference coordination (eICIC) during these tests.

5. Small cells can offload macro cell traffic. The total data volume carried by the small cells and macro cells typically increases compared to the macro-only case while the data volume and the connected UEs in the macro cells decrease.

6. If the operator can use a dedicated frequency for the small cell, it can provide higher throughput since there is no co-channel interference from the macro cell.

1) Selection of small cell location is important	• Small cells can collect traffic (only) if it can be located close to the hot spot users
2) Signal level and quality improve with small cells	• Small cells increase reference signal received power (RSRP) and quality (RSRQ)
3) UE data rates improve with small cells	• Higher signal levels and quality turn into higher user data rates
4) Co-channel integration to macro network is smooth	• Co-channel deployment is generally fine including mobility performance
5) Small cells can offload macro cell traffic	• Total traffic increases with the small cell • Macro cell traffic decreases
6) Dedicated frequency small cell gives higher throughput	• If the small cell can utilize different frequency than macro, the throughput can be maximized
7) Indoor small cells are less impacted by macro cells	• Macro cell interference has less impact on the indoor cells due to wall isolation
8) Higher small cell power gives higher capacity	• 5 W output power increases coverage and capacity of the small cell compared to lower power
9) Product performance and stability is excellent	• Very good cell availability and success rates
10) Site acquisition and backhaul challenges	• Process improvements, flexible site solutions and flexible backhaul products needed

Figure 9.3   Main learnings from small cell deployments

7. The co-channel interference is relatively simple to manage in indoor small cells because the wall isolation minimizes the interference from the outdoor macro cells to the indoor small cells.
8. The small cell can benefit from high transmission power: more power leads to larger coverage area providing more capacity and more offloading from the macro cells.
9. The product performance was very robust in terms of cell availability and in terms of key performance indicators (KPIs). One reason may be that most hardware and software components are shared with macro cells.
10. The main challenge with the small cells was the site acquisition, backhaul and power. It does not help to have low-cost product if the site and backhaul cost more than 10 kEUR.

The following figures show a few examples from the small cell measurements. Figure 9.4 illustrates drive test measurement results from co-channel small cell deployment at 2.6 GHz band. The small cells provide a considerable improvement in the signal level and also in the SNR. The median signal level improvement in this case is 7 dB and SNR improvement 3 dB,

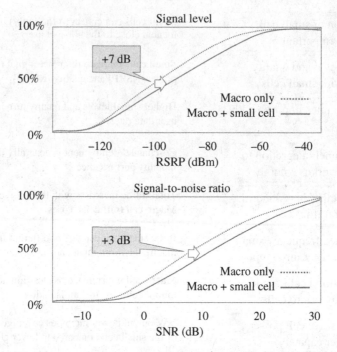

Figure 9.4 Improvement in signal level and signal-to-noise level with small cells

indicating that the typical small cell deployment improves the radio network quality which turns into higher user data rates.

The higher signal level and quality leads to higher user data rate. The data rate can be illustrated with the used modulation and coding scheme (MCS). Figure 9.5 shows MCS in five different small cells and the average value in the macro cells. We can observe that the

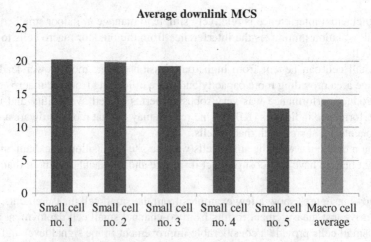

Figure 9.5 Modulation and coding scheme (MCS) in small cells and in macro cells

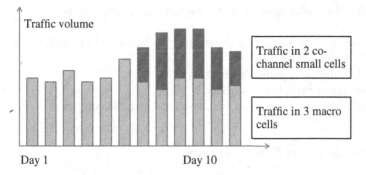

Figure 9.6 Offloading with micro base station

small cells can provide very high MCS values and, respectively, high data rates. We can also observe that there is large variance in the small cell MCS values depending on the small cell location illustrating the importance of the small cell antenna location.

Figure 9.6 illustrates the data volume carried by the macro cells and by the small cells. The total data volume increases while the data volume in the macro cell decreases, which illustrates efficient traffic offloading to the small cell.

9.5   Installation Considerations

The small cell installation requires different considerations compared to the macro cells. The macro cell installation procedures have been optimized over many years while the small cell installations are still in the early phase. The following installation aspects should be taken into account:

- Site aesthetics are important. The small cells are generally located at lower heights and closer to the users than the macro cells. Therefore, the appearance of the site needs to be considered including the number of separate boxes and the cabling. The small cell form factor is important including the street furniture camouflage, like small cell colour.
- Cost of the small cell installation should be lowered which requires minimization of the costly civil works. Simple installation and simple configuration are important. Cranes or street closures may be needed for the installation which can increase the cost.
- Always-on power supply is needed. If the street lights get power only during the night hours, it is not useful for the small cells.
- The remote access to the small cell unit needs to be considered possibly with, for example, Bluetooth.
- The backhaul design considerations including the additional boxes and RF requirements for the wireless backhaul.
- The negotiation with landlord or city council required.
- In general, the street pole deployments are more spectrally efficient than rooftop-mounted small cells because of less interference.

The small cell installations need to consider also the safety aspects:

- Safety distance from the users to the small cell which depends on the transmission power and on the antenna gain
- Structural stability of the installation posts
- Protection against vandalism and attack
- Staff training required to work on lamp poles and on radio equipment

9.6 Example Small Cell Case Study

This section shows an example small cell case study on a live network in a major city in the United States. The small cell product is Nokia Flexi Zone micro base station, see Chapter 6 for more details on the product. The small cell uses advanced wireless services (AWS) 1.7/2.1 GHz frequency. The spectrum is shared with the macro cell. The bandwidth is 10 MHz. The synchronization is based on GPS.

9.6.1 Site Solution and Backhaul

The small cell site solution is shown in Figure 9.7. The small cell is installed in the 10-m light pole. The small cell antenna is located on top of the light pole. The antenna is omni-directional with 5 dBi gain. The backhaul connection uses microwave radio at 58 GHz frequency. The small cell product itself is located below the microwave radio. This micro base station provides 5 + 5 W output power and weights 5 kg. The GPS antenna is integrated into the micro base

Figure 9.7 Site solution (1 = micro cell antenna, 2 = microwave transport, 3 = micro base station, 4 = power panel)

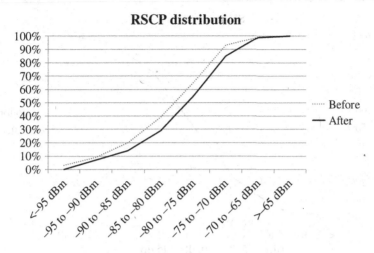

Figure 9.8   RSRP levels in drive testing before and after small cell installation

station. The power panel is also located at the pole. The total weight of the installation is approximately 23 kg.

9.6.2   Coverage and User Data Rates

The main target of the small cell installation was to improve the coverage and quality in the busy areas in the city where it is difficult to install a new macro base station. In downtown urban areas small cells have the added benefit of improving coverage along the sidewalks and roadway areas that form urban canyons between the tall skyscraper buildings and are difficult to cover by macro sites. Figure 9.8 shows the RSRP levels and Figure 9.9 the SINR levels measured by drive testing before and after the small cell installation inside the coverage area of the small cell. The median RSRP and SINR improves by 2 dB. The drive testing, however, does not show the small cell coverage benefit for the indoor locations. Another way of illustrating the coverage benefit is to analyse the UE power headroom. UEs report the available transmission power which indicates the available power resources for the uplink transmission. The distribution of the power headroom reports is shown in Figure 9.10. The UE power headroom increased considerably with the deployment of the small cell indicating clearly better coverage in the area.

The user data rate measurements in selected indoor locations are shown in Figure 9.11. The coverage was weak in those spots before the small cell installation and the data rate was just a few megabits per second. The small cell installation boosted the data rates considerably with uplink data rates exceeding 6 Mbps and downlink even higher. The ability to improve indoor data rates with outdoor small cells is incredibly important. Studies have shown that close to 80% of mobile data usage is generated indoors. Small cells are designed to better address modern mobile data traffic that is more nomadic in nature compared to more voice-centric mobile networks. These indoor measurement locations are in areas you would expect to see mobile data usage such as popular coffee shops that contain large lounge areas where

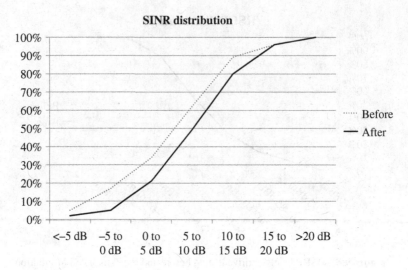

Figure 9.9 SINR levels in drive testing before and after small cell installation

mobile data is consumed. Enhancing the coverage and user data rates in these locations can vastly improve the customers' perception of the quality and reliability of their wireless connection.

9.6.3 Macro Cell Offloading and Capacity

The total data volume carried by the macro cell and by the small cell is shown in Figure 9.12. The statistics shows that the total data volume carried together by the macro and the small cell has increased while the data volume by the macro cell has decreased. The results indicated

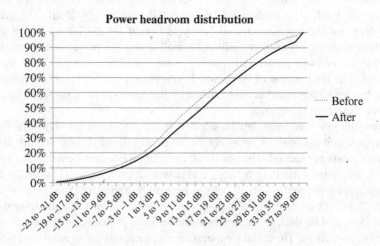

Figure 9.10 Distribution of UE power headroom reports

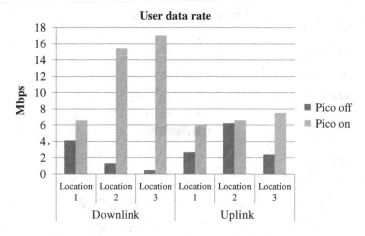

Figure 9.11   User data rate in selected indoor locations

that the small cell was able to improve the network quality leading to higher data volumes. The results also indicated that the small cell was able to offload traffic from the macro cell. Figure 9.13 shows the physical resource block (PRB) utilization in the macro cell before and after small cell installation. The result shows that the macro cell PRB utilization has decreased with the small cell which leaves more room for other users in the macro cell.

9.6.4   KPIs in Network Statistics

The network quality in terms of setup success rate and drop rate must be maintained after the small cell installation. The accessibility is shown in Figure 9.14 and the drop rate in Figure 9.15. Both performance indicators are on very good level in the macro cell and in the

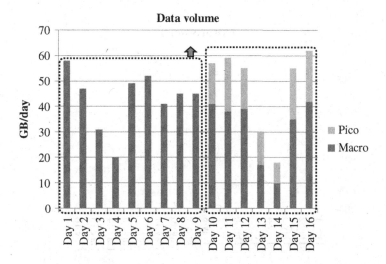

Figure 9.12   Data volumes before and after small cell installation

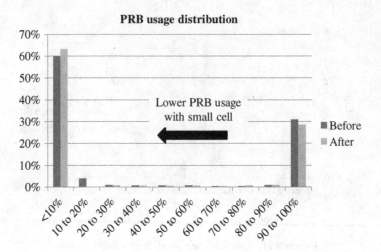

Figure 9.13   Physical resource block (PRB) utilization in the macro cell

small cell. The accessibility is clearly more than 99% and the drop rate below 0.1%. The macro and the small cells have approximately similar values.

9.6.5   Mobility Performance

The small cell covers only a limited geographical area which can cause potential issues with the mobility performance at high velocities. If the small cell signal fades too rapidly, it may lead to radio link failure before the handover has been completed to the macro cell. Figure 9.16 shows inter-eNodeB handover success rate in the small cell and it is more than 98%. In this case, the small cell provides even higher success rate than the macro cell. More than 50% of

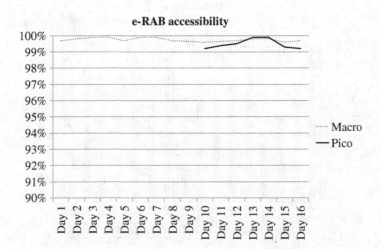

Figure 9.14   Enhanced radio access bearer (eRAB) accessibility

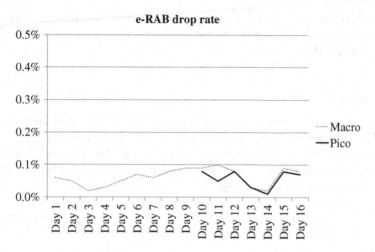

Figure 9.15 Enhanced radio access bearer (eRAB) drop rate

the outgoing handovers from the small cell were directed on one macro cell which is typical that the small cell has only a low number of macro cells neighbours.

9.6.6 Parameter and RF Optimization

No specific parameter nor RF optimization was done in this small cell case. A few example optimization ideas could be considered for this case:

- Offset parameter for idle and connected modes to push more users from macro to the small cell

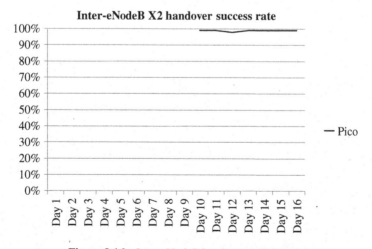

Figure 9.16 Inter-eNodeB handover success rate

- Macro cell RF tuning to shift overlapping macro coverage away from the small cell area
- Small cell RF tuning, for example, by moving the micro base station closer to the antenna to minimize the cable loss

9.7 Summary

This chapter illustrated the main learnings from the small cell deployments in live networks. The small cell performance in general looks very promising in improving the network quality and data rates and offloading the macro cells. The co-channel usage of small cells and macro cells performs fine in practical cases assuming that the small cell antenna location is properly selected. The important factors in the small cell installation are related to the site selection and to the backhaul connection. Good RF planning and placing small cells in areas with poorer macro signal coverage increase the effectiveness of the small cell to offload traffic. Multiple options exist for the backhaul connection including wireless line-of-sight and non line-of-sight radios.

10

LTE Unlicensed

Antti Toskala and Harri Holma

10.1 Introduction

This chapter presents principles of the LTE unlicensed operation which is being worked in 3GPP under the name license assisted access (LAA) for Release 13. The study phase was finalized in June 2015 and the actual work item scheduled being finalized by end of 2015. The use of LAA allows benefiting from the additional capacity available from the unlicensed spectrum. This is possible especially in the public hot spot and corporate type of environment to provide additional capacity for the licensed band operation. This chapter introduces first the 5 GHz frequency band under consideration with the key regulatory requirements to be considered and then covers the motivation for using LTE in unlicensed spectrum and the LTE unlicensed operation principles. The unlicensed band-specific coexistence issues are addressed as well as the relative performance in terms of achievable capacity and coverage of LAA for

LTE Small Cell Optimization: 3GPP Evolution to Release 13, First Edition.
Edited by Harri Holma, Antti Toskala and Jussi Reunanen.
© 2016 John Wiley & Sons, Ltd. Published 2016 by John Wiley & Sons, Ltd.

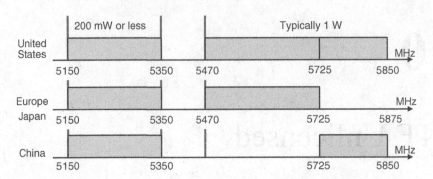

Figure 10.1   Unlicensed spectrum availability in different regions

LTE compared to using Wi-Fi in the 5 GHz band, including the performance when coexisting with Wi-Fi on the same channel. 3GPP standardization plans and expected schedule are also covered before concluding this chapter.

10.2   Unlicensed Spectrum

The frequency band of main interest in 3GPP is the 5 GHz band, which has a lot of unlicensed spectrum available globally, much more than the 2.4 GHz frequency band. Most markets offer large amount of spectrum in the 5 GHz band, for example, in Europe there is 455 MHz of spectrum available in the band as shown in the Figure 10.1. Allocations are also available in countries like Korea and Japan in 5 GHz band. In the United States, there is more than 600 MHz spectrum available on this band and the upper part of the 5 GHz band (US 5.725–5.850 GHz) differs from other regions as the regulation mainly restricts only items such as maximum transmission power there. However, for the global marketplace it is desirable to have such a standard for unlicensed band which can be applied globally and thus it is the intention of 3GPP to develop a solution which will be usable in many regions. There can be some parameterization options which may not be applicable always to all regions but basic functionality is expected to be the same.

The use of unlicensed spectrum has typically several regulatory requirements to facilitate the sharing of the band with other technologies as well as to ensure efficient use of spectrum. The European key regulatory requirements can be found from [1]. The following requirements can be identified as being necessary for 3GPP to cover in the design:

- Requirements for dynamic frequency selection (DFS) for avoiding radars operating in the 5 GHz band. Generally a system operating in the 5 GHz band (known exception being the US upper part of the 5 GHz band) needs to have the ability to detect whether a radar system is using the band (or part of the band). If a radar signal is detected then the band (or part of it) must be vacated within a given time limit. LTE needs only to copy with the detection requirements (and existing test cases which are not technology dependent) and to be able to stop the LTE operation when detection conditions are met but otherwise impacts from such a principle are rather minor for the LTE specification side.
- For the efficient use of the band there are requirements for minimum bandwidth occupancy. This prevents operating, for example, an LTE in such a way that only 180 kHz would

be allocated for a very low data rate transmission. Such an approach would unnecessary block the 20 MHz channel from other users as they would detect the energy level from the narrowband transmission which is not utilizing the full band. The regulation is Europe requires to use at least 80% of the nominal bandwidth but temporary allows going to as low as 4 MHz bandwidth. There has been some discussion to adjust this in Europe due to the needs of IEE 802.1.ac; thus some more flexibility could be potentially provided here. While the LTE downlink is well suited for this type of requirement, the uplink structure is likely to require some more changes to satisfy the minimum bandwidth occupancy requirements.

- Adaptive channel access is best known as listen-before-talk (LBT) or clear channel assessment (CCA). The key principle is to determine first if the channel is used by other users of the band before transmitting on the channel. The European regulation defines energy threshold, which is dependent on the transmit power as well, to be met before the channel can be considered 'free'. For the operation, there is need to stop the transmission and to check the channel status after the maximum channel occupancy time has been exceeded. This allows other systems to coexist with other users (other than radar) on the band. This may result in a situation that the particular channel used earlier may not be available as expected due to other increased activity of the band. Of course, all the other systems will need to follow the same LBT/CCA principles but by nature the unlicensed band may have varying availability due to the load from other system. This aspect has clear impacts on the LTE specifications and is analyzed more in Section 10.5.
- The allowed transmission power also varies depending on the part of the band. Typically, the lower portion of the 5 GHz band is often determined and allowed for indoor-only use with 200 mW or less transmit power while the upper part of the spectrum enables higher transmission power, typically 1 W at the antenna connector.

10.3 Operation Environment

The use of LTE on unlicensed bands has been generally considered for public indoor cells or outdoor hot spots, places where there is coverage from licensed band LTE operation but additional capacity would be beneficial, as shown in Figure 10.2. The intention of the 3GPP ongoing work is not to define a stand-alone system that could be used, for example, at residential/home environments since there are existing solutions (e.g. femto cell or Wi-Fi) for that case. Another typical use case would be corporate environment that would benefit from

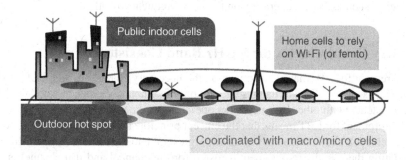

Figure 10.2   LTE-unlicensed environment

the use of high-capacity LTE radio technology. In the home environment, the use of Wi-Fi is foreseen to offer sufficient capacity and thus unlicensed band LTE operation is not targeted by mobile network operators for such an environment.

10.4   Motivation for the Use of Unlicensed Spectrum with LTE

The use of LTE with unlicensed spectrum is driven by the increase in the traffic volumes and the number of mobile broadband users globally. As mentioned previously, the 5 GHz spectrum offers a lot of available channels/bandwidth. With the LTE technology the following could be achieved:

- Better spectrum efficiency than the current technologies in use with the 5 GHz band. Since LTE radio technology is based on the state-of-the-art technology, it can achieve both *high data rates and at the same time high spectral efficiency* when operating in the unlicensed band. Besides the capacity, the coverage of LTE technology is also superior, especially when combined with the use of licensed band operation.
- From the network management point of view using the unlicensed band with LTE, instead of with an alternative radio technology, provides *an integrated solution to the existing operator radio network setup*, avoiding multiple solutions for network management, security or authentication. Having only a single technology simplifies the overall network maintenance. Finally, the use of LAA is fully transparent to the LTE core network, avoiding the need to upgrade any of the evolved packet core (EPC) elements.
- The combined use of licensed and unlicensed spectrum provides the end user opportunistic possibilities to enjoy higher data rates and overall better performance when the unlicensed band operation is available. On the other hand, the dynamic operation allows to *ensure service quality* with 1 ms resolution with the use of licensed spectrum should the unlicensed spectrum become unusable due to any reason, such as smaller coverage, interference from another systems or avoidance, for example, of a radar operating in the band. The use of carrier aggregation allows automated and extremely dynamic selection between licensed and unlicensed bands for data, thus allowing the network to always provide sufficient service quality for the end user.

Based on the studies conducted, the LTE technology can meet the regulatory requirements for the unlicensed band *and allow the coexistence with the other LTE systems as well as other technologies*, such as Wi-Fi, operating on the same frequency band.

10.5   Key Requirements for 5 GHz Band Coexistence

The requirement for LBT has clear impact on the necessary changes to adapt LTE to be used on the unlicensed band. These changes will not only facilitate coexistence with Wi-Fi but also allow coexistence with other LTE networks operating on the same band.

The requirement for LBT means that the current principle in use for licensed band operation (from the eNodeB) of continuously transmitting every 1 ms cannot be applied. In line with the regulation there is the requirement to listen for the channel and if a channel is detected busy then not to transmit. Also if the channel is determined to be free (the detected energy

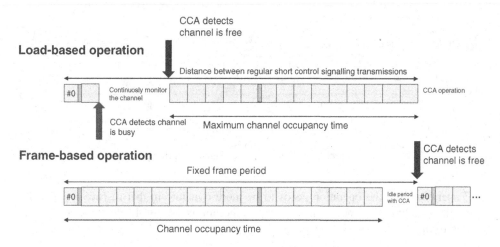

Figure 10.3 Load-based and frame-based operation principles

level is below threshold) one cannot continue transmitting forever but there is a maximum channel occupancy time after which one has to listen. This can be done either with a frame-based operation where the moment for checking the channel being available is fixed or with the load-based operation where the channel is being monitored for longer period of time (if the channel was not free at start of the monitoring), with example shown in Figure 10.3. The example in Figure 10.3 assumes in any case some elements being fixed in time domain for the load-based operation though the detailed design is still to be determined by 3GPP.

For facilitating the cell search (or discovery), the regulation allows the so-called 'short control signalling' (SCS) to be used. Within ETSI specifications [1] this is limited to be 5% channel occupancy over 50 ms measurement interval. This allows periodic signals such as the current primary synchronization signal (PSS) and secondary synchronization signal (SSS) to be transmitted periodically. As with the operation on 5 GHz band, one is first going to be synchronized to the network via the licensed band carrier and the expected mobility with 5 GHz band is foreseen to be less than 350 km/h, the requirements are not as stringent for the structure of the synchronization/discovery signals as with licensed band LTE. Thus, it is not likely to be needed to send them with the same internal (every 5 ms) as with the current LTE frame structure (as covered in Chapter 2 and in Reference 2), but rather sending them every 10 or 20 ms or even less seldom should be sufficient. SCS can be transmitted regardless of the LBT considerations, as shown in Figure 10.4. Note that the timing in Figure 10.4 is only for illustration purposes and assumes the same periodicity as the current Release 8-based frame structure. The Release 8 TDD-based frame structure is expected to be modified accordingly to fit the 5 GHz band operation.

The regulation does not determine detailed use for the SCS operation, but allows considering different uses, such as discovery and HARQ feedback. In LTE design, one may always count on the existence of the licensed band carrier, thus only such control signalling, foreseen necessary, specific for the unlicensed band operation is to be transmitted using the 5 GHz band. For example, the Release 8-based broadcast channel is not necessary to support over the 5 GHz band as it can be always obtained via the primary cell on licensed band. The common reference signals (CRS) cannot be transmitted continuously like in Release 8-based

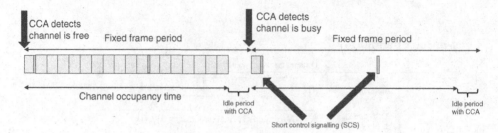

Figure 10.4   Short control signalling principle

LTE operation, but the operation needs to be more dynamic in that respect as well. Similarly, the sounding reference signals (SRS) in the uplink are not likely to be used in their current form.

10.6   LTE Principle on Unlicensed Band

The LTE operation on the unlicensed band is built on top of the LTE-Advanced carrier aggregation, which has been deployed commercially since 2013. The simplest form of LTE-Unlicensed would be to use the unlicensed band with downlink-only carrier aggregation, while the use of uplink would also be possible in-line with 3GPP carrier aggregation principles, as illustrated in Figure 10.5. This is similar to the first-phase LTE-Advanced carrier aggregation in the commercial networks which have started with the downlink-only aggregation. The primary cell ensuring connection maintenance is always located on the licensed band carrier.

When operating with downlink only on the unlicensed band (a.k.a. supplemental downlink), the LTE eNodeB can perform most of the necessary operations to ensure reliable communications, including checking whether the intended unlicensed channel is free from other use. The LTE terminal capable of operating on the unlicensed band needs to be able to make the necessary measurements for support of the unlicensed band operation, including providing the feedback when the terminal is in the coverage area of an LTE eNodeB transmitting with

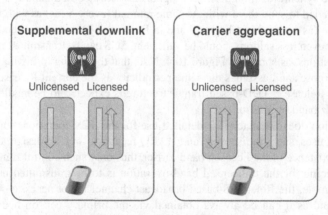

Figure 10.5   LTE-unlicensed operation modes

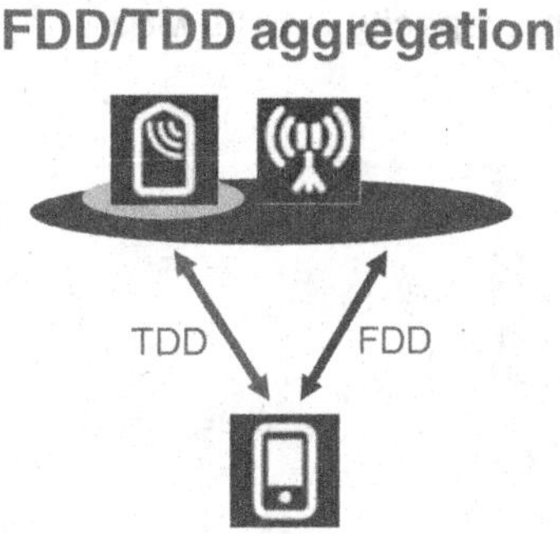

FDD/TDD aggregation

Figure 10.6   LTE-advanced aggregation between FDD and TDD bands

the unlicensed spectrum. Once the connection is activated to use the unlicensed band also, the existing channel quality information (CQI) feedback will allow the eNodeB to determine what kind of quality could be achieved on unlicensed band compared to the licensed band. The downlink-only mode is especially suited for such a scenario where the data volumes are dominated by the downlink traffic.

The uplink transmission (full TDD operation) from a terminal operating on the unlicensed band requires more features both in the terminal side and in the LTE eNodeB compared to the existing licensed band operation. Those extra features are needed to meet especially the specific requirements on the type of transmission on the unlicensed band, including enabling the LBT feature and radar detection in the terminal side. In the actual specification work the support for LAA shall be phased in such a way that first only downlink aggregation (supplemental downlink) with 5 GHz band will supported in Release 13 and in Release 14 the full TDD operation will be enabled including uplink functionality.

The LTE-Advanced carrier aggregation allows aggregating between FDD bands as well as between TDD bands from Release 10 onwards. With the Release 12 version of LTE-Advanced specifications, aggregation between FDD and TDD bands is also possible as shown in Figure 10.6, thus providing further flexibility for selecting the band that should be used together with the unlicensed band with LAA operation.

10.7   LTE Performance on the Unlicensed Band

The first-phase LTE networks provide up to 150 Mbps data rate and the latest chip using LTE-Advanced support up to 300 and 450 Mbps peak data rate. The LTE capabilities are evolving continuously to later enable higher data rates of up to 1 Gbps or even more, but the next steps after 450 Mbps is the support of 600 Mbps downlink peak rate. These rates are achieved when devices supporting more than three aggregated LTE downlink carriers become available. The LTE performance study done in an office environment was found to be roughly twice the comparable Wi-Fi network data rate, as shown in Figure 10.7, indicating only the extra performance achievable with a single 20 MHz carrier on unlicensed spectrum when there were no other interfering networks present and with 30 Mbps offered load per access node, as presented also in [2]. The office environment contained for both LTE and Wi-Fi networks six access nodes each operating on the same 20 MHz carrier without interference from other

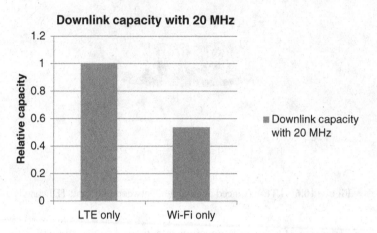

Figure 10.7 LTE capacity compared to the Wi-Fi capacity in the office environment

networks. The relative capacity of LTE network would be even higher when the load and number of users are further increased as Wi-Fi capacity will not increase more or even goes down while LTE network can still reach higher throughout. Before the work started in 3GPP, the issue of use of unlicensed spectrum with LTE technology had also been addressed in relevant conferences in the field, such as presented in [3], raising different possibilities for coexistence. Especially in the markets where regulatory requirements do not define specific methods to be used, one can of course consider solutions such as fixed amount of DTX to enable coexistence in addition to the more evolved solutions.

If one would consider single access node with large number of traffic (hot spot), the LTE design allows the system to stay robust with very large number of users while capacity of a Wi-Fi access would start to drop sharply with the increased number of traffic. The advanced features for handling the load in LTE can also be applied when operating in the unlicensed band thus facilitating high capacity also in the presence of large number of users.

From the coverage point of view the link budget with LTE is clearly better and allows having fewer nodes than Wi-Fi for a given area for reaching the same capacity that a Wi-Fi network would provide. This allows considering trade-off in deployment between the total network capacity and the number of LTE nodes being deployed for the unlicensed band.

10.8   Coexistence Performance

Important part of the deployment is coexistence with Wi-Fi and other LTE networks on the unlicensed bands. When running the LTE and Wi-Fi on the same 20 MHz carrier, both networks should use LBT solution to ensure that both systems can coexist smoothly. There is always an impact when additional interference is created with additional transmitters, regardless of whether those additional transmitters are using LTE or Wi-Fi or some other technology. Thanks to the advanced radio features, the use of LTE in a co-channel scenario with another LTE network results in smaller overall capacity loss than would be the case with two Wi-Fi networks. The performance was studied in the office environment with two corridors and

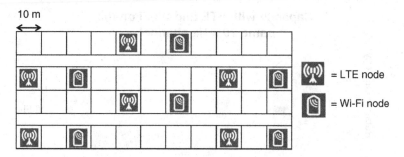

Figure 10.8   Example environment used in the simulations

considering a single floor only. The impact and additional capacity from the licensed band LTE network were not considered in the environment provided in Figure 10.8.

In Figure 10.9, the performance is illustrated with two different networks operating on the same 20 MHz channel. In the first case, the two different networks are using LTE technology while on the other case the two different networks are using Wi-Fi. For higher loads the relative performance of LTE is better.

Especially for the first-phase LTE deployments on unlicensed bands it is also important to have good performance when the other network on the same channel is a Wi-Fi network. In Figure 10.10 the downlink performance is shown with two independent networks of different technology sharing the same 20 MHz channel. As shown in Figure 10.10, when placed on the same 20 MHz channel without specific considerations other than the regulatory requirements Wi-Fi suffers more as LTE is more robust to co-channel interference. Both networks experience the degradation due to increased interference but LTE network can still maintain good performance. When adding additional fairness algorithm in the LTE side it is possible to reduce the impacts to the Wi-Fi network such that the degradation on the Wi-Fi network is of the same order compared to the degradation due to the interference from another Wi-Fi network. Thus when implementing proper fairness solution in the LTE side to control the impact

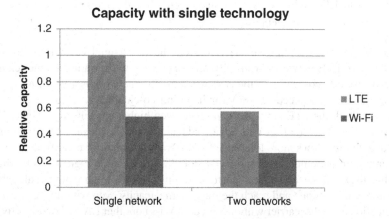

Figure 10.9   Performance per access node with two LTE or two Wi-Fi networks on the same channel

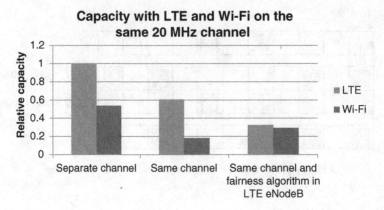

Figure 10.10 Coexistence performance between LTE and Wi-Fi

to the existing Wi-Fi network performance, it does not really matter if the network causing interference is another Wi-Fi or LTE network. This shows that LTE technology can also be implemented in such a way that it not only meets the requirements of the unlicensed band operation, but can provide even extra fairness to compensate the lower interference tolerance of Wi-Fi networks. The higher capacity of LTE technology will also reduce the needed amount of nodes to what would need to be installed to serve a given amount of traffic compared to a Wi-Fi network.

The coexistence performance presented in Figure 10.10 is actually the worst-case scenario. When considering on which channel to place the LAA transmission, one should at first set a priority to select a frequency which is not used by a high-power and high-activity Wi-Fi network or another LAA network. Selecting a channel which has either low activity or low level of interference from other users of the channel will allow reaching the performance equivalent to a single LTE network while also minimizing the interference for any of the other networks on 5 GHz band. Since LTE radio resource management is a dynamic process, it allows avoiding static worst-case interference situation by continuously monitoring the environment. LTE radio resource management should have as the first priority avoiding the use of channels overlapping with a closeby Wi-Fi or another LTE network on the unlicensed band.

LTE itself can tolerate more interference on the co-channel case; thus, as seen from Figure 10.9, two LTE networks from different operators on the same channel cause clearly smaller capacity loss compared to the case when another network is a Wi-Fi network as both LTE networks have advanced features for handling co-channel interference. In the studied example case, both the LTE networks can reach even higher capacity than single Wi-Fi network. The use of LTE technology with advanced features such as physical layer re-transmission and link adaptation allows more tolerance for different interference conditions.

An example measurement from the 5 GHz band is presented in Figure 10.11, which shows clearly that there are several places in the example European shopping mall where LAA transmission could operate and reach the maximum capacity even with multiple carriers.

The selection of such a carrier without detected Wi-Fi or other LAA network activity allows to maximize the performance. The eNodeB (or UE) needs only to have the necessary idle

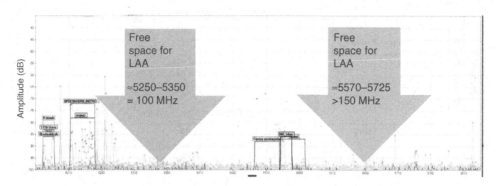

Figure 10.11 Example measurement for lower part of 5 GHz band

periods as required by the regulation but otherwise the channel may be occupied constantly due to lack of any other traffic. Similarly when looking at the full 5 GHz band, as shown in Figure 10.12, there are several channels free. Please note that the band above 5.725 GHz is not available in Europe at the moment and was thus not considered in the example measurement. The upper part of the band is more attractive as transmit power of 1 W is allowed there than what is possible with below 5.350 GHz.

The simulation result presented assumes fully loaded Wi-Fi network and corresponds thus for the worst-case results. If a Wi-Fi network is only slightly loaded, or in the extreme case sending only the periodic beacon signal, then the use of the same 20 MHz channel provides hardly any reduction to LTE capacity. Even with strong Wi-Fi signal LTE can reach more or less the maximum capacity when Wi-Fi is sending data not too often. An example utilization measurement is shown in Figure 10.13, in which network utilization of the lower part of 5 GHz band in time domain is also observed. When the utilization is below 20%, then there is clearly room for other networks to share the same channel if fully free channel cannot be found. With such low utilization, the LTE capacity would not be impacted too much even if the Wi-Fi signal is stronger than the energy detection threshold used.

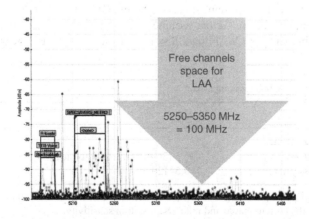

Figure 10.12 Example measurement over the full 5 GHz band in Europe

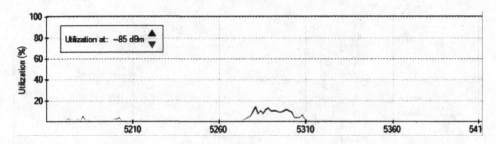

Figure 10.13 Example network utilization at lower part of 5 GHz band

10.9   Coverage with LTE in 5 GHz Band

One of the aspects with 5 GHz band is the limits for the maximum transmission power. As such this is not a new element for LTE as there has been lot of work done for optimizations for small cells operation with different eNodeB power classes being defined. The use of 5 GHz band itself causes naturally higher penetration loss due to walls and windows than lower frequency bands, but not necessarily drastically different from the 3.5 GHz band, especially for line-of-sight scenarios.

The link budget shows the maximum path loss between the terminal antenna and the base station antenna. The link budget can be used to estimate the maximum cell range with suitable propagation models. The link budgets can also be used to estimate the relative coverage of LAA and Wi-Fi. This section shows the coverage benchmarking between LAA and Wi-Fi technologies. The uplink link budget is shown in Table 10.1 and the downlink one in Table 10.2. The transmit power of LTE UE is 23 dBm while Wi-Fi terminal output power is 15–20 dBm. LTE sensitivity requirement for the medium range base station is –96.5 dBm for the Reference channel A1–3 which gives approximately 2 Mbps. The typical sensitivity is better – in the existing macro base stations even 10 dB better. Wi-Fi sensitivity is assumed to be –93 dBm for the lowest modulation and coding scheme of MCS0. The maximum uplink path loss is then 119.5–129.5 dB for LTE and 108–113 dB for Wi-Fi. The LAA downlink path is 129 dB with 30 dBm base station power (the same is assumed for Wi-Fi) and –9 dBm UE sensitivity. The sensitivity requirement is taken from 3GPP minimum requirement for Band 1 with 20 MHz bandwidth. We further assume that the typical UE sensitivity is 5 dB better than 3GPP requirement. The Wi-Fi path loss is estimated at 123 dB. The reason for the link

Table 10.1   Uplink link budget comparison

	LAA	Wi-Fi
UE transmit power (dBm)	23	15 to 20
BTS sensitivity (dBm)	–96.5 (3GPP)* to 106.5 (typical)[†]	–93 (typical)[‡]
Maximum path loss (dB)	119.5–129.5	108–113

*Reference channel A1–3 providing 2 Mbps. 3GPP requirements for medium range base station.
[†]Typical value for the current macro and micro base station sensitivity.
[‡]MCS0 providing 6 Mbps with 20 MHz.

Table 10.2 Downlink link budget comparison

	LAA	Wi-Fi
BTS transmit power (dBm)	30	30
UE sensitivity (dBm)	−94 (3GPP)* to 99 (typical)†	−93 (typical)‡
Maximum path loss (dB)	129	123

*Sensitivity requirement with 20 MHz bandwidth for Band 1.
†Typical UE sensitivity assumed to be 5 dB better than 3GPP requirement.
‡MCS0 providing 6 Mbps with 20 MHz.

budget differences between LAA and Wi-Fi comes from the radio layer optimization including channel coding (Turbo coding), fast retransmissions and soft combining, fast link adaptation, optimized modulation for RF power amplifier (SC-FDMA) and well-defined RF requirements in 3GPP.

Figure 10.14 shows the relative cell ranges of Wi-Fi and LAA for outdoor small cell. We assume that Wi-Fi with 108 dB path loss at 5 GHz band can provide 50 meter cell range. The other values are then estimated based on the assumption of path loss exponent 4.0. LAA cell range is at least double, and the cell area four times larger, compared to Wi-Fi. The larger cell area makes it possible to collect traffic from larger area.

An important factor with LAA is the use of licensed band operation with aggregation with the unlicensed band. This makes the 5% cell edge throughout less critical as one may always complement the resulting QoS with the licensed band operation from another frequency band. This allows ensuring sufficient data rate even with cell edge conditions and also getting the most from the capacity with 5 GHz LTE as one is not forced to serve a user in very poor conditions with 5 GHz band but can focus on users with good link budget and then getting higher throughout with less overhead from channel coding. Such a user selection was not implemented in the simulations in the previous section.

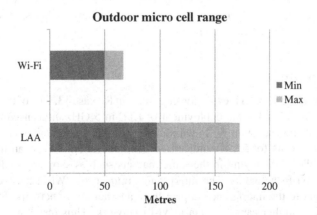

Figure 10.14 Outdoor small cell range at 5 GHz band

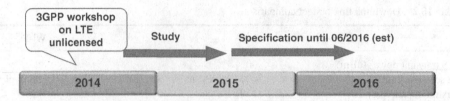

Figure 10.15 Expected 3GPP timeline for the work on LTE for unlicensed band

10.10 Standardization

3GPP Release 13 includes support of the LTE operation on the unlicensed band following the LAA for LTE study and work items finalization. 3GPP started the work with a workshop on LTE unlicensed in June 2014, with the formal study started in September 2014 [4]. The study was completed in June 2015, and Release 13 specifications are expected to be finalized (frozen) by June 2016 as visible in Figure 10.15. The study covered the necessary mechanisms for coexistence as well as evaluation of the performance when operating in the 5 GHz band. After the study was completed, the work item continued to finalize the detailed specification including the necessary band combinations to enable the LTE operation with 5 GHz band aggregated with another licensed frequency band. The latest 3GPP study status can be found from [5]. The assumption with 3GPP work is that LTE is not operated as stand-alone system on the unlicensed band but will be used in conjunction with a PCell which is operated always in the licensed band. Once the basis is specified with the Release 13 work item [6], then 3GPP will define the necessary bands and band combinations to be used with 5 GHz band, which can be done as Release independent on top of Release 13. Release 14 will then finalize support for the uplink direction, enabling full TDD operation in the 5 GHz band. Dual-connectivity is also foreseen to be addressed after Release 13.

There is also related work on-going to cover using more than five carriers in total with LTE carrier aggregation. Enabling up to 10 carriers or more would then enable similar to Wi-Fi, to aggregate up to eight carriers, that is, 160 MHz on 5 GHz band together with one or two carriers on the licensed band. 3GPP on-going work item for defining the support of up to 32 aggregated carriers is also scheduled to be finalized for Release 13 [7]. One further possibility is standalone LTE deployment on unlicensed band without any anchor carrier on the licensed band. This has not been addressed so far in 3GPP.

10.11 Conclusions

The strong momentum with LTE evolution continues in Release 13. One of the new technology components in Release 13 is the deployment of LTE in 5 GHz unlicensed band. LAA with LTE will allow coexistence with Wi-Fi without any specific coordination and will meet all the regulatory requirements for 5 GHz unlicensed band operation. This is an important feature of LAA which allows deploying in the same markets such as shopping malls and corporate environment as Wi-Fi networks. The interference impact to a Wi-Fi network from an LTE network is similar to the interference impact from another Wi-Fi network. Still such an LTE network can reach higher capacity than a Wi-Fi network. Thus installing unlicensed band-capable LTE eNodeB to be operated in aggregation with an existing LTE eNodeB in licensed band is not expected to require any more site permissions from installation point of view than a Wi-Fi access point or any other system operating on the 5 GHz unlicensed band currently would require. Especially in an environment where the traffic density is high, LAA is an attractive solution to tap the unused potential of the 5 GHz spectrum. Especially when there is possibility to control what kind of systems are installed, such as in a corporate environment, finding fully empty channels from 5 GHz band will be relatively easy, allowing LTE to reach full potential of the performance.

LTE for unlicensed band will rely on the existing LTE core network and will use the existing LTE security and authentication framework thus not requiring any changes in the core network domain. The use of LTE unlicensed together with the licensed band operation (LAA for LTE) allows having a major capacity boost from unlicensed band while still ensuring end-user quality of service regardless of the interference situation in the unlicensed band.

References

[1] ETSI EN 301 893 V1.7.1 (2012–06), 'Broadband Radio Access Networks (BRAN); 5 GHz High Performance RLAN; Harmonized EN Covering the Essential Requirements of Article 3.2 of the R&TTE Directive', European Telecommunications Standards Institute (ETSI).

[2] 3GPP Tdoc RWS-140002, 'LTE in Unlicensed Spectrum: European Regulation and Co-existence Considerations', Nokia 3GPP RAN Workshop Presentation, June 2014.

[3] T. Nihtilä, V. Tykhomyrov, O. Alanen, M. Uusitalo, A. Sorri, M. Moisio, S. Iraji, R. Ratasuk, and N. Mangalvedhe, 'System Performance of LTE and IEEE 802.11 Coexisting on a Shared Frequency Band,' Proc. WCNC'2013, IEEE Wireless Communications and Networking Conf., Shanghai, China, April 2013, pp. 1056–1061.

[4] 3GPP Tdoc RP-141664, Study Item Description, 'Study on Licensed-Assisted Access Using LTE', September 2014.

[5] 3GPP TR 36.889, 'Study on Licensed Assisted Access Using LTE', version 13.0.0, June 2015.

[6] 3GPP Tdoc RP-151045, Work Item Description, 'Work Item on Licensed-Assisted Access to Unlicensed Spectrum', June 2015.

[7] 3GPP Tdoc RP-142286, Work Item Description, 'LTE Carrier Aggregation Enhancements Beyond 5 Carriers', December 2014.

11

LTE Macro Cell Evolution

Mihai Enescu, Amitava Ghosh, Bishwarup Mondal and Antti Toskala

11.1 Introduction

This chapter covers the evolution in Release 12 and Release 13 for the macro cell layer, which is important from an operator point of view as it basically addresses how to get most of the gains out of the existing cell sites. A good part of the existing network topology has been in many cases installed originally for 3G, 2G or even analog systems, and perhaps part of the spectrum has been even refarmed to newer technologies more than once. The existing sites offer, depending on the terms and conditions of site owner of course, cost-efficient approach as infrastructure is already there in terms of power and backhaul connectivity from earlier systems. This chapter is looking into system evolution solutions at both the eNodeB and UE sides. We are first addressing the latest receiver improvement technology, on the UE side, as has been specified in 3GPP in terms of network-assisted interference cancellation and suppression (NAICS) as well as common reference signal interference cancellation (CRS-IC) in homogeneous scenarios. In the following sections, we discuss eNodeB solutions such as 3D-beamforming, also including vertical sectorization, UE-specific beamforming and

LTE Small Cell Optimization: 3GPP Evolution to Release 13, First Edition.
Edited by Harri Holma, Antti Toskala and Jussi Reunanen.
© 2016 John Wiley & Sons, Ltd. Published 2016 by John Wiley & Sons, Ltd.

DL MIMO improvements. Finally, we are elaborating on the Release 13 study on the use of non-orthogonal multiple access (NOMA) with LTE. Some of the technologies covered in Chapter 3 are usable in macro cell layer as well, such as improvements in the M2M connectivity and related coverage improvements.

11.2   Network-Assisted Interference Cancellation

Release 12 introduced in LTE more powerful receivers on the UE side, relying on non-linear processing and performing interference cancellation. Several forms of interference exist in an LTE system: *inter-stream* interference due to the non-orthogonality of spatial MIMO streams dedicated to the same UE (the so-called SU-MIMO interference) and *inter-user* interference which can arise between users spatially multiplexed in the same cell, for example, sharing the same time and frequency resources (the so-called MU-MIMO interference). Interference happens between UEs located in different cells, called *inter-cell* interference. UEs located at the cell edge are suffering in particular by the inter-cell interference.

NAICS is a downlink Release 12 technology enhancing the PDSCH interference cancellation (IC)/interference suppression (IS) capability of the UEs by providing signalling support from the network side, as shown in Figure 11.1, with the network indicating the transmission parameters used by the interfering eNodeB(s). Hence, NAICS is targeting the cancellation of inter-cell interference from the previous-mentioned interference scenarios. NAICS aims to be applicable through the whole system, in conjunction with various network configurations and technologies. At first glance, it might look that NAICS is addressing the same interference mitigation problem as FeICIC and CoMP; however, the main goal is to have a versatile utilization of NAICS, complementing the operation of FeICIC and CoMP, if these technologies are already deployed.

One of the main targets of this technology is that the NAICS UE is able to cancel the inter-cell PDSCH interference. In order to perform PDSCH cancellation, the NAICS UE

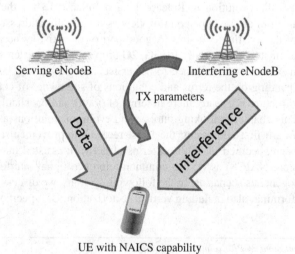

Figure 11.1   NAICS principle

needs to identify and estimate the PDSCH characteristics of the dominant interfering PDSCH coming from neighbouring cells. This is not an easy task due to the system dynamics and design flexibility. Due to the UE limitations in terms of complexity, the NAICS technology can handle the cancellation of only one dominant interferer while also only a maximum of three layers may be processed. This means that NAICS processing may happen in several combinations of desired streams and interfering streams; more specifically, NAICS processing is expected in the following configurations: one desired layer and one dominant interfering layer; one desired stream and two dominant interfering streams originating from the same interfering UE; and two desired layers and one dominant interfering layer.

Before we address the NAICS-specific operation, it is worth giving a brief description on candidate receiver structures which qualify for the task of interference cancellation and suppression. The main differentiator between these receiver structures is related to the degree of knowledge required by the receiver with respect to the signal which needs to be suppressed/ cancelled. We need to make a clear differentiation between the receiver structure as such and the mechanism used by the terminal in order to blindly estimate the parametrization of the interfering signal. In the following, we will address the receiver structures while later we discuss aspects related to the interference characteristics.

The UE receiver structures may be classified as linear interference suppression receivers and non-linear interference cancellation receivers. In order to achieve advanced interference suppression/cancellation, any such receiver structure needs to identify the dominant interferer to a certain extent. Hence, we can consider the receivers herein as interference-aware receivers. The linear receivers are based on the principle of enhanced linear minimum mean squared error interference rejection combining (LMMSE-IRC). Such receivers explicitly consider the interferer's channel estimates in the process of interference suppression. If precoding is used by the interferer, then the effective channel knowledge is utilized, that is the precoded/effective channel. Such a receiver structure reconstructs the effective channel of the interferer and suppresses the interference.

A more powerful receiver category is based on interference cancellation and such receivers are non-linear by nature. Their processing is based on maximum likelihood (ML) principle or iterative/successive cancellation. The ML receivers consist of joint detection of desired and interfering layers and are as such complex, especially if the joint detection operation is applied to a large number of layers. As a consequence, reduced complexity ML (RML) receivers are more implementation friendly and they may make use of sphere decoding, QR-MLD (a combination of ML detection and QR decomposition) or other complexity reduction principles. In terms of knowledge of the interference structure, the RML needs, in addition to the effective channel of the dominant interferer (also needed by the enhanced LMMSE-IRC mentioned above), also information of the modulation used by the interferer. The RML receiver operates on symbol level, that is, decoding of the interferer is not needed.

Another class of non-linear receivers is based on successive interference cancellation and since they operate on a modulation symbol basis, this category is abbreviated as SLIC (symbol level IC). Successive cancellation receivers are utilizing successive application of linear detection, reconstruction of the interfering signal and cancellation. In terms of interference knowledge, their operation is similar to RML; however, they may be more robust in face of imperfect interference reconstruction. A more advanced and complex version of symbol-based IC is the codeword SIC (CWIC). This receiver structure, in addition to detecting the interfering signal, also decodes it. In other words, the main difference between SLIC and CWIC is whether the subtracted signal is reconstructed based on the output of the LMMSE filter, that is, modulation symbol based, or based on the output of the turbo decoder, that is, codeword based, respectively.

Having covered some basic aspects of the advanced receiver structures, in the following we will focus on the interference structure which is needed by these receivers in the process of reconstructing it prior to the IC stage. Note that CWIC is dropped from this discussion due to the high complexity and more detailed information such receiver needs regarding the interference, as decoding the interferer requires information of the modulation and coding scheme, not only on modulation type, as needed by the symbol-based receivers. The interfering PDSCH is characterized by more *static, network topology*-oriented parameters (such as system bandwidth, cell ID, number of CRS antenna ports, MBSFN configuration) and by more *dynamic* parameters which are related to the *spatial structure* and *link adaptation* of the interference, such as transmission mode (including rank and PMI), PDCCH length, resource allocation type, modulation order and EPRE (energy per resource element). The NAICS UE would not be able to blindly estimate all the interfering PDSCH characteristics described above as there are practical UE limitations in terms of UE complexity and power consumption. On the other hand, the network cannot signal all these interference characteristics either as the signalling payload would be high while the flexibility of the network operation needs to be preserved.

The trade-off is that the NAICS UE is doing blind detection for the more dynamic parameters. The state-of-the-art NAICS receiver can blindly detect the presence of interference, transmission mode, modulation, PMI, rank and P_A offset. The network is providing, in a semi-static manner, assistance with respect to the more static parameters such as cell ID, number of CRS ports, MBSFN pattern, P_B and a subset of up to P_A values (defining the EPRE), implemented transmission modes (from the set of TM 1, 2, 3, 4, 6, 8, 9, 10) and the resource allocation and precoding granularity. Even if TM10 is signalled to the UE, there is PDSCH cancellation as such expected for a TM10 UE. While in theory possible, the cancellation of TM10 has been found to be a rather complex operation for Release 12 due to the fact that it was difficult to find network assistance versus UE blind detection common ground in terms of complexity and network flexibility. It is worth mentioning that cancelling a TM10 PDSCH would imply the need to know the virtual cell ID and scrambling information characterizing the interferer's DMRS while in addition the quasi-collocation information would be needed as well. Synchronization of CP, slot, SFN, subframe and common system bandwidth for the serving cell and interfering cells is not signalled, but rather implicitly assumed by the UE when NAICS signalling is configured (hence synchronized operation, aligned CP, same bandwidth).

The blind detection of interfering PDSCH is certainly a sensitive process as any parameter misdetection leads to the incorrect identification of the interfering PDSCH and might even bring a performance penalty in practice, as cancelling the wrong information might have the effect of noise amplification. Consequently, the power of the dominant interferer plays an important role in its (blind) estimation, bringing an interesting paradigm; the stronger the interferer is, the more reliable the NAICS UE is to detect its PDSCH characteristics.

In a bursty FTP traffic, the interference structure is changing in both time and frequency. The spatial properties of the dominant interferer would be impact (and be available for cancellation) of the NAICS UE depending on the interferers' packet duration. However, from a NAICS UE perspective, the dominant interference can change in frequency as well. At one end, we

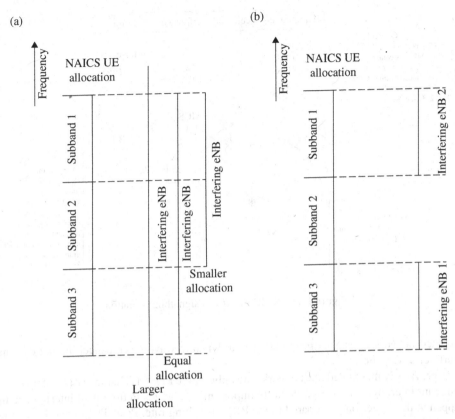

Figure 11.2 Frequency domain allocations of NAICS UE and dominant interference. (a) One interferer with smaller or equal frequency allocation as NAICS UE. (b) Multiple interferers with frequency overlap allocations as NAICS UE

have the simpler situation when a single dominant interferer allocation is larger, equal or smaller than the NAICS UE allocation, as shown in Figure 11.2a. A more complex situation is when multiple interferers overlap with the allocation of one NAICS UE, if, for example, the NAICS UE is scheduled wideband while the interferers are scheduled frequency selective (Figure 11.2b). The interferers can have same or different point of origins; however, this is not a problem from the eNodeB-to-UE signalling perspective if the assumption is that inter-eNodeB signalling is available at the serving eNodeB at the same time.

The basics of NAICS operation modes are depicted in Figure 11.3. Based on the indicated RSRP from the NAICS UE, the serving eNodeB is providing network assistance for up to eight interfering cells. The NAICS semi-static signalling is formed by two parts: the inter-eNodeB signalling and the eNodeB-to-UE signalling. The NAICS information of cell ID, number of CRS ports, MBSFN configuration, P_A subset, PB and implemented TMs is transferred via X2 between the eNodeBs and further signalled to the UE via RRC. In addition to these parameters, the RRC message may contain information on the resource allocation and precoding granularity utilized by the dominant interferer. The NAICS UE further utilizes the

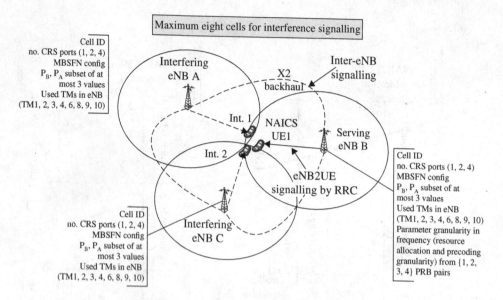

Figure 11.3 NAICS inter-cell signalling operation

indicated network assistance information in a dynamic way, hence cancelling the dominant interferer it identifies.

As previously mentioned, the network may indicate the PDSCH characteristics of up to eight dominant interferers. This is particularly important as changes in dominant interference may happen with a faster rate compared to the RRC signalling rate. Providing network assistance for multiple potential dominant interferers may be seen as a proactive signalling approach. As the network does not have the dynamic information on the dominant interferer seen by NAICS UE at a particular time, the network assistance consists of interference characteristics as point-specific information, not UE specific. Note that the signalling request is based on the indicated dominant interferers by the UE. The UE would utilize the IC if the network assistance is available at that point in time, otherwise a fallback mode to no IC should be utilized by the UE. Proactive network assistance would facilitate PDSCH IC also in the situation when there is a change in the point of origin in time of the dominant interferer.

Table 11.1 shows the NAICS system performance in a homogeneous deployment. The technology is applicable in heterogeneous networks as well. In this particular example, bursty traffic has been used while the baseline is SU MIMO operation. In these simulations, TM9 interference is cancelled by NAICS UEs also configured in TM9; however, NAICS UEs

Table 11.1 NAICS system performance in homogeneous scenarios

Rel-11 baseline, SU-MIMO, bursty traffic		
	Cell edge	Average
NAICS SLIC receiver (Rel-11 IRC CQI)	+11.7%	+7.4%

have the ability to cancel a vast amount of TMs such as TM1, 2, 3, 4, 6, 8 and 9. Further enhancements in LTE interference cancellation are being addressed in Release 13 [1, 2] as covered later in this chapter.

11.3   Evolution of Antenna Array Technology

Antenna array technology for cellular communications is continuously focusing on reducing the hardware footprint and improving the energy efficiency and cost-effectiveness of the base-station installations. Conventional base-station architecture can be viewed as a set of tower-top passive antenna panels served with coaxial cables from a BTS unit at the ground level housing the radio and transceivers (TXRUs). The radio and the transceiver unit includes up/down converters, power amplifiers, sample rate converters, power conditioning circuitry, and so on. This model is gradually evolving to a remote-radio-head (RRH) solution where the radio and the transceivers are positioned much closer to the set of passive antenna panels and could be inside the antenna casing as in the case of antenna-integrated radio (AIR) or integrated antenna system (IAS) [3]. This solution is both compact and more efficient in terms of power transfer due to reduced cable losses. An RRH installation is connected to a BBU via a high-speed link such as a CPRI or OBSAI. An active antenna system (AAS) defines the next evolution of this technology where the transceiver and the radio are integrated and distributed with the antenna elements and this system connects with the BBU via a high-speed link. Physically, this provides an attractive solution with a reduced and clean footprint due to the elimination of tower-top hardware components (such as cables, connectors, mast-head amplifiers and RET network). It also eliminates cable losses and can deliver power more efficiently to the antenna elements.

The most distinguishing feature of an AAS is the ability to provide electronic beam steering in the elevation dimension due to the distributed nature of the transceivers. In one architecture example, each antenna element can be connected to a transceiver providing baseband amplitude and phase control to every antenna element. Alternatively the number of transceivers could be smaller than the number of antenna elements with each transceiver driving a subset of antenna elements [4]. Figures 11.4 and 11.5 show two typical AAS architectures where the left-slanted bars correspond to elements with −45° polarization and the right-slanted bars correspond to elements with +45° polarization.

A conventional passive antenna system for macro cells would typically have a single transceiver feeding eight antenna elements per polarization arranged vertically in a column of cross-polarized antennas (as in Figure 11.6). As shown in Figure 11.7, an AAS can potentially allow two, four or even eight distributed transceivers feeding into the eight vertical antenna elements per polarization. The significance of a transceiver is that it allows precise baseband control of the amplitude and phase of the signals feeding into the antenna elements. Without delving into the details of AAS architectures we can simply equate the number of transceivers to the number of antenna ports in LTE terminology. Therefore, a 64-element passive antenna array comprised of 4 columns with a total of 8 transceivers each driving a set of 8 co-polarized antenna elements (vertically) can represent a typical 8 Tx macro configuration (Figure 11.6). This is comparable to an AAS offering 64 transceivers with a similar form factor thereby transitioning an 8 Tx conventional base-station to a 64 Tx Massive-MIMO base-station (Figure 11.7). The 'Massive' in Massive-MIMO, therefore, does not correspond to physical size but to the increased number of transceivers with baseband control ability.

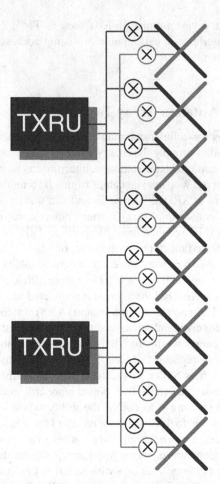

Figure 11.4 An example AAS architecture where a TXRU is connected to a sub-array of co-polarized antenna elements contained in a single column [4]

It is worth noting that conventional tower-top passive antenna systems as well as RRHs can provide a limited ability to steer beams in the elevation dimension using a remote electrical tilt (RET) feeder network. An AAS, without the need of a separate RET network, allows beamforming in the elevation and the azimuth dimension that is dynamic and more flexible (3D-beamforming). It is also evident that AAS architecture is critical in determining the number of antenna ports as well as the flexibility available for controlling the antenna array – thereby directly impacting the system performance as shown later.

11.4 Deployment Scenarios for Antenna Arrays

The spectral efficiency of a system is maximized when the antenna array is able to focus a beam on a particular user in the system achieving user-specific beamforming gains and also able to separate multiple users in the spatial domain via beamforming achieving multi-user

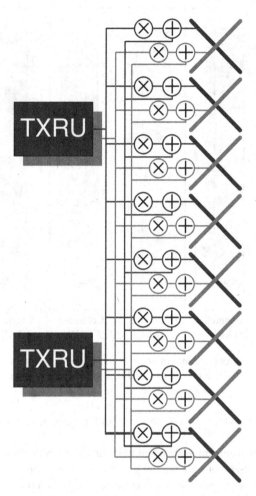

Figure 11.5 An example AAS architecture where a TXRU is connected to all the co-polarized antenna elements contained in a single column [4]

transmission gains. Conventional passive antenna systems including RRH installations are able to take advantage of user-specific beamforming and multi-user transmission gains in the azimuth (2D horizontal plane) dimension. This ability has been well studied in the literature and it is understood that the user-specific beamforming opportunity is not biased in any direction (within the sector pattern) which is a result of a uniform distribution of users within the sector in the 2D plane. This is no longer true in the elevation dimension. Figure 11.8 shows four users in the horizontal plane in NLOS condition (blocked by buildings) and their hypothetical mean elevation angle of departures. Simply from a geometrical point of view it is clear that the users closer to the base-station are more separable in the elevation dimension than the users that are far away. It is, therefore, natural to expect that the user-specific beamforming opportunity in the elevation dimension is biased favourably towards the users that are closer to the base-station. This shows that the angular distribution of the users in the elevation dimension in a particular

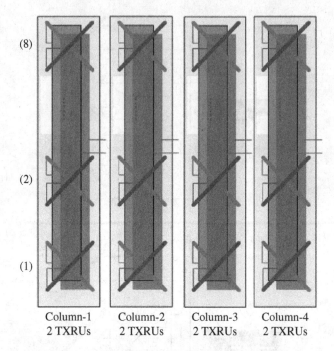

Column-1   Column-2   Column-3   Column-4
2 TXRUs   2 TXRUs   2 TXRUs   2 TXRUs

Figure 11.6 Schematic of four columns of a conventional passive antenna array, each column connecting to one transceiver per polarization via a radio distribution network (RDN)

deployment scenario provides a good indication of the gains achievable by 3D beamforming (compared to 2D).

The geometrical environment as depicted in Figure 11.8 occurs in ultra-dense urban macro deployments. The base-stations are typically installed above rooftops, the users are mostly in NLOS conditions and distributed at the street level and within a few floors of the buildings to a certain extent. The cell sizes vary from ~150 to over 500 m on average inter-site distances (ISD). Simulated system performance in Figure 11.9 shows that the gains due to 3D beamforming compared to 2D beamforming are significantly higher in smaller cell sizes in such environments and the reasoning follows naturally from the geometrical considerations above. Figure 11.10 shows that due to the better separability of users in the elevation domain in a denser deployment, multi-user transmission is more prevalent – this directly improves the gains due to 3D beamforming.

Therefore, ultra-dense macro scenarios with above rooftop deployments and small to medium cell sizes are attractive for Massive-MIMO installations. Another deployment scenario naturally fit for Massive-MIMO antenna arrays is an urban macro deployment with high-rise buildings – in this case an antenna array installed outdoors can utilize both up-tilting and down-tilting to serve users located inside high-rise buildings [5]. This is especially relevant in situations where alternative indoor solutions such as small cells or distributed antenna systems are logistically expensive. In certain cases due to constraints on antenna array installation sizes as well as due to the expense of replacing existing antenna systems Massive-MIMO deployment opportunities in the macro cells can be limited. In such cases, however, Massive-MIMO

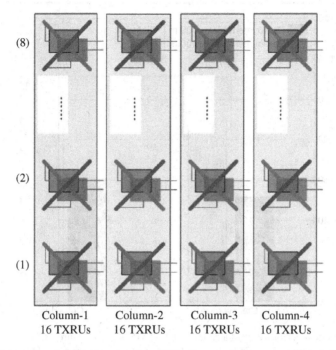

Figure 11.7 Schematic of four columns of an AAS where each antenna element is integrated with a radio and a transceiver

arrays deployed below rooftops or mounted on walls targeting capacity hot-spots like train stations can be the solution of choice [11]. These can be small cells targeted for capacity and a compact power-efficient AAS installation fits the bill. In such instances, Massive-MIMO systems with directive patch antennas operating at higher frequencies (like 3.5 GHz band) can be attractive.

The discussion above considers the performance gains due to 3D beamforming ability of AAS that is achieved by precisely controlling the gains and phases of the antenna elements dynamically at baseband. In addition, the ability to steer beams in the elevation dimension in a relatively slow fashion and the ability to form multiple beams in the elevation dimension can also be applied for different deployment needs. Such functionalities do not require baseband control and can be achieved by adapting the RF configuration of the AAS. Examples of such

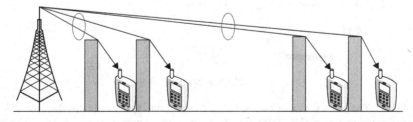

Figure 11.8 Users closer to the base-station allows better angular separability in the elevation dimension

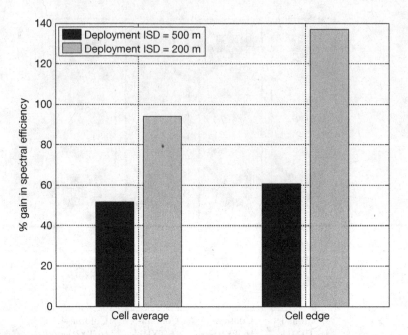

Figure 11.9 System performance of a column of 16 TXRUs (as in Figure 11.7) compared to a column of 2 TXRUs (as in Figure 11.6) – in both cases 16 antenna elements are used. This shows that the gains due to 3D beamforming compared to 2D beamforming are higher for denser deployments

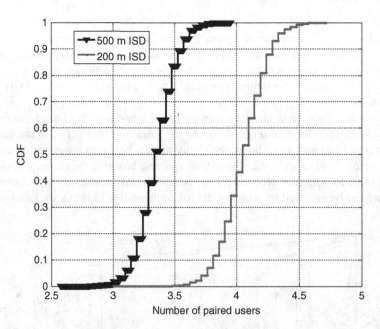

Figure 11.10 The gains shown in Figure 11.9 are in part due to the ability of the AAS base-station to transmit to multiple users (MU-MIMO) more easily in the denser scenario. This figure shows more users are paired for MU-MIMO transmission in the denser scenario as also expected from Figure 11.8

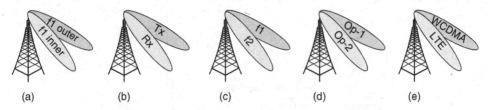

| (a) | (b) | (c) | (d) | (e) |

Figure 11.11 Application of AAS systems capable of steering multiple elevation beams: (a) cell splitting (vertical), (b) Tx/Rx tilting, (c) carrier-specific tilting, (d) operator-specific tilting and (e) technology-specific tilting

applications, as shown in Figure 11.11, include cell splitting (vertical), separate Tx/Rx tilting, carrier-specific tilting, operator-specific tilting and technology-specific tilting [3]. Cell splitting (vertical) allows spectral efficiency gains by reusing system resources; carrier-specific tilting allows operators to manage the load in each carrier; operator-specific tilting allows network sharing among multiple operators; technology-specific tilting allows managing link budgets, load and so on according to the different technologies.

11.5 Massive-MIMO Supported by LTE

The support of multiple antenna techniques (MIMO) has been one of the cornerstones of the LTE physical layer specifications since the first release (Rel-8). LTE Rel-8 was designed to provide MIMO support of up to four antenna ports for downlink transmission. This was extended to eight antenna ports in Rel-10 which is also the maximum number of ports supported in Rel-12. Massive-MIMO arrays as described above may offer the flexibility of achieving baseband control of over 16–64 antenna ports for a base-station. In this section, we describe the opportunities for Massive-MIMO deployments that are enabled with the existing Rel-8–Rel-12 specifications. This is important both from an early deployment perspective as well as from the perspective of supporting legacy mobile devices in the network.

LTE Rel-8–Rel-12 specifications can be used to support a Massive-MIMO base-station with an arbitrary number of ports even though it is not designed or optimized for such operation. In the following, we focus on the capacity performance of the shared downlink data channel which is of primary interest in the deployment scenarios described above. The restriction to four or eight antenna ports in the specifications limits the mobile devices to the measurement and feedback of channel state information (CSI) corresponding to four or eight transmit antenna ports. The CSI feedback is used at the base-station for precisely controlling the amplitude and phases at the transceivers driving the antenna array in order to adapt the data transmission to the particular user. In order to get around this antenna port limitation, there are two main approaches for supporting a Massive-MIMO base-station with Rel-8–Rel-12 specifications – sectorization-based approaches and reciprocity-based approaches.

11.5.1 Sectorization (Vertical)-Based Approaches

In the sectorization-based approaches, a mobile device is assigned a subset of four or eight antenna ports which are chosen from the set of antenna ports available at a Massive-MIMO base-station. An example would be to configure a 16 antenna port Massive-MIMO array to

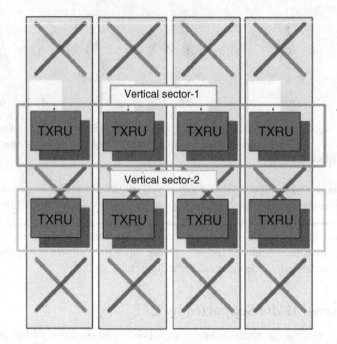

Figure 11.12 An AAS with 16 transceivers configured for vertical sectorization. Each vertical sector is comprised of eight azimuth cross-polarized ports

create two vertical sectors with 8 azimuth antenna ports in each (see Figure 11.12). A possibility is to deploy the two vertical sectors as two different cells. Alternatively Rel-11 onwards allows the possibility of deploying the two vertical sectors as two virtual cells enabling some flexibility of switching a mobile device from one vertical sector to the other without a laborious handover procedure. Rel-11 also allows dynamic coordination of scheduling between the vertical sectors.

Figure 11.13 shows the simulated performance of the shared data channel with vertical sectorization achieved with AAS compared to 8 Tx performance of a conventional base-station. These results do not assume dynamic coordination across the vertical sectors. Note that both the AAS and the conventional base-station are using a cross-polarized array of 64 antenna elements arranged in 8 rows and 4 columns with the same spacing. The AAS is equipped with 16 transceivers (8 azimuth × 2 elevation) driving the array while the conventional base-station is employing 8 transceivers in azimuth only. It shows that within a similar form factor doubling the number of transceivers can provide a significant performance improvement in certain scenarios. This performance is achievable with the existing LTE-A specifications. As shown in Figure 11.13 the AAS architecture has some impact on system performance but this alone should not be construed as a well-rounded evaluation of architecture options.

11.5.2   Reciprocity-Based Approaches

The reciprocity-based approaches do not rely on CSI feedback for controlling the amplitude and phases at the transceivers driving the transmit antenna array. Instead they depend on the

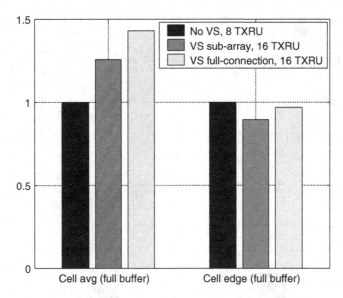

Figure 11.13 Vertical sectorization system performance gains (16 TXRU AAS) comparing to 8 TXRU passive antenna array for sub-array and full-connection architectures

uplink transmissions from the mobile devices to determine the amplitude and phases required at the transceivers for directing the downlink data transmission to the mobile device. Therefore, an arbitrary number of transmit antenna ports can be supported using this approach. An instantaneous channel between a base-station antenna array and a mobile device is always reciprocal but several practical challenges remain – for example, a mobile device may have a single transmit antenna but multiple receive antennas, the transmit and the receive RF chain responses may not be equalized perfectly at the base-station and there will be difficulty in predicting other transmission parameters (rank, MCS). Nevertheless, reciprocity remains a practical and feasible deployment option for especially TDD. Figure 11.14 shows the improvement of data channel performance obtained by increasing the number of transceivers from 8 to 64. In all cases, a 2D rectangular cross-polarized array with eight rows and four columns is assumed.

11.6   Further LTE Multi-antenna Standardization

3GPP RAN WG1 is considering the potential benefits of extending MIMO support from eight antenna ports to 16, 32 or 64 antenna ports in Rel-13. This includes investigation of enabling Rel-13 mobile devices with the ability to measure and feed back CSI corresponding to 16, 32 or 64 transmit antenna ports. The potential benefits of going from 8 to 64 antenna ports come from the increased beamforming gains and multiplexing gains. The beamforming gains are due to the increased focusing of the energy (in three dimensions) with a larger number of antenna ports towards a single user. The multiplexing gains come from the ability (of a larger number of antenna ports) to form non-interfering beams to different users in the 3D-spatial domain that allows for efficient multi-user transmission (MU-MIMO). The polarization dimension

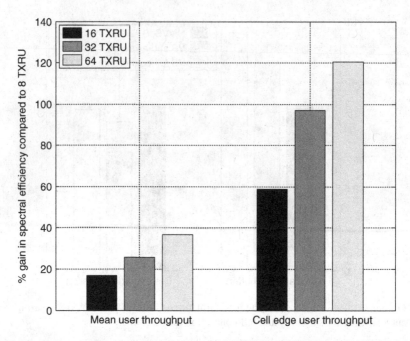

Figure 11.14 System performance comparison of 16 TXRU, 32 TXRU and 64 TXRU AAS base-stations with 8 TXRU conventional base-station. In all cases, a 64-element cross-polarized antenna array arranged in 8 rows and 4 columns is used. The results assume reciprocity-based operation in TDD

particularly helps in multiplexing multiple streams to a single user (SU-MIMO). A critical factor in achieving these gains is the accuracy of the feedback CSI. This effect is shown in Figure 11.15. The system performance attained with Rel-12 mobile devices that are capable of 8-port CSI feedback is compared with the performance attained with hypothetical Rel-13 mobile devices capable of 16-port CSI feedback. In both cases, a Massive-MIMO base-station with 16 TXRUs driving an array of 64 cross-polarized antenna elements (arranged in eight rows and four columns) is considered. The achievable performance is a strong function of the fidelity of CSI feedback and if CSI feedback is restrictive the performance improvement may not be significant.

The SU-MIMO transmissions are, for the most practical purposes, limited to two streams because a majority of mobile devices are enabled with two Rx chains. The realizable SINR at the UE is limited by Rx impairments, and the transmission MCS is limited to 64QAM. All of these imply that a Massive-MIMO deployment can offer a limited improvement with only SU-MIMO transmissions. Therefore, in addition to the CSI enhancements, enhancements focused on MU-MIMO support are also under consideration.

Regarding standardization aspects we have considered [7–9] the physical layer specifications in the above which falls under the domain of 3GPP RAN WG1. Certain aspects of an AAS-enabled base-station are also under consideration in 3GPP RAN WG4 since Rel-11. In the case of a conventional base-station, the transceiver and the antenna arrays are physically isolated and the WG4 specifications set a standard for the base-station. The integration of the radio and the transceivers with the antenna elements in an AAS affects the characteristics

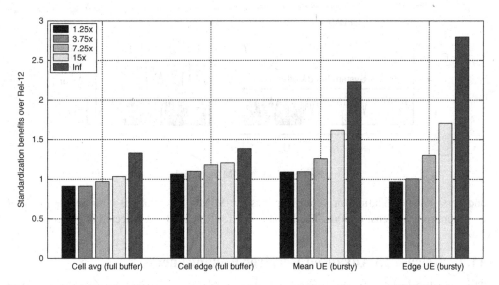

Figure 11.15 The potential benefits of standardization enhancements for supporting 16-port Massive-MIMO are shown. The improvement is a strong function of the fidelity of CSI feedback. The CSI feedback overhead per mobile device for 16-port transmission is assumed to be 1.25×, 3.75×, 7.25×, 15× the overhead due to a Rel-12 8 Tx transmission. The case of 'Inf' corresponds to ideal CSI assumption at the base-station

of the transmitted signal in the undesired channels to a certain extent. Similarly, it may also affect the ability to receive a desired signal in the presence of interference in the undesired channels. 3GPP WG4 is actively studying such aspects of deployments considering a mix of AAS base-stations and non-AAS base-stations which may lead to a revised specification for core requirements intended for AAS.

The discussion above shows the suitability of Massive-MIMO base-stations in deployment scenarios that are largely interference limited and the motivation for a Massive-MIMO base-station in such scenarios is to boost capacity by improving spectral efficiency. Although we have focused on the downlink, the capacity improvement in the uplink is also expected to be as significant as the downlink in the respective deployment scenarios. In frequency bands at about 700 MHz to 3.5 GHz where current LTE deployments are most prevalent, this seems to be the predominant role of AAS and Massive-MIMO base-stations of the future. It is worth mentioning, however, that in the context of 5G the easier availability of spectrum at higher frequencies but with unfavourable propagation characteristics is also motivating M-MIMO antenna arrays as discussed in Chapter 17. The considerations, however, at a higher band such as 30–100 GHz can be quite different. The pathloss can be higher by about 20–30 dB while at the same time a large number of antenna elements can be packed efficiently in a small area. The number of transceivers driving the array is expected to be limited due to cost and power consumption issues. The design focus may veer towards maintaining connectivity rather than optimizing for spectral efficiency. The aspect that remains relevant, however, is the ability to precisely control a Massive-MIMO array in three dimensions. The evolution of the LTE specifications for supporting Massive-MIMO is progressing as part of Release 13 supported by many operators and vendors worldwide. The focus is to provide an optimized, robust

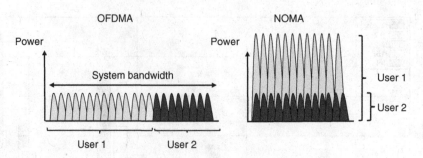

Figure 11.16   NOMA principle

and versatile specification for deploying Massive-MIMO base-stations with a wide range of configurations and cost options. The latest scope for Release 13 is reflected in [10].

11.7   Release 13 Advanced Receiver Enhancements

The cancellation of cell-specific reference signals (CRS) was studied initially in the context of Release 11 heterogeneous networks. In such scenario, pico cell UEs situated in cell range extension are impacted by high macro interference. While almost blank subframes (ABS) containing no scheduling from the macro cell are giving the opportunity to the pico users to enhance their downlink transmission, the CRS, which are always transmitted, are creating interference in subframes which otherwise should have been completely silent. Release 11 UEs are able to cancel such CRS interference in HetNet scenarios. Such CRS IC operation is extended during Release 13 by defining performance requirements [1] for using the advanced receiver at the UE to cancel the CRS interference in the homogenous deployment with multiple interference sources. Indeed, in low-load networks, the CRS transmission becomes a non-negligible source of interference. We need to highlight though that the CRS IC is limited to two antenna ports, hence the issue of four4 CRS AP IC needs to be tackled further by the advanced UEs.

Another new receiver study started in Release 13 is the NOMA [6], which looks at multiplexing far and near eNodeB users on the same downlink time/frequency resources in such a way that the far user is sent with larger power level and near users with lower power level while the UE receiving the lower power level signal would subtract the high power signal first before detecting its own transmission. This operation principle is illustrated in Figure 11.16, with the sub-carriers of two users overlapping in the frequency (and spatial domain), trusting on the advanced receiver capability of user 2 which needs to cancel high power user 1.

3GPP has started work on Release 13 to include further performance requirements related to the interference reduction for control channels, as has been proposed in [2]. The proposed approach aims to have the UE receiver to cancel also interference experienced by the downlink control channels, thus improving the downlink physical layer signaling performance in difficult to interference conditions.

11.8   Conclusions

In the previous sections, we have covered the Release 12 NAICS as well as the 3D-channel modelling which enabled further work in 3GPP for the advanced antenna technology evolution, with the work on-going in Release 13 to go beyond eight antenna ports together with the necessary enhancements for the UE feedback and other details relevant for the use of more than eight antenna ports. The use of 3D-beamforming/full-dimension MIMO-related elements is expected to be added during the work of Releases 13 and 14 of LTE-Advanced evolution. Further work on receiver will, on the other hand, extend the CRC interference cancellation capabilities in homogenous networks as well as look at the application of NOMA-type transmission/reception with two users sharing the same resources with power differences.

References

[1] 3GPP Tdoc RP-142263, 'New Work Item Proposal: Perf. Part: CRS Interference Mitigation for LTE Homogenous Deployments', December 2014.

[2] 3GPP Tdoc RP-151107, 'Interference Mitigation for Downlink Control Channels of LTE', June 2015.

[3] 3GPP Tdoc R1–133528, 'On Optional High-Rise Scenario', CMCC, RAN1 no. 74, Spain, August 2013.

[4] 3GPP TR 36.897, 'Study on Elevation Beamforming/Full-Dimension (FD) MIMO for LTE', v.13.0.0, June 2015.

[5] 3GPP Tdoc R1–144153, 'Scenarios for Elevation Beamforming and FD-MIMO', NTT DOCOMO, RAN1 no.78bis, Slovenia, October 2014.

[6] Y. Saito, Y. Kishiyama, A. Anass Benjebbour, T. Nakamura, A. Li, and K. Higuchi, 'Non-orthogonal Multiple Access (NOMA) for Cellular Future Radio Access', Proceedings of IEEE Vehicular Technology Conference, 2013.

[7] 3GPP Technical Specification, TS 36.211, 'Evolved Universal Terrestrial Radio Access (E-UTRA); Physical Channels and Modulation', 2015.

[8] 3GPP Technical Specification, TS 36.212, 'Evolved Universal Terrestrial Radio Access (E-UTRA); Multiplexing and Channel Coding', 2015.

[9] 3GPP Technical Specification, TS 36.213, 'Evolved Universal Terrestrial Radio Access (E-UTRA); Physical Layer Procedures', 2015.

[10] 3GPP Tdoc RP-151085 Elevation Beamforming/Full-Dimension (FD) MIMO for LTE, June 2015.

[11] White Paper Nokia Networks, 'Nokia Active Antenna Systems: A step-change in base station site performance', Retrieved from http://networks.nokia.com/sites/default/files/document/nokia_active_antenna_systems_white_paper.pdf (last accessed August 2015).

12

LTE Key Performance Indicator Optimization

Jussi Reunanen, Jari Salo and Riku Luostari

12.1 Introduction

Network performance is measured using a set of key performance indicators (KPIs). The selection of those KPIs should be done in such a way that they measure both the end-user-experienced performance and the resource utilization, that is, how much of the resources are used in order to deliver certain end-user-experienced performance. Sacrificing resources for signalling to maximize, for example, the call setup success rate (CSSR) is not feasible in the long run. Instead minimum resource utilization to maximize the KPI performance should be the optimization target. In this chapter the common LTE KPIs and methods to improve those are discussed. The KPI performance is further discussed in Chapter 13 where the impact of high traffic event on various KPIs is analysed.

LTE Small Cell Optimization: 3GPP Evolution to Release 13, First Edition.
Edited by Harri Holma, Antti Toskala and Jussi Reunanen.
© 2016 John Wiley & Sons, Ltd. Published 2016 by John Wiley & Sons, Ltd.

The most common call performance KPIs, based on network statistics and drive test, are introduced. Drive test based KPIs are used for a pre-launch network. However, due to increasing traffic load after network launch, network statistics can be leveraged for performance improvement in the network. Drive testing effort can then be directed towards coverage and performance verification following physical layer optimization and some parameter changes. Examples of physical layer optimization are shown together with method to pinpoint the cells which require physical layer modifications (antenna downtilts, antenna direction alignment and antenna location changes). KPIs from network statistics are analysed in details and typical optimization actions are demonstrated with corresponding, expected results on different KPIs. Higher order multiple input multiple output (MIMO) performance and ways to improve dual-stream usage are discussed and practical performance gains from higher order MIMO shown. Also high-speed train performance improvement methods are discussed in details.

All examples shown are from frequency division (FD) LTE networks but as such the optimization actions are applicable also to time division (TD) LTE networks. Also all examples in this chapter are for 10 and 20 MHz FD LTE deployment (actual carrier frequency varies between 700, 800, 900, 1800, 2100 and 2600 MHz) with 2×2 MIMO in downlink and 2Rx in uplink unless otherwise mentioned.

12.2 Key Performance Indicators

The most common KPIs which can be based on either network statistics or drive testing are listed below together with example calculation formula:

- *Random access setup success rate*: Number of received radio resource control (RRC) connection request messages (message number 3) sent by the UE (successfully acknowledged by eNodeB) divided by number of preambles received by eNodeB sent by UE.
- Call setup success rate (CSSR) can be split further as follows (see Figure 12.8):
  - *Radio resource control connection setup success rate*: Number of RRC connection setup complete messages (message 5) sent by UE and successfully received by eNodeB divided by number of received RRC connection request messages sent by UE.
  - *Evolved radio access bearer (E-RAB) setup success rate from UE point of view*: Number of sent RRC connection reconfiguration completions divided by number of sent service requests.
  - *E-RAB setup success rate from eNodeB point of view*: Number of initial context setup response messages (S1 application protocol) sent by the eNodeB to the mobility management entity (MME) divided by number of initial context setup request messages (S1 application protocol) sent by MME to the eNodeB.
- *E-RAB drop rate*: Number of abnormally released radio bearers leading to release of E-RAB divided by number of successfully setup E-RABs.
- Intra-LTE handover success rate which can be split further split as follows:
  - *Intra-frequency handover preparation success rate; inter-/intra-eNodeB from UE point of view*: Number of RRC connection reconfiguration messages with mobility info included (new cell) received by the UE divided by number of measurement reports sent by UE for corresponding event.
  - *Intra-frequency handover preparation success rate; inter-/intra-eNodeB from eNodeB point of view*: Number of received handover request acknowledgement messages divided by number of sent handover requests (over X2 or S1 for inter-eNodeB and inside eNodeB in case of intra-eNodeB).

o *Intra-frequency handover execution success rate; inter-/intra-eNodeB from UE point of view*: Number of sent RRC connection reconfiguration complete messages divided by number of received RRC connection reconfiguration messages.

o *Intra-frequency handover execution success rate; inter-/intra-eNodeB from source eNodeB point of view*: Number of received RRC connection reconfiguration messages by target cell that leads to release request message received by the source cell over X2 or received UE context release command message received over S1 in case of inter-eNodeB handover or successful handover indication delivered inside the eNodeB to source cell in case of intra-eNodeB handover divided by number of sent RRC connection reconfiguration complete messages (caused by received handover request acknowledgement over X2 or received handover command over S1).

12.3 Physical Layer Optimization

LTE spectral efficiency, limited by inter-cell inference, can be estimated based on channel quality indicator (CQI) values as specified in [1]. Optimization of cell dominance area reduces the inter-cell interference and improves CQI and therefore LTE network capacity and performance. Dominance area optimization is typically done by antenna downtilt, direction and height modifications and this is often referred to as physical layer or layer 1 optimization. The physical layer performance can be evaluated, for example, by the slope of decrease in average reported (by UE) CQI caused by increased downlink data volume per cell. The more negative this slope is the more the network suffers from inter-cell interference. Figure 12.1 shows one example of cluster-level physical layer optimization results. The slope (CQI decreased due to increased downlink data volume per cell) has improved significantly from −0.009 to −0.0035 which means that network can deliver 50% more data volume during year 2 compared to year 1 with the same CQI value (based on CQI ~ 9.5). This means better end-user experience and higher network capacity.

The analysis on cluster level can be extended to cell level (to find out the biggest interferers) by evaluating the slopes of cell X average CQI decrease versus data volume increase of cell

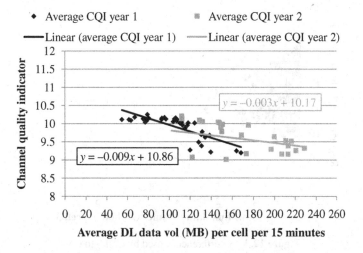

Figure 12.1 Spectral efficiency improvement by physical layer optimization

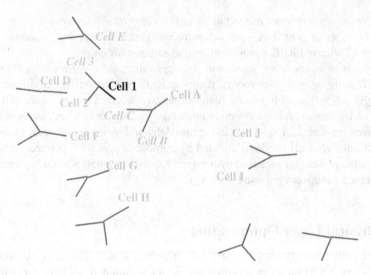

Figure 12.2 Analyzed network cluster

X's neighbours. The analysed example cluster is shown in Figure 12.2 where the cells causing biggest interference to cell 1 are indicated with italic text box.

Cell 1 is the analysed cell and cells A–I are its closest neighbours. Worst interferers are analysed based on the slopes of average reported CQI of cell 1 versus average downlink data volume for each of its neighbours. Examples are shown in Figures 12.3, 12.4, 12.5 and 12.6.

Based on neighbouring cell impact on Cell 1 CQI the physical layer optimization actions can be targeted to cells 3, B, C and E whose slope was between −0.53 to −0.95 compared to others ranging from −0.45 to −0.02. Drive testing can be conducted for coverage verification. In typical cluster optimization cases only downtilts are needed to be optimized. Example of physical layer optimization improvement (for 10 MHz FD LTE deployment) in terms of drive test is shown in Figure 12.7.

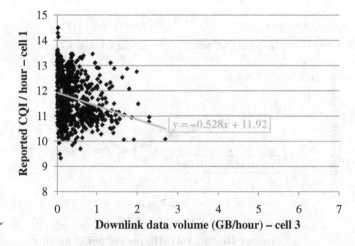

Figure 12.3 Interference caused by Cell 3 to Cell 1

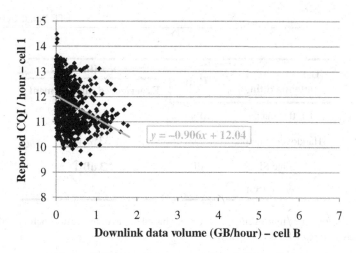

Figure 12.4 Interference caused by Cell B to Cell 1

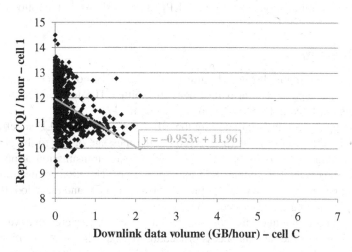

Figure 12.5 Interference caused by Cell C to Cell 1

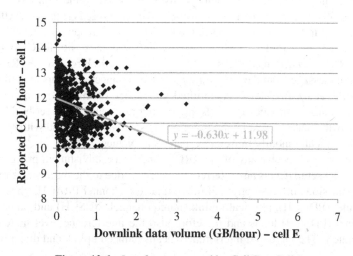

Figure 12.6 Interference caused by Cell E to Cell 1

Drive test result (antenna tilting)	Unit	Diff.
		Experienced improvement
DL throughput	Mbps	5 Mbps ↑
Handover attempts		33% ↓
Average SINR	dB	2.3 dB ↑
Average CQI		0.5 ↓

Figure 12.7 Performance improvement in a cluster based on drive test results before and after changes (10 MHz bandwidth FD LTE deployment)

Physical layer optimization impact on the KPI values is shown in the following sections.

12.4 Call Setup

The call setup signalling is shown in Figure 12.8. It should be noted that some of non-access stratum (NAS) messages and security messages are omitted. The downlink control information (DCI) formats indicated in Figure 12.8 are according to [1].

The call setup signalling is initiated by the UE that is in RRC_IDLE state (no signalling connection with the network) due to received paging message or due to the NAS procedure initiated by the higher layers in the UE. The NAS message transmission or paging response transmission triggers the RRC connection setup procedure. The RRC connection setup procedure is the signalling connection establishment between the UE and the eNodeB enabling the E-RAB setup signalling for requested service.

The RRC connection request message (layer 3 message) is delivered to lower layers within the UE and it initiates the random access procedure to establish synchronization between the UE and the eNodeB as shown in Figure 12.9. The RRC connection setup supervision timer, T300, is started in the UE layer 3 when it delivers the RRC connection request message to lower layers (T300 is stopped if RRC connection setup message is received). After successful random access setup (i.e. the UE receives the random access response), the RRC connection request message is sent to the eNodeB. The eNodeB responds with RRC connection setup message and RRC connection is successfully setup when eNodeB receives the RRC connection setup complete message from the UE. During the RRC connection setup signalling, the power control and link adaptation have very limited knowledge of the UE radio conditions. This means that the transmission power levels and coding rates need to be set by fixed parameter values given by the network (explained in following sections) combined with retry algorithms which can dynamically modify those given parameter values.

Poor success rate to receive any of the RRC connection establishment procedure messages leads to extra call setup delays and interference increase due to the retransmissions. The additional retransmissions increase physical random access channel PRACH, physical downlink control channel (PDCCH), physical uplink shared channel (PUSCH) and physical downlink shared channel (PDSCH) loads and therefore also the interference level increases for each of those channels. The increased interference means reduced uplink and downlink efficiency,

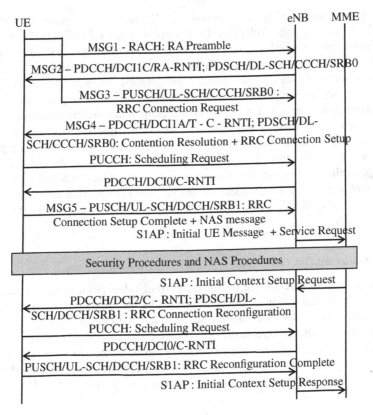

Figure 12.8   Call setup signalling

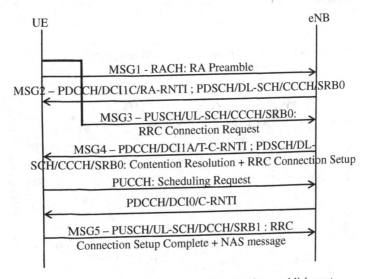

Figure 12.9   Radio resource control connection establishment

hence the optimization of initial transmission power level, number of repetitions, timer values, initial PDCCH control channel element (CCE) aggregation level and PDSCH coding rates as well as initial modulation and coding scheme (MCS) selection both in uplink and in downlink during the call setup is very important. All these aspects are discussed in the following sections.

As random access signalling is part of RRC connection setup signalling and therefore part of call setup signalling, poor random access setup success rate can cause poor overall CSSR. Also random access signalling is part of handover procedure and therefore call setup and handover procedures share large number of parameters which are used for selection of transmission power levels and capacity allocations. Hence, poor random access signalling performance causes poor call setup success performance and poor handover performance and eventually poor dropped call performance. Handover procedure is explained in Section 12.5.1.

In the following the call setup procedure analysis and optimization actions are split into the random access setup, RRC setup and E-RAB setup and each of those phases is discussed in separate sections.

12.4.1   Random Access Setup

Random access setup success rate is defined from eNodeB point of view as the number of received RRC connection request messages divided by the number of received random access preambles (see Figure 12.9). The random access setup success rate typically decreases as the traffic and inter-cell interference in the network increase as shown in Figure 12.10 but it can also be degraded because of poor random access preamble root sequence planning and/or poor physical layer design and/or poor parameter planning and/or any external interference. These most common random access setup success rate problems are briefly listed below and further discussed in later sections:

- The eNodeB receives too many preambles intended for other cells (badly overshooting cells receiving the same root sequence preambles due to root sequence reuse being not large enough).

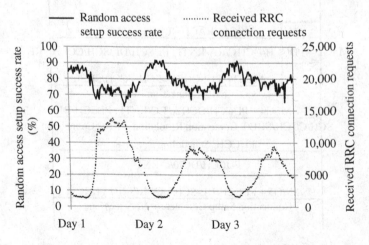

Figure 12.10   Average random access setup success rate in a small cluster of cells and average number of radio resource connection setup requests received by eNodeB per hour per cell over a period of 3 days

- The eNodeB receiver algorithms do not support high UE velocity, that is, large Doppler shift compensation is missing and therefore the PRACH, PUCCH and PUSCH performance is not adequate.
- The UE cannot receive the random access messages due to either PDCCH or PDSCH message problems caused by coverage or interference conditions and therefore random access procedure needs to be repeated several times.
- The eNodeB cannot receive the RRC connection request message because of excessive interference (caused by external or intra-LTE interference) or wrong timing detection from received preamble (due to, e.g. too small cyclic shift length for random access preambles) or too low power offset between received preamble and RRC connection request message.

Selecting the Preamble

The random access procedure is started by the UE selecting one preamble from all available preambles. A fixed amount of 64 preambles is allocated per each cell and depending on the cyclic shift length, N_{CS}, one or several random access root sequences are needed per cell to generate all 64 preambles. It should be noted that the cyclic shift length defines the cell range for eNodeB to correctly estimate the timing advance. If eNodeB receives preambles beyond the defined cell range the timing advance estimation will be wrong and random access procedure will fail causing additional reattempts from the UE. Preambles per cell are divided into two subsets:

- preambles for contention-free random access procedure used for handover procedure and for reacquiring the uplink synchronization
- preambles for contention-based random access procedure used for signalling scenarios where initial synchronization establishment is required as well as for contention-free random access procedures in case no contention-free preambles are available

For contention-based procedure the $64 - N_{cf}$, where N_{cf} is the number of preambles reserved by the eNodeB for contention-free procedure, available preambles are further divided into two sets according to the amount of bits needed to be transmitted as message 3 (msg3 in Figure 12.9), that is, preambles for small size message 3, referred to as group A preambles, and preambles for large size message 3, referred to as group B preambles. Also the path loss is taken into account by the UE to avoid large size message 3 being granted for cases where UE transmission power limits the coverage. System information block 2 contains all parameters for PRACH resource selection. The UE randomly selects one of the available preambles from either group A or group B.

Transmitting the Preamble

Once RRC connection establishment is initiated by layer 3 of the UE and the UE has selected the preamble, the UE sends the preamble at first available random access occasion at power PreambleTX given by Eq. 12.1. For the first preamble attempt, the preamble transmission counter is 1. The preambleInitialReceivedTargetPower and powerRampingStep are parameters defined by the eNodeB parameter database and sent within system information block 2. The preambleInitialReceivedTargetPower defines the targeted preamble received power level at the eNodeB and the powerRampingStep is the power increase per each retransmitted preamble.

DELTAPREAMBLE is a preamble format-specific power offset defined by [1]. Path loss is the UE-measured path loss based on filtered reference signal received power (RSRP) and reference signal transmission power indicated in system information block 2:

$$\text{Preamble Tx} = \min\{\text{PCMAX, preambleInitialReceivedTargetPower}$$
$$+ \text{DELTAPREAMBLE} + (\text{PREAMBLE\_TRANSMISSION\_COUNTER} - 1)$$
$$\times \text{powerRampingStep} + \text{PathLoss}\} \tag{12.1}$$

After the random access preamble is transmitted the UE monitors the PDCCH for random access response grant identified by the random access radio network temporary identifier (RA-RNTI). The RA-RNTI is defined by the selected subframe (within the radio frame) of PRACH resource and by the selected PRACH frequency resource. For FD LTE, there are only the time domain resources and therefore the subframe or subframes for PRACH directly define the RA-RNTI [1].

In case the UE does not receive the random access response within a specified window starting 3TTIs from the sent preamble and ending random access response window size + 4TTIs later, the UE will re-initiate the RACH procedure from the beginning (selects randomly the preamble from available preamble groups) with preamble transmission counter incremented by one per each sent preamble. Therefore, each time the random access procedure needs to be repeated the preamble is sent with higher power level, incremented as can be seen from Eq. 12.1. The random access response window size is defined by the eNodeB parameter database and sent to the UE within system information block 2. The repetition of random access procedure is done by the UE until expiry of T300.

Timer T300 supervises the RRC connection establishment procedure and is stopped at the reception of RRC Connection Setup or RRC Connection Reject message. Also cell reselection or abortion of RRC connection establishment from higher layers can cause stopping of T300. In case T300 is running, the UE can continue to send random access preambles and PREAMBLE_TRANSMISSION_COUNTER = preambleTransMax + 1 does not necessarily stop the preamble transmissions. Setting the T300 too short may result in better statistics in terms of random access setup success rate, but poor accessibility from end-user point of view due to users under poor radio conditions might not have enough power ramp-ups to compensate the large coverage area (especially in case of varying interference conditions). Therefore, the RACH procedure parameter tuning results should be always verified by drive testing to detect possible problems where the UE transmission power does not reach the maximum value and therefore the preambles cannot be detected by the eNodeB.

Message 2, the Random Access Response Message

When the eNodeB receives the preamble, it prepares to send random access response within a window specified in the previous section. The eNodeB sends the random access response on PDSCH and corresponding capacity grant, that is, DCI, on PDCCH. The UE monitors the RA-RNTI on PDCCH and if correct RA-RNTI is detected the UE decodes the actual random access message from PDSCH based on the information given by the DCI. The random access response message contains information on the received preamble identity based on which the

UE can detect if the random access response was for the preamble identity selected by it or not. If there is a match the UE decodes the rest of the random access response, otherwise the random access procedure is considered failure and UE starts the random access procedure from the beginning but with higher power (incrementing PREAMBLE_TRANSMISSION_ COUNTER). The random access response contains initial timing advance that is calculated by the eNodeB based on the received preamble. The eNodeB allocates the temporary cell-radio network temporary identifier (C-RNTI) and the uplink grant for message 3 transmission (and possible retransmissions) and includes those in the random access response message. Also the UE transmission power level for message 3 given as power increase compared to the last transmitted (by the UE) preamble transmission power is given to the UEs within the random access response message.

In case the UE does not receive any response (as random access response) to the sent preamble with selected RA-RNTI or the preamble identity decoded from random access response does not match the one selected by the UE, the UE assumes that the eNodeB did not hear the preamble and increases the preamble transmission power (incrementing PREAMBLE_ TRANSMISSION_COUNTER). This can also happen in case the problem is in downlink direction (the UE does not hear the random access response message or the corresponding DCI) and therefore the random access response transmission parameters for PDCCH and PDSCH must be properly set. For PDCCH this means that the coding rate (as CCE aggregation level) for the used DCI must be selected according to the coverage conditions of the cell. The DCI for random access response is transmitted in common search space which means that it can only have aggregation level 4 or 8. The aggregation level can be either fixed by parameters or allocated automatically by eNodeB algorithms depending on the coverage conditions. For PDSCH the coding rate for actual random access message delivery needs to be set so that it corresponds to the selected DCI PDCCH aggregation level; otherwise, the UE might be able to hear the DCI from PDCCH but not the actual random access response message on PDSCH causing unnecessary repetitions of random access procedures.

Message 3, the Radio Resource Control Connection Request Message, Transmission

Once the UE has successfully received the random access response, it will send RRC connection request message (message 3) to the eNodeB on PUSCH with the power according to Eq. 12.2 [1] (assuming PUSCH transmission and no simultaneous PUSCH and PUCCH transmission) using the resources indicated in the random access response message:

$$P_{\text{PUSCH}}\text{Msg3} = \min \left\{ \begin{array}{l} P_{\text{CMAX}}, \\ 10\log(M_{\text{PUSCH}}(i)) + \text{preambleInitialReceivedTargetPower} + \\ \Delta_{\text{PREAMBLE\_Msg3}} + \alpha(j) \cdot \text{PL} + \Delta_{\text{TF}}(i) + \Delta P_{\text{rampup}} + \Delta_{\text{Msg2}} \end{array} \right\} \quad (12.2)$$

where

- P_{CMAX} is the configured UE transmit power defined in [2] in subframe i for serving cell c;
- i is the subframe number;
- $M_{\text{PUSCH}}(i)$ is the number of physical resource blocks (PRBs) for PUSCH transmission;

- preambleInitialReceivedTargetPower is the target power for preamble reception at the eNodeB as shown in Eq. 12.1;
- $\Delta_{PREAMBLE\_Msg3}$ = DeltaPreambleMsg3 given in system information block type 2;
- $\Delta_{TF}(i)$ is the MCS-dependent component as defined in [1];
- PL is the downlink path-loss estimate calculated in the UE for serving cell c in decibels and is calculated as: eNodeB-transmitted reference signal transmission power (delivered to the UE in system information block type 2) – RSRP (measured and filtered by UE);
- $\alpha(j)$ is the path-loss compensation and is set to 1 for message 3 transmission, that is, full path-loss compensation;
- ΔP_{rampup} = the total power ramp-up from the first to the last preamble, that is, the total power ramping until the preamble is acknowledged by the RA response;
- Δ_{msg2} is given by the eNodeB in the random access response message;
- $\Delta_{msg2} + \Delta P_{rampup}$ forms the closed-loop part of the power control formula above and it should be noted that $f(0)$, that is, initial value of the closed-loop part is defined as $\Delta_{msg2} + \Delta P_{rampup}$.

Message 3 power level, therefore, depends on the preamble transmissions in terms of ΔP_{rampup} and power offsets given by the eNodeB ($\Delta_{PREAMBLE\_Msg3}$ and Δ_{msg2}). If the eNodeB fails to decode message 3 (and sends NACK) because selected MCS and number of PRBs were too high compared to the allocated UE transmission power, the UE retransmits message 3 (hybrid automatic repeat request, HARQ retransmissions). Retransmissions are done using MCS and number of PRBs given in uplink grant in random access response. Message 3 HARQ retransmissions are done until maximum allowed number is reached (given in system information block type 2) after which the UE starts the random access procedure from the beginning by selecting a new preamble at increased power level. Also in case the eNodeB does not hear message 3 at all and there is no NACK (nor ACK) or message 4 received by the UE until contention resolution timer expiry, the UE starts the random access procedure from the beginning by selecting a new preamble at increased power level. These preamble retransmissions further increase the UE PUSCH transmission power for message 3 and in case of cumulative closed-loop power control, for all coming PUSCH transmissions. Also additional PDCCH and PDSCH resources are needed for message 2 transmissions. Hence, it is very important that as many as possible preambles lead to successfully decoded message 3. It should be noted that message 3 HARQ retransmissions are typically non-adaptive HARQ retransmissions (to save the PDCCH capacity) which means that all message 3 retransmissions use exactly the same resources (number of PRBs, PRB location, MCS and transport block size) as the initial transmission and therefore there is no frequency-selective scheduling gain between the initial transmission and HARQ retransmissions. Figure 12.11 shows an example of cell suffering from high level of uplink interference and how RACH success rate gets impacted by it. As mentioned above, the eNodeB impacts on message 3 transmission power based on the parameters DeltaPreambleMsg3 and Δ_{msg2} but it is not recommended to maximize the UE transmission power for message 3 to maximize the RACH success rate due to the increased inter-cell interference which would then degrade the performance during high traffic (as shown in Figure 12.11). It is better to optimize message 3 capacity grant parameters such as the number of used PRBs, which is defined based on the size of message 3 and therefore depends also on the terminal selected preamble group, and MCS together with the transmission power parameters. It should be noted that minimum amount of bits that should

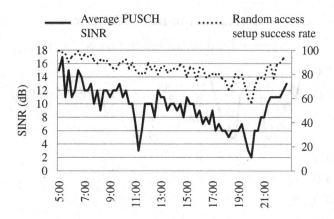

Figure 12.11 Random access setup success rate and PUSCH signal to interference and noise ratio in a cell

be assigned to message 3 is 56, which is large enough for RRC connection request message. Figure 12.12 shows improvement in RACH success rate by optimized uplink grant parameters and transmission power parameters for message 3. It should be noted that optimization of uplink grant was started in day 16 and finished on day 45. The improvement after day 45 is explained later in Section 12.4.2.

With physical layer optimization, it is possible to reduce the total amount of preambles the eNodeB receives. The random access setup success rate is improved as the overlap between the cells that allocate the preambles from the same random access preamble root sequence index is reduced and hence the amount of preambles detected by neighbouring cells with wrong timing advance (TA) estimation is reduced. Wrongly estimated TA causes message 3 reception failures and therefore reduced random access setup success rate and increased

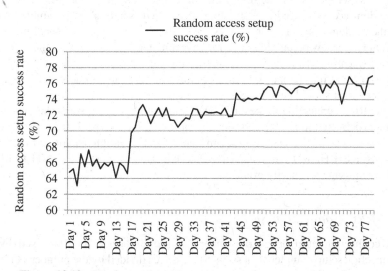

Figure 12.12 Random access setup success rate improvement in a cluster

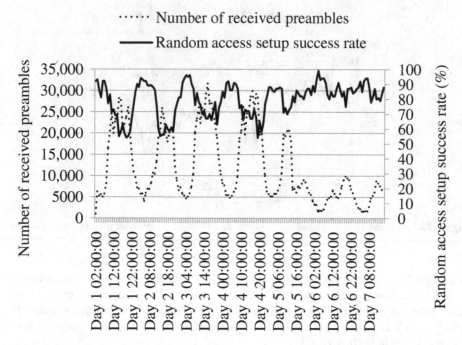

Figure 12.13   Physical layer optimization impact on random access setup success rate in a cell

inter-cell interference. An example of physical layer optimization impact on random access setup success rate and number of received preambles is given in Figure 12.13. The example in Figure 12.13 shows that after some antenna redirection and antenna downtilt tuning done for several cells, one of the cells inside the cluster received three times less preambles and also random access setup success rate improved during the peak times from ~60% to ~80%.

The random access procedure signalling, that is, UE sending the preamble and eNodeB sending the random access response, does not contain any retransmissions of those messages but rather always a new preamble transmission is initiated. Therefore, the random access setup success rate (measured from network statistics) typically does not exceed 90%, and due to the resource usage and interference increase optimization the random access setup success rate performance target should be set to 80–90%.

Once the RRC connection request has been transmitted, the UE starts the contention resolution timer (delivered to the UE in system information block type 2) to supervise message 4 reception as a response to message 3 transmission. The contention resolution timer is restarted at each message 3 HARQ retransmission. The UE uses temporary C-RNTI to monitor the PDCCH for message 4 capacity grant.

12.4.2   RRC Connection Setup

The RRC connection setup success rate can be measured as the number of received (by eNodeB) RRC connection setup complete (message 5) messages divided by the number of received (by eNodeB) RRC connection setup request (message 3) messages. Poor performance of RRC

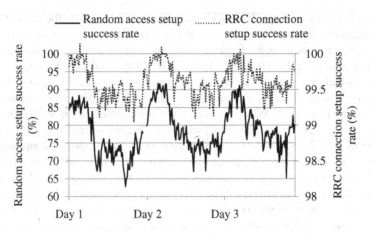

Figure 12.14 Average random access setup success rate and average RRC connection setup success rate in a small cluster of sites over a period of 3 days

connection setup success rate can be caused by message 5 reception problems at eNodeB or message 4 (RRC connection setup message) reception problems at the UE. Problems in receiving the respective message capacity grant by the UE or scheduling request (SR) reception (for message 5) problems at the eNodeB can cause problems in RRC connection setup success rate. Any capacity limitation in terms of number of RRC-connected users reduces RRC connection setup success rate as shown in Chapter 13. Figure 12.14 shows one example of traffic increase impact on the random access setup success rate and RRC connection setup success rate. Typically when the traffic increases the inter-cell interference also increases (caused by more users in general and more users under poor radio conditions) and therefore more failures in signalling message reception at the eNodeB and at the UE occur and performance decreases.

Message 4, Contention Resolution and RRC Connection Setup Message Reception

Once the eNodeB receives message 3, it prepares to send message 4, which is concatenated message including contention resolution and RRC connection setup messages, to the UE. For message 4, containing the contention resolution message, downlink grant is addressed by temporary C-RNTI, allocated to the UE in random access response message, on PDCCH. The contention resolution is simply the copy of the 40 bit UE identity (allocated by the UE itself based on SAE Temporary Mobile Station Identifier [4] or in case the UE was never previously attached to the network, random value) from message 3 contents based on which the UEs can identify the correct message 4. Several UEs can select the same random access resource, that is, use the same RA-RNTI, and select randomly the same random access preamble and therefore several UEs can receive random access response that is not intended for them and send message 3. Then finally only message 4 reception (the contention resolution part of it) can identify the correct UE for which the random access response was intended to and for which the RRC connection is being setup. Message 4 uses adaptive HARQ and the HARQ feedback is sent only by the UE which identifies its own UE identity, used in message 3, from contention resolution message. If coincidentally any other UE chose the same RA-RNTI and the same preamble, it would now understand that there was a collision and it would start the random access process from the beginning by transmitting a new preamble with increased transmission power (i.e. PREAMBLE_TRANSMISSION_COUNTER is incremented by 1). The UE for which message 4 was intended to responds with HARQ acknowledgement (in case correct decoding of the whole message 4) and stops the contention resolution timer and prepares to send message 5. While the contention resolution timer is still running in the UE (started when message 3 was transmitted) and the UE receives negative acknowledgement for transmitted messag 3, the UE will retransmit message 3 and restart the contention resolution timer at each retransmission of message 3. After transmission (or retransmission) of message 3, the UE monitors the PDCCH resources identified by temporary C-RNTI (given in the random access response message) for message 4 capacity grant until the expiry of contention resolution timer. If the contention resolution timer expires, contention resolution is considered unsuccessful and the UE is expected to start the random access procedure from the beginning, that is, select the preamble from preamble group and send the selected new preamble with increased power as PREAMBLE_TRANSMISSION_COUNTER is incremented by 1.

The coding rate for message 4 delivery on PDSCH and CCE aggregation level for DCI transmission on PDCCH for message 4 are either fixed by parameters or adjusted by the eNodeB algorithm. Similar to random access response those parameters should be set so that good enough RRC connection setup success rate is achieved. It should be noted that in case the UE cannot receive message 4 it cannot send message 5 and therefore from statistics point of view message 5 reception problems are detected. The RRC connection setup success rate improvement with decreased message 4 coding rate on PDSCH and increased CCE level for DCI delivery on PDCCH of one test cluster during busy hour is shown in Table 12.1.

It should be noted that other parameters impacting on message 4 reception success rate that should be tuned as well are

- HARQ retransmission times for message 4.
- Related to above the contention resolution time should be long enough so that terminal can receive all specified transmissions of message 4. For example, if 4 HARQ transmissions are allowed for message 4 then contention resolution timer should be minimum 4×8 ms = 32 ms.

Table 12.1 Radio resource control connection setup success rate optimization depending on message 4 parameters

	PDSCH coding rate = 0.39	PDSCH coding rate = 0.12	PDSCH coding rate = 0.05
	PDCCH CCE aggregation level = 4	PDCCH CCE aggregation level = 4	PDCCH CCE aggregation level = 8
Total number of received RRC connection request messages (message 3)	10,577	8625	9488
RRC connection setup success rate (%)	99.41	99.9	99.96

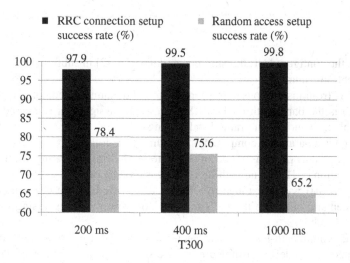

Figure 12.15 RRC connection setup success rate and random access setup success rate as a function of T300

- T300 is started by layer 3 transmission of RRC connection request to lower layers within the UE and stopped when layer 3 receives either RRC connection reject or RRC connection setup message. Therefore, T300 should be set long enough to cover all message 3 and message 4 transmissions. However, it should be noted that in case T300 is increased the terminal can transmit more preambles (not restricted by PREAMBLE_TRANSMISSION_COUNTER = preambleTransMax + 1 as described earlier) causing degradation of random access setup success rate. Example of T300 tuning and its impact on RRC connection setup success rate and random access setup success rate is shown in Figure 12.15.

Message 5, Radio Resource Control Connection Setup Complete Message Transmission

The UE does not have any uplink grant for RRC connection setup complete (message 5) message transmission so before message 5 can be transmitted the UE needs to send scheduling request (SR) on physical uplink control channel (PUCCH) and wait for uplink capacity grant on PDCCH. Successful message 5 transmission (and reception by eNodeB) is impacted by PUCCH power control, PUSCH power control and PDCCH DCI decoding (to send the uplink capacity grant with high enough CCE aggregation level) and the resource allocation on PUSCH for the actual message 5 delivery (selected MCS and number of PRBs). Power control for both PUCCH and PUSCH must be set so that high enough transmission power from the UE is possible when transmitting message 5 and scheduling request.

Message 5 is the first message sent by the UE that is following the given power control strategy and uses dynamic PUSCH scheduling. The UE power for PUSCH is given by Eq. 12.3 according to [1] (assuming no simultaneous PUSCH and PUCCH transmission):

$$P_{\text{PUSCH}}(i) = \min \left\{ \begin{array}{l} P_{\text{CMAX}}(i), \\ 10 \log_{10}(M_{\text{PUSCH}}(i)) + P_{\text{0\_PUSCH}}(j) + \alpha(j) \cdot \text{PL} + \Delta_{\text{TF}}(i) + f(i) \end{array} \right\} \quad (12.3)$$

where

- P_{CMAX} is the configured UE transmit power defined in [2] in subframe i for serving cell c;
- i is the subframe number;
- For PUSCH (re)transmissions corresponding to the dynamic scheduling j equals 1;
- $M_{PUSCH}(i)$ is the bandwidth of the PUSCH resource assignment expressed in number of resource blocks valid for subframe i and serving cell c;
- $P_{0\_PUSCH}(j)$ is a parameter composed of the sum of the following:
  - $P_{0\_NOMINAL\_PUSCH}(j)$ has range of -126 to $+24$ dBm indicating from full path-loss compensation (-126) to no path-loss compensation ($+24$).
  - $P_{0\_UE\_PUSCH}(j)$ with the range of -8 to 7 dB is used by the eNodeB to compensate any systematic offsets in the UE's transmission power settings arising from a wrongly estimated path loss.
- PL is the downlink path-loss estimate calculated in the UE for serving cell c in decibels and is calculated as eNodeB-transmitted reference signal transmission power (delivered to the UE in system information block type 2) – RSRP (measured and filtered by the UE);
- $\alpha(j)$, alpha, is used as path-loss compensation factor as a trade-off between total uplink capacity and cell-edge data rate:
  - Full path-loss compensation maximizes fairness for cell-edge UE's;
  - Partial path-loss compensation may increase total system capacity, as less resources are spent ensuring the success of transmissions from cell-edge UEs and less inter-cell interference is caused to neighbouring cells;
  - For $j = 1$, α can be 0, 0.4, 0.5, 0.6, 0.7, 0.8, 0.9 and 1.0, where 0.7 or 0.8 gives a close-to maximum system capacity by providing an acceptable cell-edge performance;
- $\Delta_{TF}(i)$ is the MCS-dependent component as defined in [1];
- The closed-loop part $f(i)$ is adjusted by the eNodeB according to power control strategy. In case of accumulation type the previous value of $f(i)$, that is, $f(i-1)$ is summed together with the new transmission power control command from eNodeB and in case of absolute type only the new transmission power control command from eNodeB is used to update $f(i)$. The accumulation type transmission power control command is selected from $\{-1, 0, 1, 3\}$ dB and absolute type transmission power control command is selected from $\{-4, 1, 1, 4\}$ dB. The accumulation type closed-loop power control starts from the initial value, $f(0)$, and adjusts that initial value based on eNodeB transmission power control command. The $f(0)$ value is the one given for message 3 transmission power as part of Eq. 12.2. The absolute closed-loop power control type only uses the current transmission power control command from the eNodeB as the new closed-loop power control, $f(i)$, value.

The UE transmission power for PUCCH, given by Eq. 12.4 according to [1] (assuming PUCCH transmission), follows the same principle as transmission power for PUSCH:

$$P_{PUCCH}(i) = \min \begin{cases} P_{CMAX,c}(i), \\ P_{0\_PUCCH} + PL_c + h(n_{CQI}, n_{HARQ}, n_{SR}) + \Delta_{F\_PUCCH}(F) + \\ \Delta_{TxD}(F') + g(i) \end{cases} \quad (12.4)$$

where the main parameters are explained below:

- P_{CMAX} is the configured UE transmit power defined in [2] in subframe i for serving cell c;
- P_{0_PUCCH} is a parameter composed of the sum of the following:
  - $P_{0\_NOMINAL\_PUCCH}$ is $P_{0\_NOMINAL\_PUSCH}$ equivalent for PUCCH and has a typical range of -127 to -96 dBm;
  - $P_{0\_UE\_PUCCH}$ is $P_{0\_UE\_PUSCH}$ equivalent for PUCCH and has a typical range of -8 to 7 dB;
- PL is the downlink path-loss estimate calculated in the UE for serving cell c in decibels and is calculated as eNodeB-transmitted reference signal transmission power (delivered to the UE in system information block type 2) – RSRP (measured and filtered by UE);
- $\Delta_{F\_PUCCH}(F)$ corresponds to a PUCCH format (F) power relative to PUCCH format 1a, where each PUCCH format (F) refers to all other PUCCH formats: 1, 1b, 2, 2a, 2b and 3) delivered in system information block type 2 by the eNodeB;
- $\Delta_{TxD}(F')$ is assumed to be 0 indicating no PUCCH transmission from two antenna ports by the UE as defined in [1];
- $h(n_{CQI}, n_{HARQ}, n_{SR})$ is a PUCCH format-dependent value explained in [1];
- Note that there is no fractional power control for PUCCH, that is, alpha equals 1;
- The closed part $g(i)$ is adjusted by eNodeB according to the power control strategy. The transmit power control command, δ_{PUCCH}, included into the PDCCH format 1,1A, 1B, 1D, 2, 2A, 2B or 2C can have values $\{-1, 0, 1, 3\}$. The transmit power control command, δ_{PUCCH}, included into the PDCCH format 3 or 3A can have values $\{-1, 1\}$. The transmit power control command is included into the previous value of $g(i)$ according to $g(i) = g(i-1) + \sum_{m=0}^{M-1} \delta(i - k_m)$, where for FD mode $M = 1$ and $k_0 = 4$ and for TD as defined in [1].

The power control settings for PUSCH and PUCCH must follow the same strategy, that is, in case PUSCH power control is aggressive then PUCCH power control should also be aggressive. Similarly in case PUSCH power control is conservative then PUCCH power control should also be conservative. This way the UE transmission power has less fluctuation between PUSCH and PUCCH and therefore power differences per allocated physical resource block are smaller and signalling on PUCCH is improved (CQI and ACK/NACK). UE transmission power control for PUCCH is further analysed in Section 12.5.

At the point of message 5 transmission the closed-loop power control does not cause any increase or decrease of the UE transmission power as there are no received signal strength indicator and signal to noise and interference measurement results available from previous PUSCH dynamic scheduling allocations. Therefore, message 5 power is set by open-loop power control part (alpha, $P_{0\_NOMINAL\_PUSCH}(j) + P_{0\_UE\_PUSCH}(j)$, path loss measured by the UE and in case of cumulative closed-loop power control, message 3 closed-loop part explained in Section 12.4.1). Similarly for scheduling request transmission the UE transmission power is determined by the open-loop power control part $(P_{0\_NOMINAL\_PUCCH} + P_{0\_UE\_PUCCH})$ as there are not many PUCCH transmissions based on which the closed-loop part could be adjusted. The open-loop parameter values for PUSCH and PUCCH are network specific due to the different coverage and interference (inter-cell and load based) conditions and should be adjusted together with resource allocation optimization. The PDCCH CCE aggregation level (i.e. coding rate) for DCI is determined by the received CQI reports (parameters given in RRC connection setup message) and therefore based on link adaptation or in case of no CQI reports received the aggregation level is decided based on fixed parameter value. The actual resource allocation for message 5, that is, MCS and number of PRBs should be optimized together with the transmission power control in order to maximize the RRC connection setup success rate while minimizing the transmission power from the UE.

The link adaptation can adjust the MCS (and number of PRBs) during retransmissions as message 5 is the very first message sent on dedicated control channel (DCCH) using acknowledged mode (AM) radio link control (RLC) protocol on signalling radio bearer one (SRB1). In addition to HARQ retransmissions, there are also RLC retransmissions and in between RLC transmissions the link adaptation can change the MCS and number of PRBs (in case of HARQ transmissions, the MCS and number of PRBs cannot change from the initial transmission). Therefore, the link adaptation operation must also be considered when optimizing message 5 decoding performance. Message 5 typically contains piggybacked NAS message; for example, in case of RRC connection setup initiated for the attach request, the attach request message is combined together with the RRC connection setup complete, hence the resource allocation for message 5 should be large enough for both RRC connection setup complete and possible NAS message. Figure 12.16 shows one example of RRC connection setup success rate improvement by parameter optimization to improve message 4 decoding at the UE and message 5 decoding at the eNodeB in coverage-limited cluster.

The random access setup success rate and RRC connection setup success rate are highly correlated as shown in Figure 12.17. In high loaded cells (PRB utilization >50%) the RRC connection setup success rate quickly degrades when random access setup success rate drops below 80%. Therefore, in order to maintain high RRC connection setup success rate during high load the random access setup success rate should be maintained above 80%. Both random access setup success rate and RRC connection setup success rate are impacted by the increase of inter-cell interference and therefore when traffic grows and inter-cell interference grows parameter optimization becomes essential. The RRC setup success rate optimization also

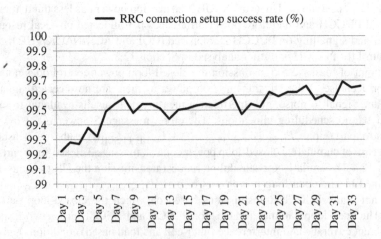

Figure 12.16 RRC connection setup success rate improvement after parameter optimization in coverage-limited cluster, day-level statistics

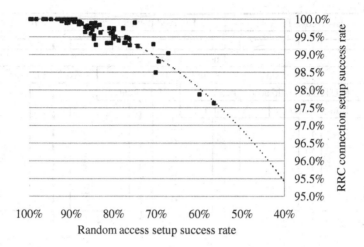

Figure 12.17   Random access setup success rate versus RRC connection setup success rate in a cluster during moderate traffic, 15 minutes statistics collection period

improved the random access success rate, that is, the improvements after day 5 in Figure 12.16 are also visible as an improvement in Figure 12.12 after day 45. This is caused by the less amount of repetitions of preambles due to message 4 problems and less amount of RRC connection setup attempt procedures due to the improved RRC connection setup complete detection.

12.4.3   E-RAB Setup

After successful RRC connection setup procedure the call setup continues with the E-RAB setup procedure as shown in Figure 12.8. The E-RAB setup success rate itself can be defined based on network statistics from eNodeB point of view:

- Number of initial context setup response messages (S1 application protocol) sent by the eNodeB to the mobility management entity (MME) divided by number of initial context setup request (S1 application protocol) sent by MME to the eNodeB or
- Number of E-RAB setup response messages (S1 application protocol) sent by the eNodeB to the MME divided by number of E-RAB setup request messages (S1 application protocol) sent by MME to the eNodeB.

As shown in Figures 12.8 and 12.18, the E-RAB setup covers radio bearer (RB) setup between the UE and the eNodeB and the S1 bearer setup for the UE between eNodeB and the evolved packet core (EPC) serving gateway (S-GW). It should be noted that here the E-RAB setup success rate analysis contains only the RB setup part, that is, S1 bearer setup challenges are not considered. The E-RAB setup success rate is only marginally impacted by the increase in traffic (and inter-cell interference) compared to random access setup success rate and RRC connection setup success rate as can be seen from Figure 12.19. The RB setup

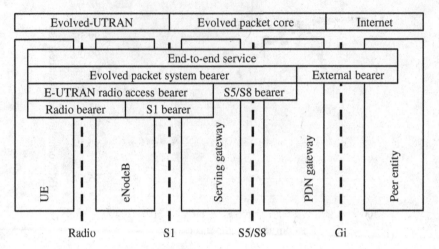

Figure 12.18   LTE bearer architecture [3]

success rate in Figure 12.19 also includes incoming handovers (HO) for which the RB is established in the target cell whereas the E-RAB setup success rate does not contain any HO scenarios hence the difference. The increased inter-cell interference causes RRC connection reconfiguration message decoding challenges in the UE and RRC connection reconfiguration complete message decoding challenges in the eNodeB. E-RAB setup success rate improvement after parameter optimization (explained below) is shown in Figure 12.20.

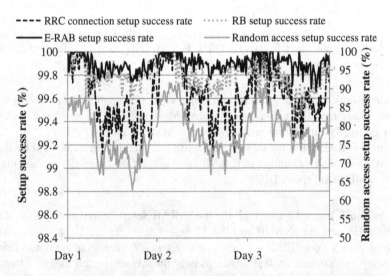

Figure 12.19   Average E-RAB setup success rate, radio bearer (RB) setup success rate, RRC connection setup success rate and random access setup success rate per cell in a cluster during moderate traffic (15 minutes statistics collection period)

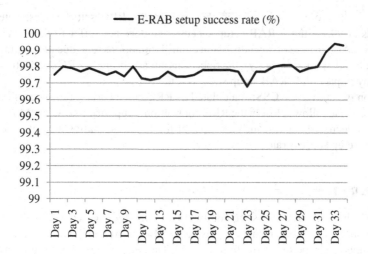

Figure 12.20 Average E-RAB setup success rate improvement after parameter optimization for coverage-limited cluster, day-level statistics

The improvement in E-RAB setup success rate after day 31 comes from link adaptation parameter optimization. This includes minimum resource allocation size optimization together with UE transmission power control for coverage-limited scenarios, that is, minimizing the number of PRBs and MCS in order to minimize the UE transmission power while maximizing the E-RAB setup success rate.

Figure 12.21 shows the impact of moderate traffic on E-RAB setup success rate and RRC connection setup success rate. During highest peak of signalling traffic (RRC connection setup attempts increase to around four times the lowest level) the RRC connection setup success rate decreases significantly to 99.6% level but the E-RAB setup success rate decreases only

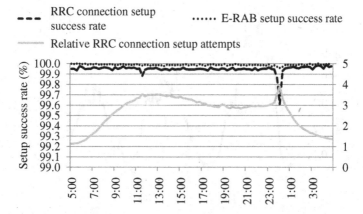

Figure 12.21 Average RRC connection setup success rate, average E-RAB setup success rate and average relative RRC connection setup attempts in cluster during moderate traffic, 15 minutes statistics collection period

marginally. Therefore, it can be said that in case the RRC setup success rate remains at high levels of 99.x% the E-RAB setup success rate does not suffer from increased inter-cell interference. Also in case of any admission control limitations (e.g. due to reaching the number of RRC-connected users capacity), the rejections already occur at RRC connection setup phase without any impact on E-RAB setup success rate.

Based on the above, the CSSR calculated as RRC connection setup success rate × E-RAB setup success rate follows closely the RRC connection setup success rate. E-RAB setup success rate seldom decreases significantly as most of the failure scenarios are covered by the RRC connection setup success rate.

12.5   E-RAB Drop

The E-RAB drop rate defined as the number of abnormally released RBs leading to release of E-RAB divided by number of successfully setup RBs is correlated with the traffic and E-RAB setup attempts as can be seen from Figure 12.22. The setup success rates (random access, RRC connection and E-RAB) and E-RAB drop rate are having similar trend with traffic (inter-cell interference) increase. This is due to similarities between signalling procedures for setups and handovers and due to the fact that E-RAB drop rate is highly correlated with handover success rate (in this case intra-frequency handover success rate) as shown in Figures 12.23 and 12.24.

12.5.1   Handover Performance

The intra-LTE, intra-frequency handover execution success rate is defined from source eNodeB point of view as number of received RRC connection reconfiguration complete messages by

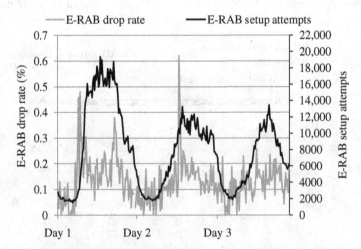

Figure 12.22   Average E-RAB drop rate and total amount of E-RAB setup attempts in a cluster during moderate traffic (15 minutes statistics collection period)

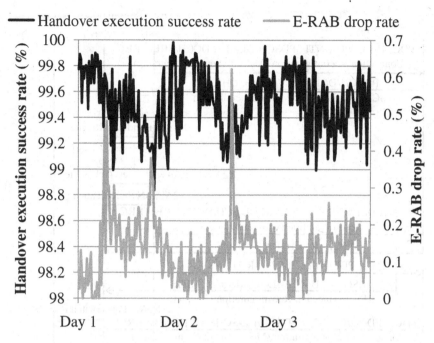

Figure 12.23 Average intra-frequency handover execution success rate and average E-RAB drop rate in a cluster during moderate traffic (15 minutes statistics collection period)

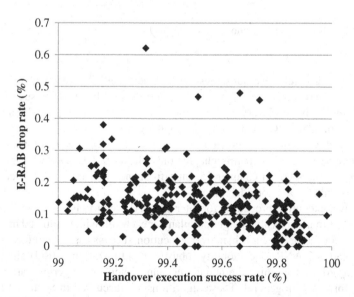

Figure 12.24 E-RAB drop rate as a function of intra-frequency handover success rate averaged over cluster of cells during moderate traffic (15 minutes statistics collection period)

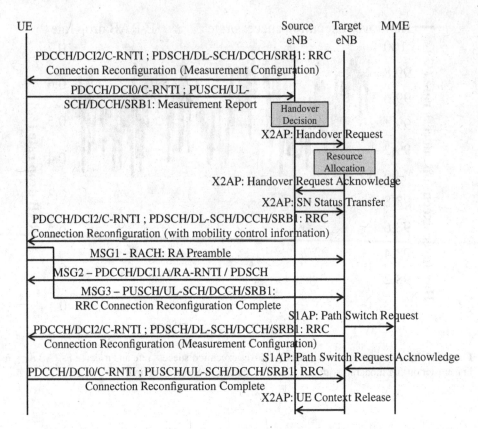

Figure 12.25 Inter-eNodeB handover signalling using contention-free random access procedure

target cell that leads to release request message received over X2 or received UE context release command message received over S1 in case of inter-eNodeB handover or successful handover indication delivered inside the eNodeB to source cell in case of intra-eNodeB handover divided by the number of sent RRC connection reconfiguration messages by source eNodeB (caused by received handover request acknowledgement over X2 or received handover command over S1). It should be noted that in this chapter only the intra-LTE, intra-frequency handover performance is discussed. In Chapter 15, inter-frequency and inter-radio access technology handover performance are considered.

The handover execution procedure for inter-eNodeB scenario is shown in Figure 12.25.

Having as optimal as possible neighbour plan per each cell with optimized handover parameters is the main factor for high handover execution success rate. Therefore, the automatic neighbour relations (ANR) and mobility robustness optimization (MRO) algorithms should be utilized in order to optimize the neighbour plan, handover triggering parameters and UE mobility performance in general. These algorithms are discussed in Section 12.6 and in this section the handover signalling and resource allocation for messages are analysed.

During the resource allocation phase the target eNodeB allocates the following parameters which are given to the source eNodeB within the X2AP Handover Request Acknowledge message and to the UE by the source eNodeB within the RRC Connection Reconfiguration message:

- The target cell physical cell identity (PCI).
- T304 supervises the handover procedure from the UE point of view: started when the UE receives RRC Connection Reconfiguration message containing mobility control information and stopped once the UE has successfully completed the random access procedure and sent the RRC Connection Reconfiguration Complete message to the target cell (eNodeB).
- New C-RNTI, which is used by the UE right after the random access procedure for PDCCH DCI monitoring under the target cell.
- In case of contention-free random access procedure the target eNodeB allocates contention-free preamble and all preamble transmission parameters. In case the UE does not receive any preamble identification, it knows that it should use contention-based random access procedure for the handover procedure. Also contention-based preamble parameters are given to the UE for contention-based random access procedure (to be used in case contention-free preamble index is not given). Use of contention-based random access procedure instead of contention-free procedure for the handover procedure would increase the handover execution delay (the time between the UE receiving the RRC connection reconfiguration message from the source eNodeB/cell and the target eNodeB/cell receiving the RRC connection reconfiguration complete message).
- Message 3 HARQ transmission amount.
- Reference signal (RS) transmission power parameters.
- PUSCH parameters for supported hopping mode and uplink RS configuration.
- Target cell antenna configuration.
- UE maximum transmission power limit.
- Discontinuous reception (DRX) parameters.
- PUCCH resources for CQI, RI and SR reporting.
- Transmission mode and related possible code book restriction parameters.
- Ciphering and integrity protection algorithms and related parameters.

After the UE has received the RRC Connection Reconfiguration message it releases the physical layer connection to the source cell (source eNodeB) and sends RRC Connection Reconfiguration Complete message to the target cell which in turn initiates the contention-free random access procedure towards the target cell (target eNodeB). It should be noted that according to [4] the UE does not need to acknowledge (HARQ or ARQ) the reception of RRC Connection Reconfiguration message (including mobility control information) but perform the handover without any delay.

The preamble transmission power for handover signalling is defined by the UE in the same way as the preamble transmission power for RRC Connection Setup procedure. After transmitting the preamble the UE monitors the PDCCH DCIs using the selected RA-RNTI in order to receive message 2. The message 2 PDCCH and PDSCH transmission parameters for handover procedure are defined in the same way as for the RRC Connection Setup procedure. Message 2 defines the PUSCH resource allocation for the initial transmission of message 3. The resource allocation for message 3 retransmissions (in case of contention-free random access procedure) is done following adaptive HARQ retransmissions and the allocation informed to the UE by PDCCH DCI0. Message 3 resource allocation should be large enough for RRC Connection

Reconfiguration Complete message and long buffer status report (BSR) and power head room report (PHR). The number of PRBs and the MCS for handover procedure (for message 3 transmission) should therefore be optimized in a similar way as for RRC Connection Setup procedure (minimizing the transmission power needed to send the whole message 3 in single PUSCH transmission). It should be noted that in case the resource allocation is too small message 3 is segmented causing additional delay in handover procedure completion. When message 3 resource allocation is optimized in a similar manner as for RRC Connection Setup procedure, the handover procedure failure rate in poor coverage conditions in terms of failed message 3 receptions by the eNodeB can be reduced significantly (up to 20% in one example cluster). In case of challenging radio conditions the handover procedure failure can also be caused by the source cell problems, that is, the UE might not hear the RRC Connection Reconfiguration message. In case the MRO algorithms (too early or too late handovers) cannot improve this problem then the source cell link adaptation in terms of sending the RRC Connection Reconfiguration message (PDCCH and PDSCH) to the UE needs to be made more robust.

12.5.2   UE-Triggered RRC Connection Re-establishments

UEs monitor the radio link performance and in case of problems RRC connection re-establishment can be triggered unless the problem can be recovered. Successful RRC connection re-establishment requires that the target eNodeB has the UE context (basically all the information about the UE's current call parameters) of the incoming UE. Therefore, in LTE special arrangements are needed for the UE context delivery from the source eNodeB to the target eNodeB in order to achieve as high RRC connection re-establishment success rate as in wideband code division multiple access (WCDMA) networks. In WCDMA networks, the radio network controller (RNC) holds the UE context and therefore only in case of inter-RNC RRC connection re-establishment some type of context fetching by target RNC from source RNC is needed. In LTE, the UE context fetching can be enabled using MRO algorithms as shown in Chapter 14. Therefore, in order to minimize the dropped call rate in LTE networks the triggering of re-establishments should be minimized and on the other hand the success rate of the RRC connection re-establishment should be maximized. The UE handling of radio link problems is based on [4], [5] and [6] as shown below (it should be noted that additional triggers, on top of 3GPP-defined ones, for RRC connection re-establishments can exist in the UEs):

- Timer T310 expiry
- Maximum number of RLC retransmissions in uplink is reached
- Handover failure and timer T304 expiry
- Non-handover-related random access problem

Timer T310 Expiry

While in RRC-connected state the physical layer of the UE monitors the radio link quality based on the cell-specific reference signals of the primary cell (Pcell) as defined in [6]. The UE compares the downlink radio link quality estimation against the out-of-sync and in-sync thresholds (defined in [6]). The out-of-sync threshold, Qout, is defined as the level at which the downlink radio link cannot be reliably received corresponding to 10% block error rate (BLER) of a hypothetical PDCCH transmission taking into account the PCFICH errors. The in-synch

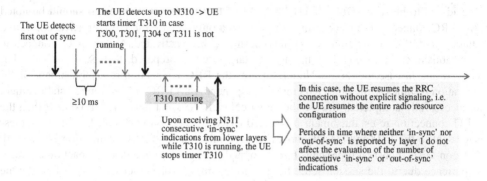

Figure 12.26   Out-of-sync and in-sync evaluation

threshold, Qin, is defined as the level at which the downlink radio link can be reliably received corresponding to 2% BLER of a hypothetical PDCCH transmission taking into account the PCFICH errors. The in-sync and out-of-sync are evaluated during time window which depends on the DRX settings as shown in [6]. In case of no DRX: if the downlink radio link quality of the Pcell estimated over the last 200 ms is worse than Qout, the physical layer sends out-of-sync indication to layer 3. In case out-of-sync is detected the UE initiates evaluation of in-sync. The UE layer 3 evaluates radio link failure based on the out-of-sync and in-sync indications after layer 3 filtering (layer 3 filtering as specified in [4]). As shown in Figure 12.26, upon receiving N310 consecutive out-of-sync indications for the Pcell from physical layers while none of the timers T300, T301, T304 or T311 is running the UE starts the timer T310. In case while T310 is running there are N311 in-sync indications from physical layer the radio link is considered to be recovered and normal operation continues. The time between two successive radio link status indications from physical layer is defined in [6] to be minimum 10 ms.

In case the UE layer 3 does not receive N311 consecutive in-sync indications while T310 is running and T310 expires, the UE initiates re-establishment procedure as shown in Figure 12.27. If the UE can find a suitable cell and successfully send the RRC connection re-establishment request message (message 3) with cause value 'other failure' within the timer T311 still running, the T311timer is stopped and timer T301 started to supervise the RRC connection re-establishment message reception from the eNodeB. Then provided that RRC connection re-establishment is completed within T301 time the RRC connection is

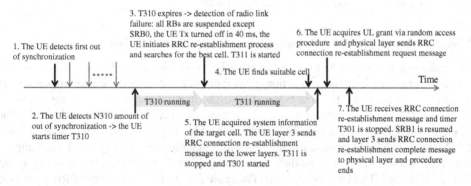

Figure 12.27   RRC connection re-establishment caused by T310 timer expiry

considered to be successfully re-established and E-RAB is not dropped. It should be noted that RRC connection re-establishment is triggered only in case the UE is in RRC-connected state (active RRC connection exists) and security has been activated. And the RRC connection re-establishment succeeds only in case the target cell is prepared, that is, it has valid UE context and in case the target eNodeB accepts the RRC connection re-establishment SRB1 operation resumes while other RBs remain suspended until RRC connection re-establishment is successfully completed. In case the target eNodeB does not have the UE context then the RRC connection re-establishment is rejected (RRC connection re-establishment reject message sent to the UE) and the UE moves to RRC idle state and RRC connection is released. For non-GBR (non-real time) calls these dropped calls do not have much impact on end-user experience due to the session setup but for, for example, voice over LTE (VoLTE) calls the end-user experiences muting as well as a dropped call. It should be noted that GBR or non-GBR calls can be re-established by new RRC and UE context setup through the target cell and if this is successful then the end users experience muting of voice or no throughput but no dropped calls. During the T311 running the UE tries to find a suitable cell according to the normal idle mode cell selection procedure [7] and if suitable cell is found the UE acquires system information messages and initiates transmission of RRC connection re-establishment message. In case the UE does not find any suitable cells and T311 expires the UE enters RRC idle state and RRC connection is released. In case timer T301 expires or selected cell is no longer considered as suitable cell, the RRC connection re-establishment procedure fails and the UE enters RRC idle state and releases the RRC connection.

The E-RAB drop rate can be easily improved by tuning the timers T310 and T311 as well as out-of-sync, N310, and in-sync, N311, indications. However, it should be noted that any optimization of the timers and indication amounts above can cause increase of silent period, or the so-called mute time for VoLTE calls, when no data can be transmitted to the UE or received from the UE. Therefore, the duration of the silent period should always be considered when optimizing the timers and indication amounts. Typically the optimization is done in such a way that quite high amount of out-of-sync indication is needed before T310 timer is started (big N310) and detection of radio link failure is done fast, that is, short T310 with very small N311 to stop timer T310. With long (longer than T310) T311 value, the UE is allowed to try to find suitable cell and send the RRC connection re-establishment message and increase the probability for successful RRC connection re-establishment procedure. It should also be noted that timer T301 that supervises the RRC connection re-establishment procedure is started and stopped in exactly the same way as T300 for RRC connection setup procedure and therefore the T301 should be set equal to or longer than T300.

Maximum Number of RLC Retransmissions in Uplink Is Reached

The radio link failure can be triggered by the UE in case maximum number of RLC retransmissions is reached in uplink and actions after radio link failure detection are the same as for T310 expiry scenario as shown in Figure 12.28. The maximum number of retransmissions is given in RRC connection reconfiguration messages for all AM RLC DRBs and in RRC connection setup message for SRB1. Typical value for AM RLC retransmission amount is 16 for DRBs and 2 for SRB1; increase of retransmissions to higher than 16 typically do not improve the dropped call rate further (as there are already $17 \times 7 = 119$ transmissions taking into account RLC and HARQ transmissions). The cause value used by the UE in RRC connection re-establishment message is the same as in the case of T310 expiry.

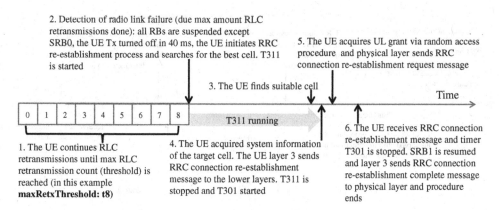

Figure 12.28 RRC connection re-establishment caused by UE reaching max number of RLC retransmissions

Handover Failure and Timer T304 Expiry

In case the UE receives the RRC connection reconfiguration message containing mobility control information element from the source cell and the UE can comply with the given configuration, the UE will start the timer T304 to supervise handover procedure completion. It should be noted that the timer T304 [4] value should be set so that it contains the random access procedure and all related reattempts time as shown in Figure 12.29.

After the T304 expiry the UE detects radio link failure and initiates the RRC connection re-establishment procedure the same way as in the case of T310 expiry. In case of T304 expiry the RRC connection re-establishment cause value is set to handover failure. As mentioned above the value of T304 should be set so that it covers the random access procedure and all related retransmissions but on the other hand the value of T304 should not be too long as then the re-establishment procedure cannot be initiated early and recovery from handover failure takes long time. Typical value of intra-LTE T304 is 1000 ms.

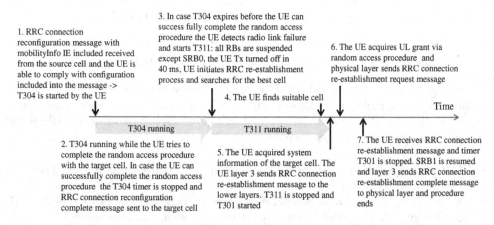

Figure 12.29 RRC connection re-establishment caused by T304 expiry

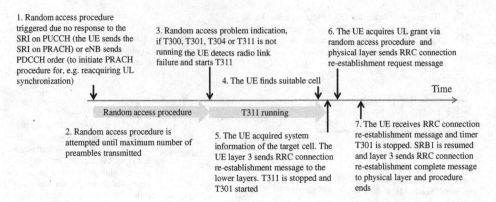

Figure 12.30 RRC connection re-establishment caused by non-handover-triggered random access procedure failure

Non-handover-Related Random Access Failure

The UE initiates non-handover-related random access procedure due to receiving PDCCH order (containing contention-free preamble identifier), for example, for re-establishment of uplink synchronization (lost due to the long discontinuous reception, DRX, cycle) for downlink data delivery from eNodeB. In case the UE is in uplink out-of-synchronization state and the UE has something to transmit, it sends the scheduling request (SR) indication on PUCCH but the eNodeB cannot receive it due to UE in out-of-synchronization state. Therefore, after retransmissions of SRs on PUCCH the UE sends the SR using random access procedure and enters into uplink in-synchronization state. The UE sends selected preambles until it successfully completes random access procedure or until the maximum amount of preambles has been sent. In case maximum number of preambles is reached the random access procedure is considered as failure and UE detects radio link failure and triggers RRC connection re-establishment procedure as shown in Figure 12.30.

Minimizing this type of re-establishment can be done by random access setup success rate improvement parameters as shown in Section 12.4.1. Also PUCCH power control should be set to compensate max pathloss while minimizing the inter-cell interference to avoid SRI being sent using random access procedure (due to the poor coverage of PUCCH).

12.5.3   eNodeB-triggered RRC Connection Re-establishments

The eNodeB has triggers for radio link problem detection and in case any of those link monitors detect the problem, the eNodeB will wait for radio link recovery until defined timer expires without any RRC connection re-establishment request received from the UE. Radio link problem detection is typically as below:

- *UL PUSCH DTX detection*: After the uplink capacity grant is given there is nothing received on the allocated PUSCH resources
- Certain amount of consecutive periodical CQI reports not received on PUCCH or PUSCH
- No uplink ACK/NACK received but only DTX for downlink data transmissions

- No allocated preamble received as a response to PDCCH order
- No allocated sounding reference signal (SRS) received but instead DTX detected on allocated resources

Each of the above-mentioned radio link problem triggers has its own controlling parameters which impact on the radio link problem detection and therefore start of radio link recovery timer. In case of expiry of the radio link recovery timer of the RRC connection, all E-RABs and S1 connection are released. The eNodeB radio link recovery timer should be longer than T310 + T311 so that the UE can perform RRC connection re-establishment to the source cell or to any other cell before connection is considered to be dropped. Figure 12.31 shows the impact of the radio link recovery timer increase for a cluster (timer change implemented at day 10).

Also possible UE context fetching time should be included into the radio link recovery time in case of advanced RRC connection re-establishment procedures are implemented. Radio link problem indication detection parameter tuning can also improve the dropped call rate. Figure 12.32 shows the number of consecutive missed periodical CQI reporting based radio link problem detection parameter tuning impact on E-RAB drop rate (note that the cluster is different compared to Figure 12.31). In case more periodical CQI reports are allowed to be missed the E-RAB drop rate is improved. However, it is not recommended to extend the radio link recovery timer too long or try to avoid any radio link problem detection, as if the problematic radio link is not released it just would reserve resources unnecessarily from the eNodeB and end user would experience poor performance.

As shown in Figure 12.32, increasing the amount of missed periodical CQI reports improves the dropped call rate. The periodical CQI report is sent typically using PUCCH and therefore the PUCCH power control (explained in Section 12.4) also plays an important role when

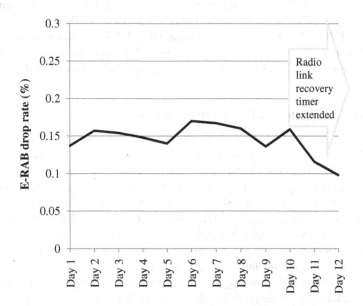

Figure 12.31   E-RAB drop rate improvement by radio link recovery timer increase

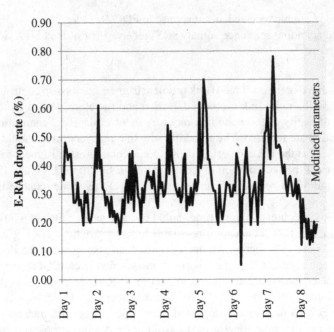

Figure 12.32 Amount of consecutive missed periodical CQI reporting impact on E-RAB drop rate

optimizing the E-RAB drop rate. As a general rule the PUCCH power control strategy should follow the PUSCH power control strategy. The PUCCH is not used as often as PUSCH and therefore the PUCCH closed-loop power control commands from eNodeB are not sent as frequently as for PUSCH. Therefore, the PUCCH closed-loop power control decision windows (RSSI and SINR) should be narrower compared to PUSCH and in general lower compared to PUSCH. Figure 12.33 shows the gain in E-RAB drop rate in terms of percentage point improvement compared to before change when RSSI and SINR decision windows are narrowed for conservative PUCCH (and PUSCH) closed-loop power control strategy.

12.6 Handover and Mobility Optimization

As mentioned in Section 12.5, the handover success rate has a strong impact on the dropped call rate. Good neighbour plan is the basis for high handover success rate. Good neighbour plan means that each cell has all the main handover target cells defined as neighbours. The most optimal neighbour plan can be achieved with the 3GPP-defined ANR feature. The ANR feature operation is introduced in [8]. The ANR feature can automatically add neighbours based on the UE measurements provided that this neighbour is good enough target cell. There can be several unnecessary neighbours added as well and therefore the ANR feature requires additional neighbour relation removal which can then remove seldom used neighbours or neighbours that have very low handover success rate.

Enabling the ANR feature shows significant improvement in dropped E-RAB rate caused by the handover execution problems (basically handover signalling problem) and improvement in handover success rate as shown in Figure 12.34.

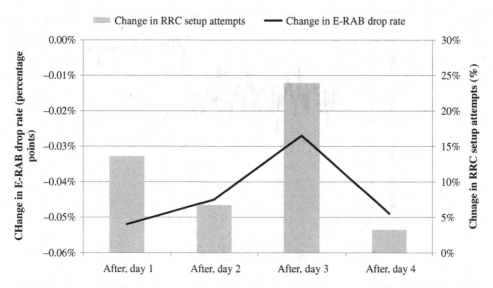

Figure 12.33   PUCCH closed-loop power control decision window tuning gain. Network-level statistics 2 days before the parameter tuning are compared to the statistics samples taken on four consecutive days after the parameter tuning

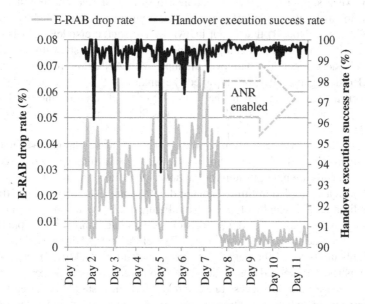

Figure 12.34   E-RAB drop rate caused by handover signalling problems and handover execution success rate before and after enabling ANR feature. Network-level hourly statistics collected for 11 days

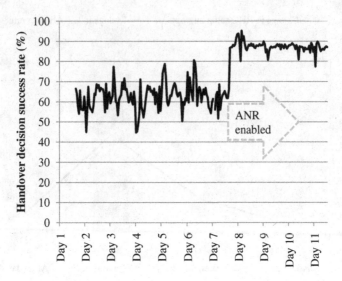

Figure 12.35 Handover decision success rate before and after enabling the ANR feature. Network-level hourly statistics collected for 11 days

Another gain from ANR is the improvement in handover decision success rate, that is, the ratio between started handover procedures to sum of started and not started handover procedures. In case source eNodeB receives measurement report that indicates handover to a target cell (target eNodeB) that is not known by the source eNodeB, that is, no mapping from reported PCI to E-UTRAN cell global identifier (ECGI) exists, without ANR activated the source eNodeB rejects the handover attempt. With ANR this mapping can be verified by eNodeB ordering the UE (that reported the unknown PCI) to decode the ECGI from the target cell and reporting that to the source eNodeB. Therefore, the handover decision success rate is improved from around 60% to below 90% when ANR is activated as shown in Figure 12.35.

Allowing ANR feature to add all measured cells by UE as neighbours will not be optimal. Having too many adjacent cells as neighbours may lead to a situation where radio conditions for handover target cells are not optimum for handovers to be successful, for example, minimum radio conditions are fulfilled in limited area or duration only. Therefore, the ANR feature use needs to be limited. This can be observed from Figure 12.36. As it can be seen right after the ANR feature was enabled the handover decision success rate improved a lot but the handover execution success rate did not improve enough. Therefore, after ANR was enabled and it added several neighbours some of the neighbours were deleted (or blacklisted) based on handover attempts statistics. This caused the handover decision success rate to decrease but at the same time the handover execution success rate improved further. It should be noted that ANR can easily add neighbours which are extremely far away from the reporting UE but just happen to provide very good coverage for certain location, for example, high-rise building, but then the handover execution can simply fail due to the random access preamble configuration (random access preamble cyclic shift length is too small). So at around day 180 further neighbour cell blacklisting was executed and handover decision success rate degraded further but at the

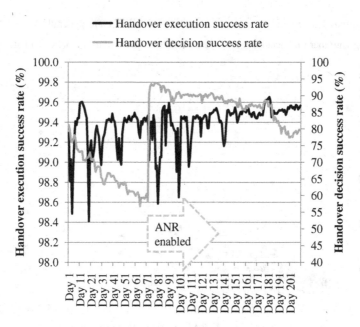

Figure 12.36 Handover decision success rate and handover execution success rate before and after enabling ANR feature. Network-level daily statistics

same time handover execution success rate improved further close to 99.6%. Therefore, ANR feature is excellent but the operation must be carefully controlled so that it does not add all neighbours measured by the UEs.

The addition of neighbours by ANR feature can be seen as increased handover attempts (comparing after ANR activation to before ANR activation) as shown in Figure 12.37.

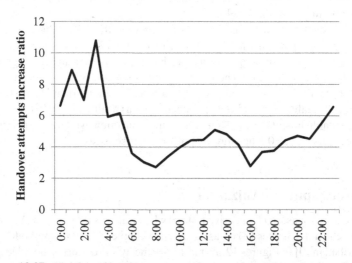

Figure 12.37 Number of handover attempts increase due to the activation of ANR

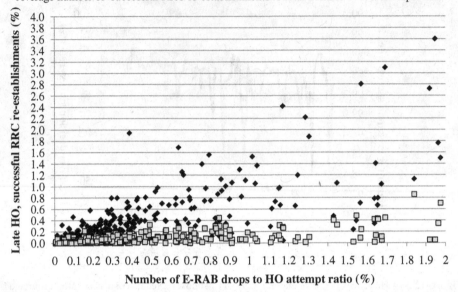

Figure 12.38 Late handover to total number of handover attempts ratio and successful RRC re-establishment to total number of handover attempts as a function of the number of E-RAB drops to total number of handover attempts

The handover attempts increased due to the ANR around four times during the highest traffic times (6–9 AM and 5–8 PM). This increase in signalling activity needs to be taken into account for the highest traffic cells and during the special events, that is, neighbour lists need to be further optimized for those cells.

The MRO algorithms can be used to further improve the UE mobility performance and therefore the E-RAB performance. As shown in Figure 12.38 the bigger the share of late handovers the higher the amount of drops per handover attempt and the higher the amount of successful re-establishments. Therefore, by enabling the MRO algorithms the too late handover share could be reduced and the end-user-experienced quality improved. The end-user experienced quality here refers to dropped calls as well as the RRC re-establishments. The RRC re-establishments cause short break in connection which is experienced as mute time for voice calls and as 0 Kbps throughput for other services.

12.7 Throughput Optimization

Throughput can be measured from drive tests or when traffic is high enough based on cell level counters (as shown in Chapter 13) or, as used in this chapter, based on methodology defined in [9]. As described in [9] in Figures 12.39 and 12.40, the 3GPP-defined scheduled IP throughput formula is slightly modified so that denominator contains also the time when the eNodeB

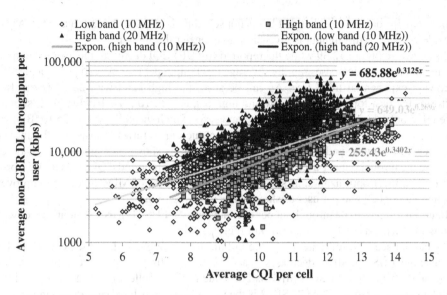

Figure 12.39 Downlink average non-GBR user throughput as a function of reported CQI. Cells with more than 250 MB downlink data volume transmitted per 15 minutes statistics sample are included into the analysis

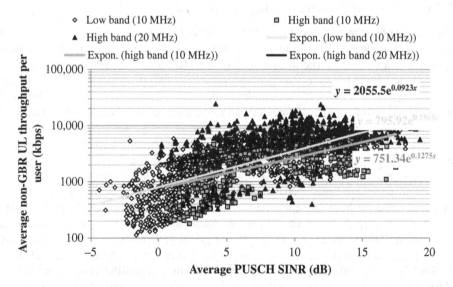

Figure 12.40 Uplink average non-GBR throughput per user as a function of average PUSCH SINR. Cells with more than 100 MB uplink data volume received per 15 minutes statistics sample are included into the analysis

buffer contains data to a UE but that UE is not scheduled. Also carrier aggregation (CA) is not considered, that is, throughput is given as single-carrier throughput.

As shown in Figure 12.39 while the reported (by UE) CQI decreases the average non-GBR throughput per user decreases rapidly. Average throughput per user per cell per band decreases following exponential trend line. As can be seen, the high band with 20 MHz bandwidth performs the best followed by the low and high bands with 10 MHz bandwidth. It should be noted that only the 15 minutes statistics samples for which the cell delivers 250 MB or more downlink data volume are included. The CQI reduction in turn is caused by the increased inter-cell interference and hence the increased traffic as shown in Figure 12.1. In order to improve the throughput per user while traffic increases the reported CQI needs to be improved by physical layer optimization as shown in Figure 12.7. Use of MIMO can also improve the throughput per cell and per user as shown later in this section.

In uplink direction the throughput per user depends on PUSCH SINR similarly as downlink throughput depends on CQI as shown in Figure 12.40.

As can be seen from Figure 12.40 the uplink user throughput decreases exponentially as the average PUSCH SINR decreases. The low band cells provide extremely good uplink throughput in case the PUSCH SINR is good due to the excellent coverage compared to the high band cells. Also the PUSCH SINR is the worst for the low band cells due to the excessive overlap of some of the cells. In case of uplink throughput the performance can be greatly improved in case the PUSCH SINR can be improved as discussed later in this section.

12.7.1 MIMO Multi-stream Usage Optimization

In the context of communication theory the term MIMO means a system that is able to transmit and receive multiple parallel information streams within the same spectrum, therefore increasing the spectral efficiency. The theoretical basis of MIMO, also called spatial multiplexing, dates back to 1950s [5] and was later popularized in the 1990s [6,7]. In radio communications, multi-stream communication requires multiple antennas at both ends of the radio link and the maximum number of parallel streams is equal to the minimum of the number of transmit and receive antennas.

In this section the focus is on understanding, from the viewpoint of radio network planning, the factors that impact the throughput gain from spatial multiplexing; for details of multi-antenna communications the reader is referred to the extensive literature. It is, however, important to understand that spatial multiplexing is only one of the many ways to utilize multiple antennas at both link ends. Also transmit/receive diversity and beam forming are techniques requiring multiple antennas at transmitter, receiver or both. While diversity and beam forming aim to improve received signal power (reliability) of a single information stream, spatial multiplexing aims to split the signal power between multiple parallel streams. Correspondingly, spatial multiplexing brings most benefits at high SINR regime while diversity and beam forming are useful at low SINR. Since in a real-world network the SINR experienced by UEs varies, practical systems implement some form of dynamic switching between multi-stream and single-stream transmission. An example of a hybrid transmission mode is the 3GPP transmission mode 8 that combines beam forming and spatial multiplexing; the details can be found in [8].

Factors Impacting Spatial Multiplexing Throughput

The classical Shannon channel information capacity for single-stream communication link without modulation constraints is

$$C = \log_2(1 + \text{SNR})$$

where C is the maximum achievable information capacity in bits per second per Hertz and SNR = S/N is the signal to noise ratio at output of the receiver, S is the signal power and N is the noise power. For a spatial multiplexing system the corresponding generalization of the classic Shannon formula reads as

$$C = \log_2(1 + \text{SNR}_1) + \log_2(1 + \text{SNR}_2) + \cdots + \log_2(1 + \text{SNR}_k)$$

where $\text{SNR}_k = S_k/N$ now denotes the SNR of the kth information substream. The SNR of a substream depends on the values of the channel impulse responses between receiver and transmitter antenna pairs and on the receiver implementation. In a closed-loop communication scheme the transmitter can use some form of precoding to improve substream SNR and thus capacity. The impulse response of a MIMO system is usually represented as an n by m matrix. For n receive ports and m transmit ports the maximum number of parallel streams is $k = \min(n, m)$; for example, a 4Tx/2Rx system can have two parallel streams. With channel fading the SNRs of the substreams also experience fading and hence the instantaneous link capacity is also time varying. Some form of link and rank adaptation has to be implemented in practical implementations. The adaptation algorithm is vendor dependent, which makes it difficult to give any guideline for parameter optimization.

In multi-stream communication the signal power is shared between information streams, the sharing ratio is defined by the squared singular values of the $n \times m$ channel matrix. As mentioned, the signal powers of the individual streams fade as the UE moves, and thus it is not feasible to transmit multiple streams at every time instant. An example is shown in Figure 12.41 for a 2×2 narrow-band Rayleigh fading antenna system. The uppermost curve is the SNR of a 2×2 diversity system. The two other curves show the SNR of the first and second information stream. It can be seen that the SNR of the second stream is always less than the first one. Another point worth noting is that the amount of fading of the second stream is much higher than the first stream, which results in the second stream SNR occasionally dipping below 0 dB. At such points where the channel rank effectively drops to one, it is not possible for the receiver to successfully decode both information streams and, instead of splitting transmit power to two streams, it is more spectrally efficient to switch to single-stream diversity transmission with higher modulation order. In the downlink of the LTE system each subcarrier is subjected to fading as shown in Figure 12.41.

To measure the suitability of the radio channel for MIMO transmission, one possible figure of merit is the condition number CN which in the present context can be defined as the ratio of stream signal powers, or for a dual-stream system

$$\text{CN} = \frac{S_1}{S_2}$$

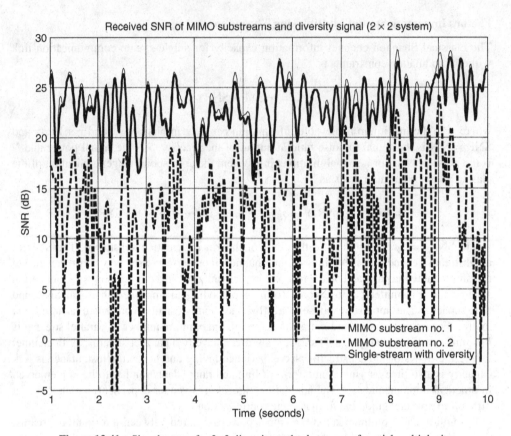

Figure 12.41   Signal power for 2×2 diversity and substreams of spatial multiplexing

From Figure 12.41 it can be seen that the condition number varies between 3 and 30 dB, depending on the instantaneous radio channel fading conditions. In case of more than two streams, some other figure of merit could be substituted, for example, the so-called ellipticity statistic which is the ratio of geometrical and arithmetical means of the stream signal powers [10].

To reach maximum spectral efficiency the ideal radio channel would have equal power for all streams, in other words $S_1 = S_2 = \cdots = S_k$. In this case, the information streams are completely orthogonal and do not interfere with each other. The reason why this is not possible in real-world conditions is mainly threefold: instantaneous channel fading, long-term antenna correlation and antenna power imbalance. All of these factors increase the condition number and thus increase the SNR difference between streams. Even when the entries of the channel matrix have no correlation or power imbalance on average, the random channel fading results with high probability in non-orthogonal information streams at any given time instant.

Figure 12.42 shows a measurement example with a real base station, 20 MHz FD LTE and a real UE where the average SNR = 25 dB is constant and antenna correlation is artificially altered with a fading simulator. The terms low, medium and high correlation refer to the standardized 3GPP channel models in 3GPP TS 36.101. At the beginning of the measurement, antenna

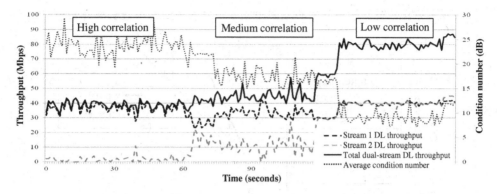

Figure 12.42 The 2×2 spatial multiplexing throughput for different correlations, SNR is 25 dB throughout the measurement. Carrier frequency is 2.6 GHz and bandwidth is 20 MHz

correlation is high at both receive and transmit ends of the 2×2 system, and correspondingly one stream carries almost all of the 40 Mbps throughput. At 60 seconds time the correlation is switched to 'medium' and a jump in second stream throughput can be seen with the first stream still carrying most of the data. At about 120 seconds time, the correlation is switched to 'low' and both streams are now carrying 40 Mbps with the total dual-stream throughput of 80 Mbps. Thus, depending on the level of antenna correlation throughput can vary between 40 to 80 Mbps, at fixed average SNR of 25 dB. From the condition number,[1] shown on the right-hand side axis, it can be seen that when the correlation is high the power difference between the first and second substreams is in the order of 25 dB which indicates that the SINR of second substream is about 0 dB since average SNR is 25 dB. For low correlation, the average condition number drops to below 10 dB; in this case the higher signal power of the second substream allows sending and detecting two streams simultaneously.

A logical question for practical deployments is that of how to minimize antenna correlation. While this is in general difficult, one simple rule is to use cross-polarized eNodeB and UE antennas whenever possible. A drive test example with two different antenna setups is shown in Figure 12.43 to illustrate the impact. In the example, the same drive test route in an unloaded test network is driven twice, once using two vertically polarized antennas at the rooftop of the measurement car and once using a custom-manufactured cross-polarized antenna with a near-omni-horizontal pattern. The result shows that the dual-stream usage of the cross-polarized antenna is noticeably higher improving the throughput distribution at the high-SNR regime (typically coinciding with line of sight). However, at low-SNR regime the throughput of the vertically polarized antennas is slightly better in the example measurement (the gain of the vertically polarized antennas was higher and was not compensated). Similar behaviour has been noticed in numerous other measurements and the practical conclusion for field deployments is that cross-polarized configuration provides gain over co-polarized setup especially in line of sight, and this gain can be attributed to reduced antenna correlation.

---

[1] Condition number varies between subcarriers. The condition number shown is the average over system bandwidth and time.

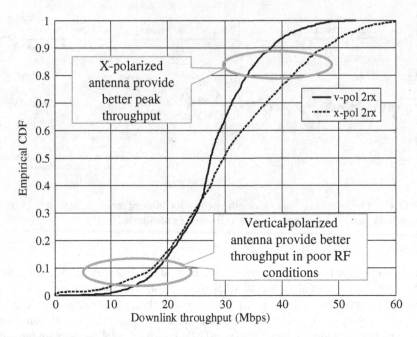

Figure 12.43 Downlink throughput distribution for the same drive test route with two different UE antenna setups

When deploying 4×2 open- or closed-loop spatial multiplexing, antenna correlation impact can be minimized by proper connection of feeders to antenna ports. To improve diversity gain, the correlation between reference signals R_0 and R_2 (also R_1 and R_3) should be minimized by connecting the corresponding antenna branches to cross-polarized or spatially separated ports.

Carrier frequency also has an impact on the antenna correlation: as the wavelength increases the antenna correlation also increases for a given physical antenna separation. Figure 12.44 shows a scanner measurement sample where a cluster of eNodeBs is measured simultaneously on two frequencies, 800 MHz and 2.6 GHz. Each cell has two cross-polarized transmit antennas and the antenna chassis and azimuth direction are the same on both the frequencies. The scanner has two omni-directional receive antennas placed on the rooftop of the measurement car at 10 cm spacing. The RSRP difference, shown in the right-hand side subfigure, between the two frequencies is in excess of 10 dB, although in measurement the 800 MHz cells have 3 dB higher reference signal transmit power. Regardless of the higher path loss, the condition number on 2.6 GHz, shown in the left-hand side subfigure, is lower indicating smaller power difference between the first and second MIMO substreams, and thus better multi-stream capability.

Power imbalance between antenna branches results in similar degradation as correlation. A rule of thumb for a 2×2 system is that if RSRP difference between UE branches is above 6 dB noticeable dual-stream throughput degradation results. With practical integrated UE antennas power imbalance is difficult to control since the UE antenna patterns are not identical. In drive test measurements, external antennas with identical patterns may be used to reduce degradation and increase repeatability of results. On the other hand, as seen in Figure 12.43, using external

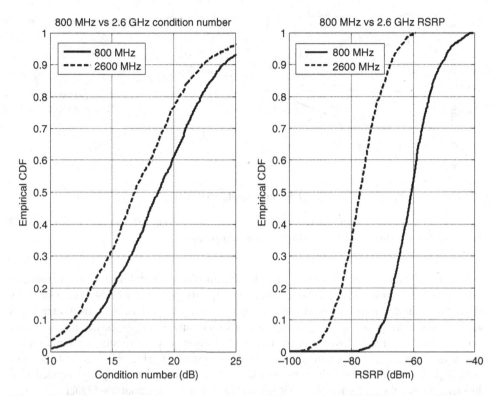

Figure 12.44 Drive test measurement example of condition number of 800 MHz versus 2.6 GHz when both frequency layers are co-sited

antennas with the same polarization can also result in higher receive correlation reducing the net throughput gain. Maximizing antenna spacing may alleviate this problem.

In indoor installations, if the eNodeB antenna branches are different physical antennas their patterns should be directed to serve the same coverage area, otherwise unnecessary power imbalance will occur, again reducing the multi-stream throughput. For dual-stream MIMO, using a single cross-polarized indoor antenna has been found to be a good field-tested solution.

Practical Throughput Gain of Some 3GPP Transmission Modes

3GPP has standardized several single-user spatial multiplexing schemes. The basic schemes include open-loop spatial multiplexing (3GPP TM3), closed-loop spatial multiplexing (3GPP TM4) and hybrid of beam forming and dual-stream spatial multiplexing (3GPP TM8). Additionally, advanced transmission modes with up to eight streams have been standardized in latest releases.

Figure 12.45 shows a field measurement example of throughput gain for basic 2×2 and 4×2 3GPP transmission schemes with stationary UE. For 2×2 schemes, there is not much difference between open- and closed-loop transmission schemes. Upgrading from two transmit antennas to four antennas produces clear link-level throughput gain especially at low to medium SNR

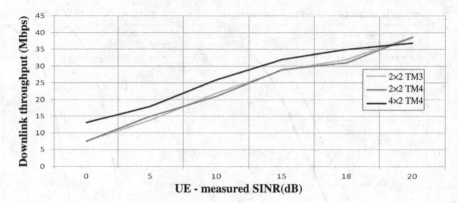

Figure 12.45 Example of field measurement of 2Tx and 4Tx downlink throughput, stationary UE in different SINR conditions. TD-LTE 10 MHz, frame configuration 1

regime. In the example measurement, the total transmit power with four transmit antennas is 3 dB higher so the result includes both power and diversity gain on the PDSCH channel.[2] The SINR shown in the horizontal axis is the value measured by the UE. As shown later in this section, even more important benefit from four eNodeB antennas is the possibility to employ four-way maximum ratio (MRC) or interference rejection (IRC) diversity combining in the uplink; indeed it is usually the uplink link budget that is the limiting factor in practical networks.

At high-SNR regime the 4Tx transmission schemes actually have slightly reduced peak throughput since the reference signal (RS) overhead from 4Tx is about 5% higher.

While the performance difference between open-loop and closed-loop transmission modes for two transmit antennas is fairly small, the gain of four eNodeB antennas over the two antenna eNodeB configuration has been demonstrated in a number of simulations and field measurements. The main reason is that for four transmit antennas the larger number (relative to two antennas) of precoding matrices in the closed-loop mode offers more granular antenna weighting which in turn results in higher gain than the open-loop transmit mode 3 at low UE speeds. This difference is even further pronounced at high antenna correlation (e.g. closely spaced eNodeB antennas and narrow angular spread) and low-SNR regime where at a given location the throughput of four antenna closed-loop system can be twice the throughput of a two transmit antenna system. On the other hand, large variation in the evaluated throughput gains has been noticed between different measurements, depending on the radio channel conditions, UE model capability, UE speeds and so on.

Performance of 4×2 MIMO

As shown in Figure 12.45 the 4×2 MIMO provides considerable throughput gain compared 2×2 MIMO across the whole UE-measured SINR range (the peak throughput is slightly decreased

---

[2] When per-antenna transmit power is kept constant, RSRP and SNR measured from reference signals do not change when changing from two to four transmit antennas.

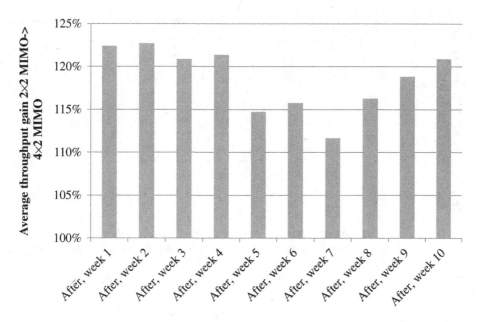

Figure 12.46   Average downlink cell throughput comparison between 4×2 MIMO and 2×2 MIMO

due to the higher RS overhead). Similar performance improvement can also be measured from the network statistics as a cell level throughput improvement as shown in Figure 12.46. The average cell throughput improvement is measured over network-level over weekly statistics. The 2×2 MIMO downlink throughput is averaged over 9 weeks and then the 4×2 MIMO weekly downlink throughput is calculated over 10 weeks. The 4×2 MIMO gain comes from power gain (i.e. 3 dB higher total power compared to 2×2 MIMO) and therefore higher usage of dual-stream transmission especially at low coverage areas (as shown in Figure 12.45). Double cross-polarized antennas were used to minimize the correlation impact and get the best possible performance gain.

On average per cell the 4×2 MIMO provides around 120% throughput compared to the 2×2 MIMO which is similar to the improvement for around 5 dB UE-measured SINR area in Figure 12.45.

In the uplink direction the four-way interference rejection combining (IRC) reception at the eNodeB improves the average PUSCH SINR. The network statistics of four-way IRC reception PUSCH SINR improvement over two-way IRC reception is shown in Figure 12.47. The PUSCH SINR improvement is calculated as the difference between daily cluster-level PUSCH SINR after activation or four-way IRC reception and average of PUSCH SINR over 10 consecutive days before activation of four-way IRC reception. The PUSCH SINR improvement is around 4 dB which provides uplink throughput improvement as shown in Figure 12.40 as well as decreases the number of UEs for which the transmission power reaches the maximum value as shown in Figure 12.48. According to Figure 12.48 after activating the four-way IRC reception the number of UEs transmitting with maximum power decreases by 20 percentage points which means longer battery life time as well as less inter-cell interference.

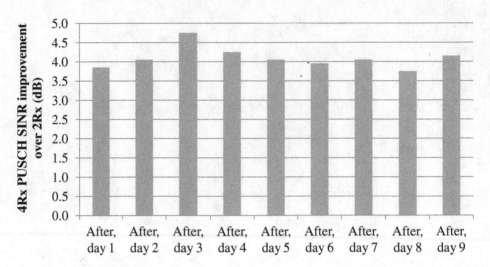

Figure 12.47 Average PUSCH SINR improvement by four-way IRC reception over two-way IRC reception

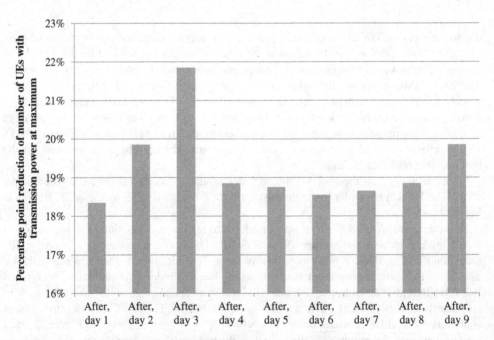

Figure 12.48 UE transmission power reduction by four-way IRC reception as a percentage point reduction in the number of UEs transmitting with maximum power

12.8 High-Speed Train Optimization

3GPP has defined in [11] high-speed and performance limits in case of up to 350 km/h UE speed for band 1. These performance requirements for eNodeB demodulation are discussed in details in [8]; here, only a short summary is presented.

When the eNodeB is activated with high-speed support then those performance requirements from [11] are met and performance of PRACH, PUCCH and PUSCH is guaranteed for up to 350 km/h depending on the high-speed scenario. There are two different scenarios defined in [11], one for open space (scenario 1) and another for multi-antenna tunnel case (scenario 3). For both scenarios, the Doppler shift is defined according to [11] as shown in Eq. 12.5. In Eq. 12.5 $f_s(t)$ is the Doppler shift, f_d is the maximum Doppler frequency given in Eq. 12.6 and cosine of the signal arrival angle $\theta(t)$ is defined in [11] and shown in Figure 12.49:

$$f_s(t) = f_d \cos \theta(t) \tag{12.5}$$

$$f_d = \frac{v f_c}{c} \tag{12.6}$$

In Eq. 12.6, v is the UE travelling speed, f_c is the carrier frequency and c is the speed of light. The eNodeB experiences double Doppler shift in the worst case as the UE synchronizes to the downlink frequency which already includes Doppler shift. Therefore, f_d in Eq. 12.5 needs to be multiplied by 2.

As shown in Figure 12.49, the highest Doppler shift is experienced by the UE that is in the middle of the two cells that are pointing towards each other (cell from eNodeB A and from eNodeB B). Table 12.2 based on [11] shows the high-speed scenario 1 and 3 parameters and the resulting required supported Doppler shift. Note that the Doppler shift requirements in Table 12.2 and Doppler shift trajectories are applicable for all bands. The UE travelling

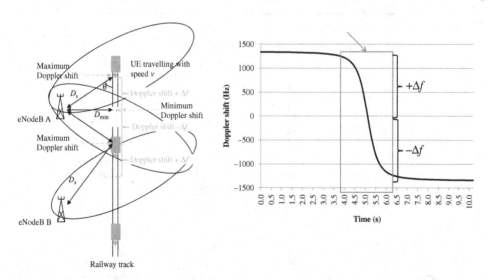

Figure 12.49 Parameters for signal arrival angle θ calculation and the Doppler shift trajectories for high-speed scenario 1

Table 12.2 Parameters for high-speed train condition

Parameter	Value	
	Scenario 1	Scenario 3
D_s (m)	1000	300
D_{min} (m)	50	2
v (km/h)	350	300
f_d (Hz)	1340	1150

speed, v, in Table 12.2 is derived for band 1 and in case lower frequency band is used then the corresponding speed needs to be calculated using Eq. 12.6. The lower the used frequency band the higher the speed for the same Doppler shift.

In case the high-speed support for physical random access channel is taken into use the length of Zadoff–Chu sequence, N_{ZC}, for preamble generation is defined according to [3] based on the restricted set of cyclic shift length, N_{CS}. The preambles generated based on restricted set of N_{CS} provide protection against Doppler shift which destroys the zero autocorrelation property of unrestricted generated preambles. The restricted set preambles are generated by masking some of the cyclic shift positions in order to retain acceptable false alarm rate while maintaining high detection performance even for very high speed UEs. Therefore, the number of cyclic shifts generated from one Zadoff–Chu root sequence is greatly limited compared to the unrestricted set. The supported cell ranges for restricted set are shown in Table 12.3.

Table 12.3 Supported cell ranges depending on restricted set of cyclic shift length

N_{cs} config index	N_{cs}, cyclic shift length	Delay spread $= 5.2$ μs	
		Length of single cyclic shift (μs)	Kilometres
0	15	9.1	1.37
1	18	12.0	1.79
2	22	15.8	2.37
3	26	19.6	2.94
4	32	25.3	3.80
5	38	31.0	4.66
6	46	38.7	5.80
7	55	47.2	7.09
8	68	59.6	8.95
9	82	73.0	10.95
10	100	90.2	13.52
11	128	116.9	17.53
12	158	145.5	21.82
13	202	187.4	28.11
14	237	220.8	33.12

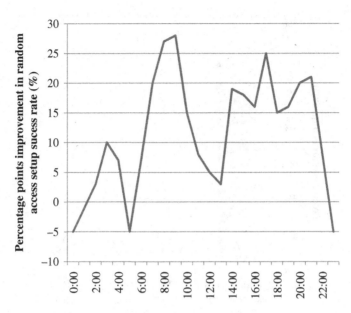

Figure 12.50 Improvement in random access setup success rate after high-speed feature activation. Statistics taken from the eNodeB next to the train tracks

The impact of using high-speed support of eNodeB in terms of random access setup success rate improvement is shown in Figure 12.50.

The handover execution success rate also improved due to the activation of high-speed support and use of restricted set of cyclic shift lengths to generate preambles. The improvement in handover execution success rate was around 1.2 percentage points due to which the dropped call rate also improved by 1 percentage points. Further optimization requires redesign of below parameter sets:

- Signalling robustness for call setup and handovers following the methods detailed in Sections 12.4, 12.5 and 12.6.
- Downlink power allocations in terms of reference signal (RS) boosting can cause unnecessary large coverage areas and handover execution problems, so typically RS boosting is avoided for the high-speed scenarios.
- Uplink power settings with full path-loss compensation while maximizing the signalling robustness and minimizing the extra inter-cell interference. Also any scheduling algorithms that can minimize the inter-cell interference should be used to improve signalling robustness.
- Speed up the cell reselection in order to have the UEs always under the best cell but ping pong cell reselections need to be avoided by, for example, speed-dependent cell reselection feature and reduced hysteresis needed to be achieved before reselection can occur.

By optimizing the above-mentioned parameters the high-speed train performance can be further improved as shown in Figure 12.51.

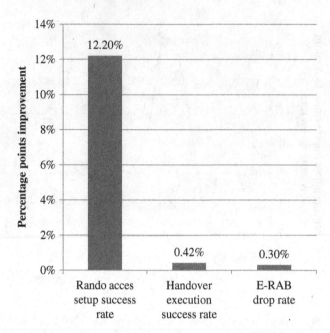

Figure 12.51 Performance improvement after parameter optimization

It should also be noted that use of repeaters especially inside the high-speed train tunnels should be carefully designed so that the transmissions from each repeater are perfectly time aligned and Doppler impact minimized by proper repeater antenna arrangement.

12.9 Network Density Benchmarking

The denser the network is the more cells there are; but just plain number of cells is not that good as a benchmark indicator due to the different stages of the network rollout (initial rollout for, e.g. hotspots only, country wide rollout and capacity increase) between the networks in different areas. The number of cells per number of subscribers indicates the absolute density of the network. In Figure 12.52, one example of network density benchmarking is shown as the estimated number of sectors per 1000 subscribers. As it can be seen, the densest networks are in Japan, South Korea, China and Scandinavia where in general the networks are two to three times denser compared to Europe, North America and APAC and up to 10 times denser compared to Latin America, Middle East and Africa. The subscriber numbers include all the technologies (2G, 3G and 4G) in Figure 12.52.

The average UE-reported CQI for each network seen in Figure 12.53 shows no big difference between networks with high or low amount of sectors per 1000 subscribers. This means that the high-density networks have good physical layer performance and therefore very high downlink and uplink throughputs can be achieved. However, the more dense the network is the more challenging it is to control the inter-cell interference and physical layer optimization is challenging but excellent results can be achieved as shown in Section 12.3. The KPI optimization actions described in this chapter become more and more important as the number

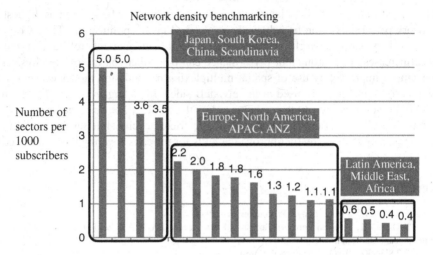

Figure 12.52   Network density benchmarking as estimated number of sectors per 1000 subscribers

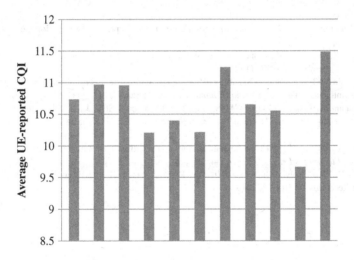

Figure 12.53   Average UE-reported CQI

of cells in the network increases as with very dense networks any excessive resource usage for signalling can cause massive inter-cell interference problems.

12.10   Summary

The LTE network performance measured by network or drive test KPIs can be perfected by parameter tuning. When the networks get more and more dense the parameter tuning must also take into account that inter-cell interference should be always minimized. Therefore, the optimization target is to have the best possible KPI values with lowest possible resource and power allocations. Physical layer optimization should never be ignored as highest gains in network performance can be achieved via physical layer optimization. This is especially the case of very dense networks and when the network loading and therefore inter-cell interference increases. The throughput performance can be further (on top of the physical layer optimization) improved by use of spatial multiplexing transmission modes especially when cross-polarized antennas are used at the eNodeB side. Upgrading from 2×2 MIMO to 4×2 MIMO improves user and cell throughputs across the whole SINR range but especially at the poor SINR area. Activation of 3GPP-specified restricted cyclic shift set for random access preamble generation can greatly improve the KPIs for the cells covering the high-speed train.

References

[1] 3GPP TS 36.213, 'Physical Layer Procedures', v.10.9.0, 2013.
[2] 3GPP TS 36.101, 'Evolved Universal Terrestrial Radio Access (E-UTRA); User Equipment (UE) Radio Transmission and Reception', v.10.9.0, 2013.
[3] 3GPP TS 36.300, 'Evolved Universal Terrestrial Radio Access (E-UTRA) and Evolved Universal Terrestrial Radio Access Network (E-UTRAN); Overall Description; Stage 2', v.10.9.0, 2012.
[4] 3GPP TS 36.331, 'Evolved Universal Terrestrial Radio Access (E-UTRA); Radio Resource Control (RRC); Protocol Specification', v.10.9.0, 2013.
[5] S. Kullback. *Information Theory and Statistics*, Dover (1968). Reprint of 1959 edition published by John Wiley & Sons, Inc.
[6] E. Telatar, 'Capacity of Multi-antenna Gaussian Channels', *European Transactions on Telecommunications*, November–December, 585–595 (1999).
[7] G. J. Foschini and M. J. Gans, 'On Limits of Wireless Communications in a Fading Environment When Using Multiple Antennas', *Wireless Personal Communications*, January, 311–335 (1998).
[8] H. Holma and A. Toskala. *LTE for UMTS*, 2nd ed., John Wiley & Sons, Ltd., Chichester (2011).
[9] 3GPP TS 36.314, 'Evolved Universal Terrestrial Radio Access (E-UTRA); Layer 2 – Measurements', v.10.2.0, 2011.
[10] J. Salo, P. Suvikunnas, H. M. El-Sallabi, and P. Vainikainen, 'Ellipticity Statistic as Measure of MIMO Multipath Richness', *Electronics Letters*, 42(3), 45–46 (2006).
[11] 3GPP TS 36.101, 'Evolved Universal Terrestrial Radio Access (E-UTRA); Base Station (BS) Radio Transmission and Reception', v.10.10.0, 2013.

13

Capacity Optimization

Jussi Reunanen, Riku Luostari and Harri Holma

13.1 Introduction

In this chapter the LTE network capacity limits are introduced and ways to improve the performance during high traffic events (mass events) are discussed in detail. Capacity analysis is done on cluster level (whole mass event area) as well as on cell level. The highest loaded cells (in terms of data volume and resource utilization) from several different mass events are used to show the process of capacity bottleneck analysis. The main factors impacting the high traffic event performance are shown in the Figure 13.1 and are all analysed in detail in the following sections.

The average user traffic profile changes greatly during the mass event compared to a normal traffic condition and typically the uplink traffic volume exceeds the downlink traffic volume during the highest loaded time. Therefore, the uplink inter-cell interference control becomes crucial from end-user performance point of view.

The downlink interference control is very important from the throughput per user point of view and due to the physical downlink control channel (PDCCH) capacity. High downlink

LTE Small Cell Optimization: 3GPP Evolution to Release 13, First Edition.
Edited by Harri Holma, Antti Toskala and Jussi Reunanen.
© 2016 John Wiley & Sons, Ltd. Published 2016 by John Wiley & Sons, Ltd.

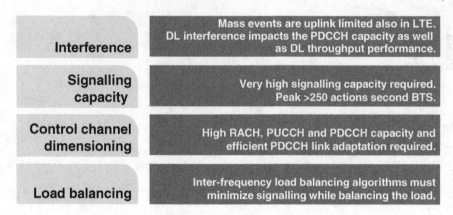

Figure 13.1 Main factors impacting the mass event performance

interference reduces the PDCCH capacity and the spectral efficiency of the physical downlink shared channel (PDSCH). The decreased PDCCH capacity in turn reduces the number of UE that can be scheduled simultaneously per transmission time interval (TTI) for UL and DL further reducing the spectral efficiency (due to the reduced frequency-selective scheduling gain).

The amount of signalling traffic per eNodeB in terms of call setups and releases, and incoming and outgoing handovers, can also limit the total traffic and therefore signalling peaks (amount per second) need to be carefully monitored.

Control channel dimensioning and maximum number of supported radio resource control (RRC)-connected users are of great importance during the mass event. The capacity of all control channels needs to match or exceed the maximum number of supported simultaneously RRC-connected and scheduled users so that each of the users has the required control channel capacity. In case control channel capacity is not adequate there can be massive amount of call setup rejections which typically cause huge increase in capacity requests (rejected capacity request is simply re-attempted by the UE until it gets the requested capacity) which in turn cause interference increase.

Load balancing can distribute the load, that is, users, evenly between different carriers and this should be done without any additional increase of signalling load, that is, preferably without handover signalling (this in case target carrier coverage measurements are not needed and handover to the target cell can be done blindly). This can be achieved via RRC connection release within which the target frequency layer is indicated to the UE. The users can also be distributed between the same frequency cells using handover thresholds but here we must keep in mind that the UE should always be connected to the cell that provides the smallest path loss. Therefore the handover thresholds-based tuning can only be used to move users from one cell to another when at the cell edge.

It should be noted that all examples in this chapter are from 10 and 20 MHz FD LTE deployments (actual carrier frequency varies between 700, 800, 900, 1800, 2100 and 2600 MHz) with 2×2 multiple-input multiple-output (MIMO) in downlink and 2Rx in uplink. Also the eNodeB transmission power per sub-carrier is the same for all carrier bandwidths unless otherwise stated.

13.2   Traffic Profiles in Mass Events

The uplink data volume increases heavily during the mass events most likely due to the social media usage to share the event experience and automatic cloud backup of photos taken. A mass event area's uplink and downlink data volumes per subscriber during normal day (2 days prior the event) and during the mass event day are shown in Figure 13.2. The uplink data volume per subscriber increases more than four times during the event from average prior to the event hours and compared to the normal day. Similarly the downlink data volume per subscriber decreases during the event to around one-third and one-fourth compared to prior the event hours and normal day, respectively. The analysed mass event starts at around sample 76 and ends at around sample 91.

Similar to the uplink and downlink data volume per subscriber, the uplink and downlink data volume per setup E-RAB changes a lot during the mass event hours. Figure 13.3 shows that the uplink data volume per setup E-RAB increases during the event hours five times compared

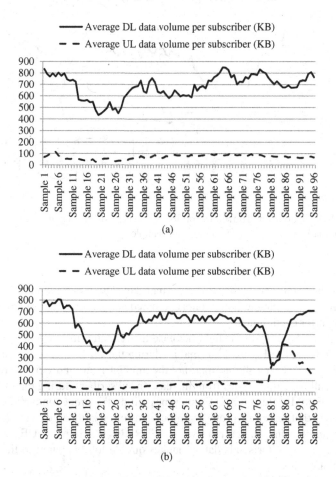

Figure 13.2   Average uplink and downlink traffic volume per subscriber in mass event area during (a) normal day and (b) mass event day per 15-minute data samples

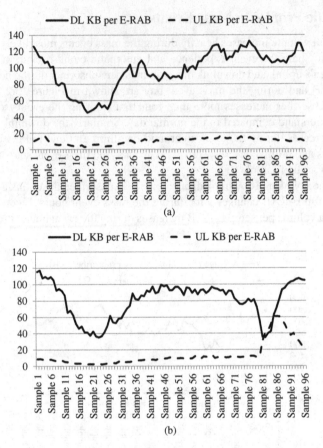

Figure 13.3 Average uplink and downlink traffic volume per E-RAB in mass event area during (a) normal day and (b) mass event day per 15-minute data samples

to hours prior the event and compared to normal day. The downlink data volume per setup E-RAB is lower throughout the event hours compared to the normal day and reduces to around one-third from prior event hours.

During the mass event there is slight increase in signalling in terms of E-RAB setups per subscriber as indicated in Figure 13.4, that is, during the event hours each subscriber makes 7 E-RAB setups and during the normal day the amount is around 6.5. Therefore, the main increase in signalling traffic comes from increase in subscribers during the mass event hours. The amount of E-RAB setups increases six times during the event compared to lowest during the event day and that increase is roughly three times more compared to the normal day.

During the mass event the session time per E-RAB (the time the UE occupies the data channel and receives and/or sends data) changes quite a lot compared to the normal day as depicted in Figure 13.5.

The QCI1 (VoLTE) E-RAB session time is reduced during the event time compared to the normal day and this is most likely due to the fact that end users seldom make any voice calls

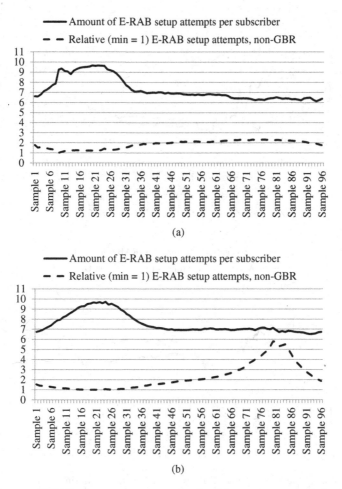

Figure 13.4 Average E-RAB setup attempts per subscriber in mass event area during (a) normal day and (b) mass event day per 15-minute data samples

during the actual mass event and rather send some social media updates. The non-GBR E-RAB session time has a sharp peak exactly at the event time. This is due to the fact that the activity is heavily uplink dominated and as the uplink is having, in general, lower spectral efficiency compared to downlink it takes longer time to deliver similar amount of data in uplink than in downlink. Also uplink is very sensitive to interference and UE power control as shown in Section 13.3.

The user behaviour, that is, amount of data calls and amount of data volume per each of the call, changes heavily during the day even if there are no special events. Figures 13.6 and 13.7 show per registered user the amount of E-RABs requested per hour versus downlink and uplink data volume per hour. Figures 13.6 and 13.7 are based on whole network level hourly statistics averaged over several normal days, that is, there are no special events occurring during the analysed days.

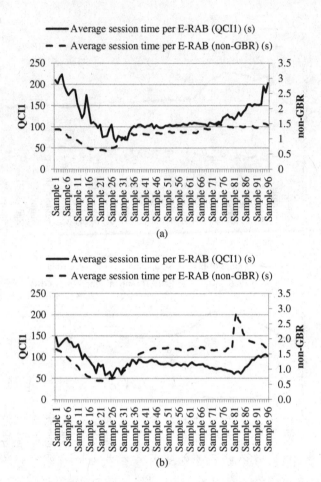

Figure 13.5 Average session time per E-RAB in mass event area during (a) normal day and (b) mass event day per 15-minute data sample

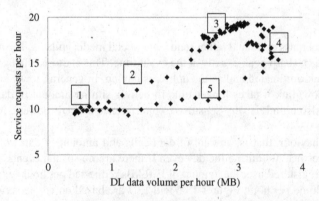

Figure 13.6 User activity during a normal working day versus downlink data volume

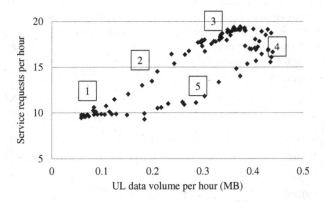

Figure 13.7 User activity during a normal working day versus uplink data volume

The user behaviour during a normal day can be analyzed based on Figures 13.6 and 13.7 as below:

1. During the night time (01.00–05:00) each of the registered users generates roughly 10 E-RABs per hour, receives roughly 1 MB of data and transmits roughly 0.1 MB per hour.
2. During the morning hours (05:00–08:00) the number of E-RABs per hour per registered user rapidly increases from 10 to 18 and data volume per hour increases to 2 MB in downlink and 0.3 MB in uplink.
3. During the working hours (08:00–18:00) each registered user generates 15–19 E-RABs per hour and data volume per hour is around 3 MB in downlink and 0.3–0.4 MB in uplink.
4. During the evening hours (18:00–23:00) the amount of E-RABs per registered user per hour decreases to around 15 while data volume per hour in downlink increases close to 4 MB and in uplink close to 0.5 MB.
5. During the late evening hours (23:00–01:00) the data volume per registered user quickly decreases to 1 MB in downlink and to 0.1 MB in uplink and the number of E-RABs decreases to 10 per registered user per hour.

During the normal working days the user activity (data volume and signalling) changes quite a lot depending on the hour of the day and this causes big changes to the capacity demand in different parts of the network in different times. This should be taken into account when predicting the capacity demand for business areas (users located during office hours) and residential areas (users located during non-office hours).

The mass event causes impact to relatively small area (limited number of cells) but the impact should be carefully considered and capacity estimation should be done taking into account the increase in the number of subscribers and signalling increase due to that as well as massive increase in uplink data volumes per E-RAB.

13.3 Uplink Interference Management

Due to the massive uplink traffic increase during the mass event compared to the normal traffic time, the uplink inter-cell interference limits the performance. Therefore, the power

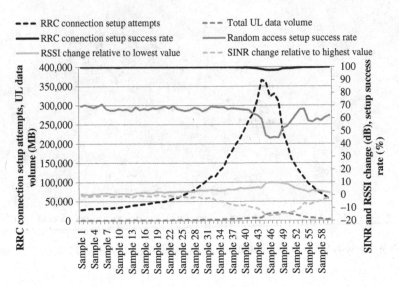

Figure 13.8 Impact of aggressive uplink power control on RRC and random access setup success rates and PUSCH RSSI and SINR. Statistics (15-minute samples) are averaged over cluster of cells for which RRC connection setup attempts are the highest during the peak time

control strategy of physical uplink shared channel (PUSCH) and physical uplink control channel (PUCCH) needs to be set so that inter-cell interference is always minimized. In case aggressive uplink power control strategy is used, to achieve maximum uplink throughput, all the UEs will transmit with extremely high power regardless of the location under the cell. This causes the uplink received signal strength indicator (RSSI) to increase and the signal to interference and noise ratio (SINR) decrease with increasing traffic (uplink data volume and/or signalling). Degraded PUSCH SINR in turn causes the degradation of random access setup success rate and RRC connection setup success rate and causes poor end-user experience as shown in Figure 13.8. Note that the RRC and random access setup success rate performance indicators used in Figure 13.8 are introduced in Chapter 12.

As it can be seen from Figure 13.8 the PUSCH RSSI increase and PUSCH SINR decrease start when signalling (in terms of number of RRC attempts) increases. The lowest PUSCH SINR value and lowest random access setup success rate occur at the highest level of RRC connection setup attempts (at sample 45). At this point also the RRC setup success rate is at lowest level and PUSCH RSSI is highest. Then RRC connection setup attempts start to decrease and that causes PUSCH SINR to improve and even the uplink data volume increases as shown in Figure 13.9. The PUSCH RSSI also decreases significantly when RRC connection setup attempts decrease. This behaviour of PUSCH SINR and RSSI compared to RRC connection setup attempts and uplink data volume means that very large portion of uplink interference is caused by signalling and can be explained by the UE transmission power control (TPC) behaviour for PUSCH for signalling messages. As explained in Chapter 12, message 1 and message 3 (i.e. preamble and RRC connection request message respectively) are following open-loop power control based on parameters given by the network plus power increase offsets for message 3 on top of message 1 power. Similarly, message 5 follows message 3

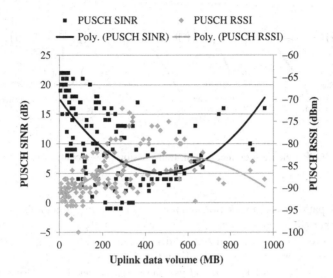

Figure 13.9 Uplink data volume versus uplink (PUSCH) signal to interference and noise ratio (SINR) and uplink (PUSCH) received signal strength indicator (RSSI) per cell per 15-minute data samples

transmission power as there are no PUSCH RSSI or SINR measurements available at the eNodeB for closed-loop power control to adjust the UE transmission power. Therefore, the UE transmission power should be minimized for message 1 and message 3 transmissions.

More examples of power control impact on different performance indicators during mass event are shown in Section 13.3.3.

13.3.1   PUSCH

Transmission Power Control

The purpose of the uplink TPC is to maintain high enough energy per transmitted bit to fulfil certain quality of service requirement and on the other hand to minimize the inter-cell interference. The uplink TPC operation is presented in [1]. The 3GPP-specified [2] uplink transmitted power control scheme is based on open-loop and closed-loop power control as depicted in Figure 13.10. Note that Chapter 12 contains more details on uplink power control scheme and therefore here only the most important factors impacting on mass event performance are considered.

The open-loop power control adjustments are based on path-loss changes, that is, slow channel variations measured by the terminals. The closed-loop power control is based on the power control commands sent by the eNodeB to try to mitigate any path-loss measurement and power amplifier errors as well as rapid changes in the inter-cell interference conditions.

In Figure 13.11 the UE transmission power as a function of path loss is shown. P_{0_PUSCH} is varied, alpha (α) is fixed to 1 and the resource block allocation is 1. As it can be seen P_{0_PUSCH} controls the UE transmission power based on the path loss and the higher the P_{0_PUSCH} is the higher is the UE transmission power for a given path loss. With alpha = 1 the UE TPC is

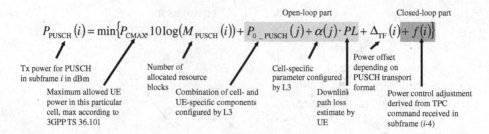

Figure 13.10 3GPP Release 8-defined uplink transmission power control for PUSCH, expressed in dBm

working as full path-loss compensation and the UE is trying to maintain P_{0_PUSCH} dBm power at the receiver per allocated resource block. Therefore with $P_{0\_PUSCH} = -60$ dBm the max UE transmission power is reached at path loss 83 dB whereas with $P_{0\_PUSCH} = -100$ dBm the max UE transmission power is reached at 40 dB higher path loss of 123 dB.

With fixed $P_{0\_PUSCH} = -100$ dBm and varying alpha values the fractional path-loss compensation can be used. Alpha < 1 means that only fraction of path loss is compensated as indicated in Figure 13.12. For example, with alpha $= 1$ and $P_{0\_PUSCH} = -100$ dBm the maximum UE transmission power is used at path loss 123 dB but with alpha 0.9 maximum UE transmission power is used at $123/0.9 = 137$ dB level. Therefore, values of alpha < 1 should be used together with higher P_{0_PUSCH} values in order to compensate the impact of fractional path-loss compensation.

The UE transmission power as a function of combined fractional path-loss compensation and different P_{0_PUSCH} values is shown in Figure 13.13.

When varying P_{0_PUSCH} and alpha (α) values the maximum UE transmission power is reached at path loss 124 dB with $P_{0\_PUSCH} = -100$ dBm and alpha (α) $= 1$ but at path loss 138 dB with $P_{0\_PUSCH} = -60$ dBm and alpha (α) $= 0.6$. This means that in case of large cells for

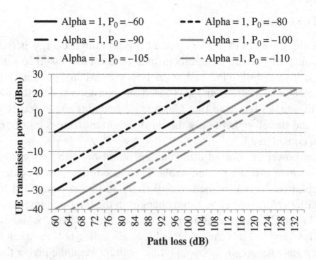

Figure 13.11 UE transmission power as a function of path loss, fixed alpha (α) and varying P_{0_PUSCH}

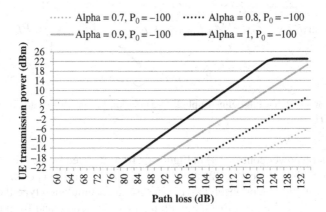

Figure 13.12 UE transmission power as a function of path loss, fixed P_{0_PUSCH} and varying alpha (α)

which the max path loss is, for example, 124 dB it is better from performance point of view to use conservative power control settings $P_{0\_PUSCH} = -100$ dBm and alpha (α) = 1 as then at the cell edge UE transmission power is at maximum (21 dBm) compared to $P_{0\_PUSCH} = -80$ dBm and alpha (α) = 0.8 with cell edge UE transmission power of 19.2 dBm. Too low UE transmission power allocation for the cell edge users can cause RRC setup failures due to missing RRC Setup Complete message (message 5), as explained in Section 13.3.2, and poor uplink throughput for cell edge users. However, open-loop power control settings where UEs close to the cell edge start to use maximum power would cause unnecessary inter-cell interference for cell edge users (many UEs transmitting with maximum power). Therefore, during mass events it is recommended to set the open-loop power control part according to the cell edge path loss so that the UE transmission power is set as low as possible without sacrificing the call setup related signalling performance. Typically this means having alpha (α) = 1 and $P_{0\_PUSCH} \leq -100$ dBm.

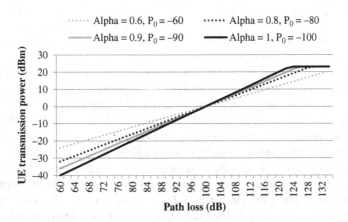

Figure 13.13 UE transmission power as a function of path loss when varying P_{0_PUSCH} and alpha (α) values

$$P_{\text{PUSCH}}(i) = \min\{P_{\text{CMAX}}, 10\log(M_{\text{PUSCH}}(i)) + P_{0\_\text{PUSCH}}(j) + \alpha(j) \cdot PL + \Delta_{\text{TF}}(i) + f(i)\}$$

$$P_{0\_\text{PUSCH}} = P_{0\_\text{PRE}} + \Delta_{\text{PREAMBLE\_Msg3}} \qquad \alpha=1 \qquad f(0) = \Delta P_{\text{rampup}} + \delta_{\text{msg2}}$$

Figure 13.14 3GPP Release 8-defined uplink transmission power control algorithm for message 3 power on PUSCH

The closed-loop power control, $f(i)$ in Figure 13.10, is updated by the UE based on the TPC command sent by the eNodeB. In [2] it is specified that closed-loop power control part, $f(i)$, can be either accumulation or absolute type. In case of accumulation type the previous value of $f(i)$, that is, $f(i-1)$ is summed together with the new TPC command from the eNodeB and in case of absolute type only the new TPC command from the eNodeB is used to update $f(i)$. The accumulation-type TPC command is $\{-1, 0, 1, 3\}$ dB and absolute type TPC command is $\{-4, 1, 1, 4\}$ dB. The accumulation-type closed-loop power control always starts from the initial value, $f(0)$, and then adjusts that initial value based on the eNodeB TPC commands from the eNodeB. The absolute closed-loop power control only uses the current TPC command from the eNodeB and replaces the closed-loop power control, $f(i)$, value with the new received one.

The initial value, regardless of closed-loop power control type selected to be used, $f(0)$, is given in [2] as a sum of TPC command given in the random access response, δ_{msg2}, and total preamble transmission power ramp-up, ΔP_{rampup}, from the first to the last (acknowledged by random access response) preamble. The RRC connection request message, message 3, sent by the UE is the first message sent on PUSCH and therefore is the transmission that uses the initial value, $f(0)$, for closed-loop power control. Message 3 total transmission power is as given in Figure 13.10 with modifications according to [2] as shown in Figure 13.14.

The preamble power ramp-up is also increasing the message 3 uplink transmission power and in case of accumulation-type closed-loop power control the preamble power ramp-up impacts the closed-loop power control part for the whole duration of the call. Similarly the TPC command given in random access response impacts the closed-loop power control part until the end of the call for the accumulation type of closed-loop power control. Therefore, from uplink inter-cell interference point of view it is extremely important that the preamble ramp-up rounds are kept down to minimum.

The eNodeB decision algorithm for TPC command, not specified by 3GPP, is typically based on uplink SINR and/or uplink RSSI target windows. As an example, in case the eNodeB-measured SINR is below the low SINR target then the TPC command is +1 or +3 dB depending on how much below the target the eNodeB-measured SINR is. In case the eNodeB-measured SINR is within the target window, that is, in between the low and high SINR targets, then TPC command is 0. In case the eNodeB-measured SINR is above the high SINR target, the TPC command is –1 dB. This way the target window defines the target range of SINR and/or RSSI for each UE. The eNodeB tries to maintain the measured PUSCH SINR and/or RSSI within the defined window by sending the TPC commands to the UE. The higher the SINR and/or RSSI target window is the higher the UE transmission power. High target window provides high uplink cell throughput while load is low, but generates high inter-cell interference that causes negative performance during high load. The inter-cell interference can be further optimized by, for example, providing the UE position in terms of inter-cell interference generation to the closed-loop power control decision algorithm and therefore lowering the transmission power of most inter-cell interference-generating UEs. For mass events the closed-loop uplink TPC should be set very conservatively so that for cell edge users the power control is decided mainly by the open-loop power control. This means large decision window size for SINR and RSSI as well as low absolute SINR/RSSI target values (starting point of the decision window).

Examples in Section 13.3.3 show the impact of power control settings on different performance indicators during the mass event.

Throughput per User and Spectral Efficiency

The uplink throughput per user depends on the bandwidth allocation, spectral efficiency and number of users simultaneously transmitting data to a cell. As explained in the previous section the traffic increase causes inter-cell interference increase which is visible as PUSCH SINR decrease and PUSCH RSSI increase. PUSCH SINR decrease in turn means that the spectral efficiency of the cell decreases and therefore the end-user throughput decreases. On the other hand, the increase in the number of connected users which are uploading at the same time causes additional throughput per user degradation as the cell radio resources are shared among all the users.

The throughput per user can be estimated based on the cell-level counters, for example, simply dividing the cell-level throughput by the number of estimated users which are having data in the UE buffer (i.e. those UEs from which the eNodeB has received buffer status report indicating that there is still data in the buffer and are therefore taken into account for scheduling of the resources in the next TTI). The cell-level throughput in turn can be estimated by dividing the received user plane bits by the number of active TTIs used to allocate resources to receive those bits. The number of bits should contain only the user plane data bits excluding the retransmissions, hybrid automatic repeat request (HARQ) or radio link control (RLC) retransmissions. Active TTI is a bit difficult to measure as the TTI is active regardless of the number of physical resource blocks (PRBs) used during the TTI. Therefore, this type of performance indicator is not very accurate when the PRB utilization (how many PRBs are used from all available) is very low (<30%). Scaling the cell throughput for, for example, 10% PRB utilization to 100% PRB utilization, simply by dividing the cell throughput by PRB utilization cannot be done. This due to the fact that such simple scaling assumes that the uplink interference conditions do not change at all when the PRB utilization increases and such assumption is not correct. Therefore, in case the cell throughput performance indicator is calculated using the simple formula described above the results should be analysed carefully taking into account the PRB utilization.

Figure 13.15 shows the example of average throughput per user (calculated as described above) as a function of the number of RRC-connected users. The cells included into the analysis are high traffic cells for which PRB utilization is above 20% per 15-minute sample. It should be noted that all the carriers are not deployed on the same sites and the optimization is performed on carrier basis trying to achieve highest possible performance per carrier. As can be seen from Figure 13.15 the low band (700, 800 or 900 MHz) 10 MHz cells provide slightly worse average user throughput compared to the high band (1800, 2100 or 2600 MHz)

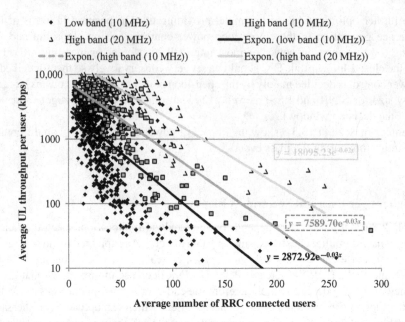

Figure 13.15 Average uplink throughput per user (kbps) as a function of the number of RRC-connected users per FD LTE cell, based on 15-minute statistics

10 MHz cells. The highest average user throughput can be achieved by the high band (1800, 2100 or 2600 MHz) 20 MHz cells.

If uplink throughput limit is set to, for example, 500 kbps per user then that means that on average the analysed cells can support the following amount of RRC-connected users:

- 20 MHz high band FD LTE can support 180 RRC-connected users
- 10 MHz high band FD LTE can support 90 RRC-connected users
- 10 MHz low band FD LTE can support 58 RRC-connected users

The throughput difference between 10 MHz low band and high band allocations comes from radio quality difference between the bands, that is, high band is in general easier to optimize and therefore high band can provide better spectral efficiency. The higher the carrier frequency the less is the PUSCH SINR decrease while the uplink data volume per cell increases as shown in Figure 13.16. Also the more the bandwidth available the more uplink data volume can be delivered with certain PUSCH SINR level. Based on Figure 13.16 the 20 MHz bandwidth allocation provides more than double uplink data volume per cell compared to the high band 10 MHz bandwidth allocation at the same PUSCH SINR level (with PUSCH SINR 5 to 10 dB).

As can be seen from Figure 13.16 there are several low band cells (grey circle) which are suffering from excessive uplink inter-cell interference. These cells are having massive overshooting coverage area and therefore PUSCH SINR performance is poor regardless of uplink data volume.

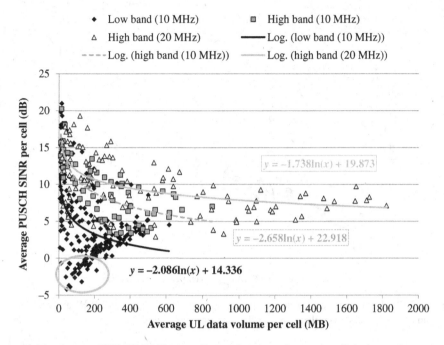

Figure 13.16   Average PUSCH SINR per cell as a function of average uplink data volume per cell. Highest loaded cells included with 15-minute statistics samples

The actual spectral efficiency in terms of bit/s/Hz per cell can be calculated based on the active cell throughput statistics and allocated bandwidth per cell. Figure 13.17 shows example per different bands and bandwidths.

As can be seen from Figure 13.17 the 3GPP-simulated spectral efficiency of 0.74 bit/s/Hz [3] per cell for 100% load can be met and exceeded provided that the PUSCH SINR is better than 6 dB. It should be also noted that this high spectral efficiency can be achieved by low band and high band with 10 and 20 MHz bandwidths. The 3GPP simulations for spectral efficiency are using full buffer traffic model which means that each cell is having 100% UL PRB utilization at the same time. The own cell UL PRB utilization and its neighbours' UL PRB utilization are hardly ever at 100% at the same time in real live networks as shown in Figure 13.18.

Figure 13.18 shows that when the own cell UL PRB utilization increases the share of its neighbours which are having the same or higher UL PRB utilization also increases. For example, when own cell UL PRB utilization reaches 90–95% then 8% of its neighbours are having UL PRB utilization less or equal than 15%. And at the same 90–95% UL PRB utilization for own cell, 55% of its neighbours have higher than 80% DL PRB utilization at the same time. This means that in real networks the spectral efficiency can exceed the 3GPP limits due to the unbalanced load between the cells i.e. all cells (own cell and it's neighbors) are not having very high load at the same time. However, it should be noted that the UL PRB utilization imbalance seems to be much less compared to the DL as seen when comparing Figures 13.18 and 13.35.

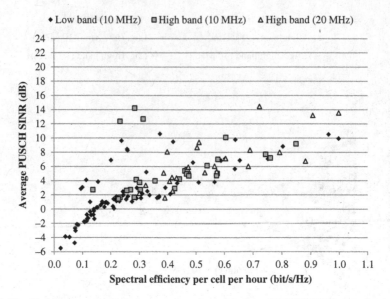

Figure 13.17 Uplink spectral efficiency as a function of average PUSCH SINR (dB) for cells having 30% and above uplink PRB utilization over the measurement period

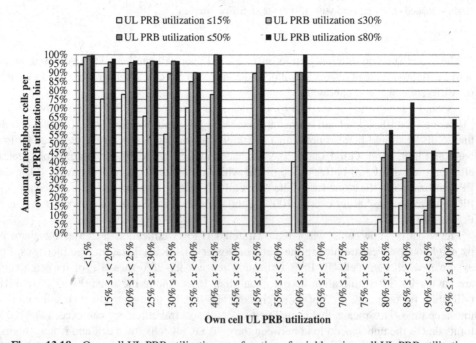

Figure 13.18 Own cell UL PRB utilization as a function of neighbouring cell UL PRB utilization

$$P_{\text{PUSCH}}(i) = \min\{P_{\text{CMAX}}, P_{\text{0\_PUCCH}} + \text{PL} + h(n_{\text{CQI}}, n_{\text{HARQ}}) + \Delta_{\text{F\_PUCCH}}(F) + g(i)\}$$

$P_{\text{0\_NOMINAL\_PUSCH}} + P_{\text{0\_UE\_PUCCH}}$

Closed-loop PC part updated
by eNodeB per UE basis

Figure 13.19   3GPP Release 8-defined uplink transmission power control algorithm for PUCCH

13.3.2   PUCCH

As explained in Chapter 12, the PUCCH power control should follow the strategy of PUSCH power control. The PUCCH power control formula is discussed more in detail in Chapter 12 and only the most important aspects in terms of mass event performance are discussed here. This means that $P_{\text{0\_NOMINAL\_PUCCH}}$ and $P_{\text{0\_UE\_PUCCH}}$ in Figure 13.19 should be set following the similar strategy as used for PUSCH. Each UE basically uses only single PRB on PUCCH and therefore the thermal noise + eNodeB noise figure gives the minimum value for $P_{\text{0\_NOMINAL\_PUCCH}}$ as around −118 dBm. Adding some safety margin (against fading) on top of that we end up having $P_{\text{0\_NOMINAL\_PUCCH}}$ as −114 to −116 dBm depending on the RF conditions (if there is plenty of fast fading then $P_{\text{0\_NOMINAL\_PUCCH}}$ should be set higher otherwise dropped call rate due to the periodical CQI reporting failure will increase). The $g(i)$ is the closed-loop power control part similar to the PUSCH. The RSSI closed-loop decision window should be then set to ±2 dB from the $P_{\text{0\_NOMINAL\_PUCCH}}$. The SINR closed-loop decision window should be set so that the minimum SINR is above the 3GPP-defined performance targets as shown in [4]. This means that SINR window should be set to e.g. −1 to 3 dB, depending on the RF conditions.

13.3.3   RACH and RRC Setup Success Rate

Figure 13.20 shows the typical performance of interference-limited cell for aggressive uplink power control scenario. Aggressive uplink power control here means that the closed-loop power control tries to maintain 18 dB SINR for each UE and open-loop power control is set conservatively ($P_{\text{0\_PUSCH}} = -100$ dBm and alpha (α) = 1) to protect the RRC setup success rate. RACH setup success rate (number of message 3 divided by number of message 1) and RRC setup success rate (number of message 5 divided by number of message 3) collapse when average PUSCH SINR reduces down to 0 dB. The increase in inter-cell interference is first compensated by aggressive closed-loop power control and when the UE has reached the max transmit power the average SINR for the cell reduces while average PUSCH RSSI increases from −90 to −82 dBm. The RACH setup success rate decreases as UEs need to do several rounds of preamble retransmissions with increased preamble transmission power to get message 3 through and receive message 4. Similarly the RRC setup success rate decreases.

The number of required preambles per each RRC attempt increases rapidly when the SINR collapses as shown in Figure 13.21. The rapid increase of required preambles also causes preamble collision probability (two UEs randomly select the same preamble) to increase and due to that increased amount of required preambles per RRC attempt and increased interference levels.

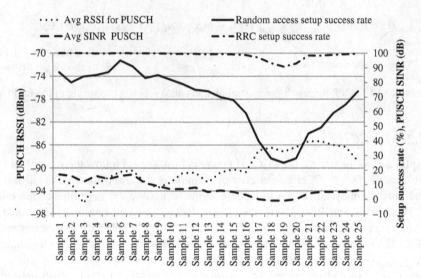

Figure 13.20 Average received signal strength indicator (RSSI), signal to interference and noise ratio (SINR), RACH setup success rate and RRC setup success rate of highest loaded cells for aggressive uplink power control; 15-minute data samples are used

The preamble collision probability as given in [5] is shown in Figure 13.22 as a function of RAO (random access opportunity, i.e. the amount of random access TTIs per radio frame) and random access intensity per second (sent preambles per second). In order to have very low collision probability and retransmissions of preambles due to that, the amount of random access TTIs per radio frame should be increased for mass events. Typically collision probability of below 1–2% is recommended in order to limit the additional interference increase.

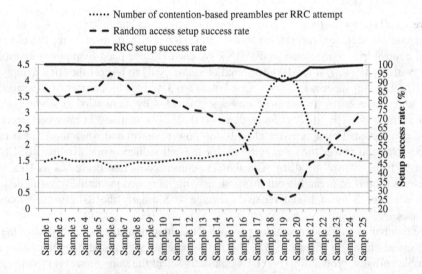

Figure 13.21 Number of required preambles per RRC setup attempt for highest loaded cells using 15-minute data samples

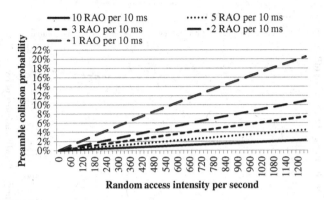

Figure 13.22 Preamble collision probability versus sent preambles per second

In order to calculate the peak random access intensity per second based on the, for example, 15-minute period statistics a peak (highest 1-second value) to average (average per second measured over 15-minute period) ratio should be used. Figure 13.23 shows one example where peak to average ratio is ~5.6 for highest loaded time for highest loaded cell for number of received RRC connection requests.

During the mass events it is recommended to use very conservative power control settings in order to reduce the inter-cell interference and protect the random access success rate and RRC setup success rate KPIs and uplink throughput performance. Conservative power control settings mean that the open-loop part of the uplink power control should be set to have $P_{0\_NOMINAL\_PUCCH}$ as around −105 dBm and alpha as 1. PUSCH closed-loop power control settings should be set to have large SINR window of 0–20 dB and fairly narrow RSSI window of −108 to −98 dBm. When this type of uplink power control strategy for PUSCH is used the performance is greatly improved as can be seen when the graphs for the same cells with aggressive power control shown in Figures 13.8 and 13.20 are compared to conservative power control shown in Figures 13.24 and 13.25, respectively. It can be seen that with conservative

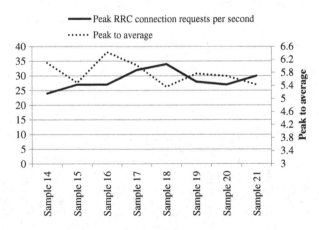

Figure 13.23 Peak RRC connection requests per second and peak to average ratio of received RRC connection requests

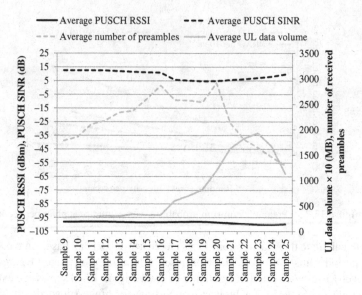

Figure 13.24 Uplink data volume, SINR, RSSI and number of received preambles averaged over highest loaded cells for conservative uplink power control strategy; 15-minute data samples are used

uplink (PUSCH and PUCCH) power control the collapse of PUSCH SINR, random access and RRC connection setup success rate can be avoided as well as the increase of PUSCH RSSI when uplink data volume or signalling increases.

In case the maximum supported RRC-connected users capacity is too low it can cause massive amount of RRC connection setup rejections and increased interference due to the re-attempts. This topic is discussed more detailed in Section 13.5.2.

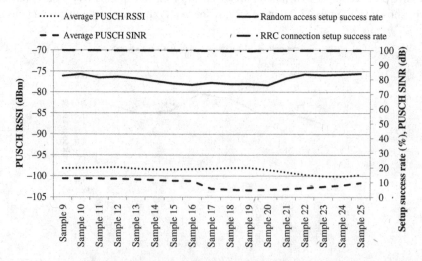

Figure 13.25 Average received signal strength indicator (RSSI), signal to interference and noise ratio (SINR), RACH setup success rate and RRC setup success rate of highest loaded cells for conservative uplink power control; 15-minute data samples are used

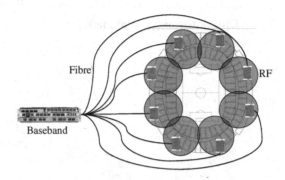

Figure 13.26   Centralized RAN solution for stadium capacity

13.3.4   *Centralized RAN*

The uplink capacity can be enhanced considerably by using coordinated multipoint (CoMP) reception in uplink. The solution for implementing CoMP is called here as centralized RAN (CRAN) because of centralized baseband location. A typical stadium installation is illustrated in Figure 13.26 where the stadium is covered by eight cells each with 2×2MIMO. The RF units are connected with fibre to the baseband unit. The traditional solution without uplink CoMP receives the UE transmission only via one cell while the UE transmission causes interference to the adjacent cells. The inter-cell interference is a problem especially in the open areas like stadiums where it is difficult to create clear dominance areas. The uplink CoMP solution can receive the UE transmission via multiple cells increasing the signal level and reducing the interference, which leads to a clear gain in the quality of the uplink signal. The traditional solution is shown in Figure 13.27 and the uplink CoMP solution in Figure 13.28. A number

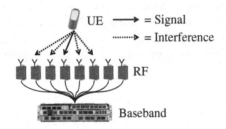

Figure 13.27   Traditional uplink reception via single cell

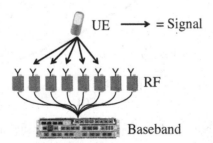

Figure 13.28   Uplink CoMP reception via multiple cells

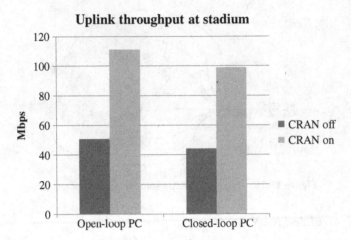

Figure 13.29   Throughput gain with uplink CoMP in centralized RAN (CRAN)

of stadiums already have such an installation. The gain of the uplink CoMP in throughput is shown in Figure 13.29, which illustrates the total uplink throughput with and without CRAN. The gain is more than 100% both with open-loop and with closed-loop power control.

The uplink CoMP needs no support from the UEs. Actually, the UEs do not even know that uplink CoMP is being utilized. Therefore, uplink CoMP works fine also with Release 8–10 UEs while the downlink CoMP requires Release 11 UEs.

13.4   Downlink Interference Management

Table 7.2.3-1 in [2], shown as a graph in Figure 13.30, specifies the connection between UE-reported downlink channel quality indicator (CQI) and spectral efficiency in a network where no reference signal power boosting is used. The higher the average reported downlink CQI value in a cluster of cells the better the downlink spectral efficiency.

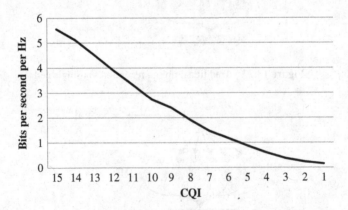

Figure 13.30   Connection between channel quality indicator per stream (CQI) and downlink spectral efficiency (bits per second per hertz)

The CQI reports from the UEs, together with other indicators such as HARQ feedback, are used by eNodeB algorithms to decide the downlink modulation and coding scheme (MCS) and number of control channel elements (CCEs) (i.e. aggregation level) used to deliver the downlink control information (DCI) (i.e. uplink and downlink capacity grants) on PDCCH. Once the radio conditions for a UE worsens, due to higher inter-cell interference or poor coverage, the eNodeB needs to use more robust MCS as well as higher aggregation level is required to deliver the DCI i.e. more CCEs are used.

13.4.1 PDSCH

Spectral Efficiency

Spectral efficiency is a factor of large number of variables, for example, layer 1 design, system bandwidth, cell density, traffic load, UE measurement and reporting accuracy and UE distribution across the network. Spectral efficiency is not constant in a network, but varies during the day. As the traffic increases and inter-cell interference increases the average reported CQI falls which means loss in spectral efficiency.

Highest spectral efficiency in any network can measured when the UEs are under favourable radio conditions (close to the eNodeB) and there is very low inter-cell interference (low traffic) from surrounding cells. Often this is visible during the night time when UEs are least active and tend to be distributed more evenly across the large geographical area covered by several cells. During the working hours traffic is higher and subscribers are concentrated within central business districts which are challenging from inter-cell interference control point of view (due to the high rise buildings with site acquisition constraints).

Figure 13.31 shows how the network efficiency varies as a function of downlink data volume per cell per hour. It can be seen how the average reported CQI decreases in linear manner as the downlink data volume per cell increases.

The starting point, indicated by point 1 in Figure 13.31 for average CQI during the hours of low data volume per cell per hour is mainly a function of cell density and dominance. During the low data volume per cell per hour the interference is caused by downlink cell-specific

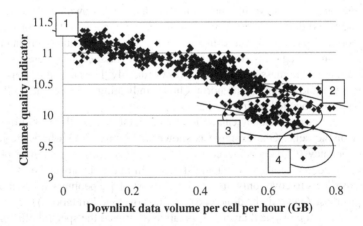

Figure 13.31 Channel quality indicator as a function of downlink data volume per cell per hour over all the cells in the network

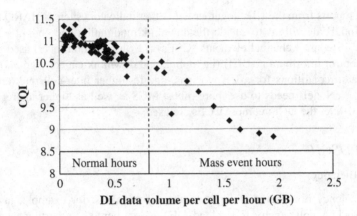

Figure 13.32 Average channel quality indicator as a function of downlink data volume per cell per hour in a cluster of 24 cells over a few days, including the mass event hours

reference signals transmitted by neighbouring cells (assuming no cancellation of cell-specific reference signals by the UEs). When the data volume per cell per hour increases, indicated by point 2 in Figure 13.31, the PDSCH usage increases and becomes the dominant source of interference from and to neighbouring cells. The slope is defined mostly by the same factors as the starting point – cell density and dominance.

There is also a separate distinctive group, indicated by point 3 in Figure 13.31, below the linearly distributed samples that is generated during the working hours as the distribution of UEs in the network has changed to more concentrated in certain areas increasing the data volume per cell per hour more than on average across all the cells in the network.

Mass event is an extreme example of radical change in how UEs are distributed in the network as a large number of UEs become concentrated within a small geographical area. In some cases a large mass event can be visible even in network level statistics as indicated by point 4 in Figure 13.31. Figure 13.32 shows the impact on the average reported CQI during the mass event on 24 cells covering the main event area. The downlink data volume per cell is almost quadrupled during the peak mass event hour compared to the data volume per cell during the normal hours. This traffic increase causes additional inter-cell interference and a big reduction in average reported CQI and therefore in spectral efficiency. The average CQI decrease during the mass event hours follows roughly the same slope depending on the downlink data volume as during the normal hours indicating that the inter-cell interference conditions have not changed during the mass event hours.

The network spectral efficiency can be improved (CQI increased) by improving the dominance areas of the cells. One example is shown in Figure 13.33 where the improvement in spectral efficiency (CQI) is achieved by physical layer optimization. The physical layer optimization included antenna downtilt adjustments (limit the cell coverage spillage), antenna direction adjustments to cover only the areas with most of the people and downlink transmission power decrease (can only be done in case of no coverage limitations).

In the case above the physical layer optimization improved the spectral efficiency so much that ~50% more downlink data volume per cell could be delivered with the same CQI compared to previous year.

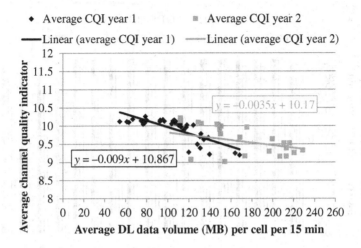

Figure 13.33   Spectral efficiency improvement by physical layer optimization

The actual spectral efficiency in terms of bit/s/Hz per cell can be calculated based on the active cell throughput statistics and allocated bandwidth per cell. Figure 13.34 shows example per different bands and bandwidths.

As can be seen from Figure 13.34 the 3GPP-simulated spectral efficiency of 1.7 bit/s/Hz [6] per cell for 100% load can be met and exceeded provided that the reported CQI is 9 or better. It should also be noted that this high spectral efficiency can be achieved by low band and high band with 10 and 20 MHz bandwidths. The 3GPP simulations for spectral efficiency are using full buffer traffic model which means that each cell is having 100% DL PRB utilization at the same time. The own cell DL PRB utilization and its neighbours' DL PRB utilization are hardly ever at 100% at the same time in real live networks as shown in the Figure 13.35.

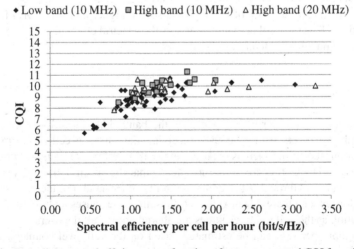

Figure 13.34   Downlink spectral efficiency as a function of average reported CQI for cells having 80% and above downlink PRB utilization over the measurement period

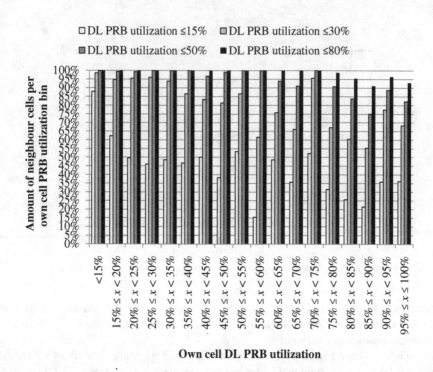

Own cell DL PRB utilization

Figure 13.35   Own cell DL PRB utilization as a function of neighbouring cell DL PRB utilization

Figure 13.35 shows that when the own cell DL PRB utilization increases the share of its neighbours which are having the same or higher DL PRB utilization slightly increases as well. For example, when own cell DL PRB utilization reaches 90–95% then 35% of its neighbours are having DL PRB utilization less than or equal to 15%. Or at the same 90–95% DL PRB utilization for own cell, only 5% of its neighbours have higher than 80% DL PRB utilization at the same time. This means that in real networks the spectral efficiency can exceed the 3GPP limits due to the unbalanced load between the cells.

Throughput per User

The downlink throughput per user depends on the bandwidth allocation, spectral efficiency and the number of users simultaneously receiving data from a cell. As explained in the previous section, the traffic increase causes inter-cell interference increase which is visible as CQI decrease. The CQI decrease in turn means that the spectral efficiency decreases and therefore the end-user throughput decreases. On the other hand, the increase in the number of connected users which are downloading at the same time causes additional throughput per user degradation as the cell radio resources are shared among all the users.

The throughput per user can be estimated based on the cell-level counters, for example, simply dividing the cell-level throughput by number of users which are having data in the eNodeB buffers (i.e. those users which are currently downloading something). The cell-level throughput in turn can be estimated by dividing the delivered user plane bits by the number of active TTIs used to transmit those bits. The number of bits should contain only the user plane data bits excluding the retransmissions, HARQ or RLC retransmissions. The active TTI is a bit difficult to measure as the active TTI is active regardless of the number of PRBs used during the TTI. Therefore, this type of performance indicator is not very accurate in case the PRB utilization (how many PRBs are used from all available) is very low (<30%). Scaling the cell throughput for, for example, 10% PRB utilization to 100% PRB utilization simply by dividing the cell throughput by the PRB utilization cannot be done. This is due to the fact that such a simple scaling assumes that the downlink interference conditions do not change at all when the PRB utilization increases and such assumption is not correct. Therefore, in case the cell throughput performance indicator is calculated using the simple formula described above the results should be analysed carefully taking into account the PRB utilization.

Figure 13.36 shows the example of average throughput per user (calculated as described above) as a function of the number of RRC-connected users. The cells included into the analysis are high traffic cells for which PRB utilization is above 20% per 15-minute sample. It should be noted that the carriers are not deployed on the same sites and optimization is performed on carrier basis trying to achieve highest possible performance per carrier. As can be seen from Figure 13.36 the low band (700, 800 or 900 MHz) 10 MHz cells provide slightly worse average user throughput compared to the high band (1800, 2100 or 2600 MHz) 10 MHz cells. The highest average user throughput can be achieved by the high band (1800, 2100 or 2600 MHz) 20 MHz cells.

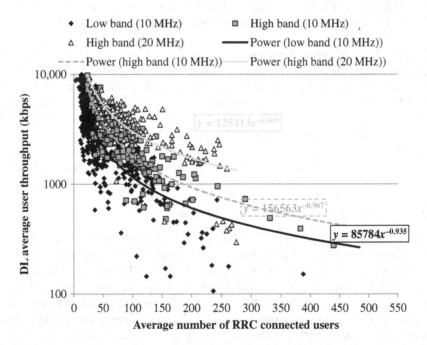

Figure 13.36 Average downlink throughput per user (kbps) as a function of the number of RRC-connected users per FD LTE cell. Highest loaded cell 15-minute statistics are used

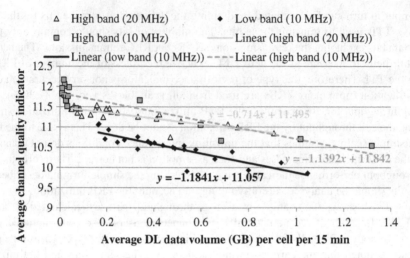

Figure 13.37 Spectral efficiency per carrier as a function of downlink data volume

If downlink throughput limit is set to, for example, 1 mbps per user then that means that on an average analysed cells can support the following amount of RRC-connected users:

- 20 MHz high band FD LTE can support 392 RRC-connected users per cell
- 10 MHz high band FD LTE can support 186 RRC-connected users
- 10 MHz low band FD LTE can support 117 RRC-connected users

The throughput difference between 10 MHz low band and high band allocations comes from radio quality difference between the bands, that is, high band is in general easier to optimize and therefore high band can provide better spectral efficiency as can be seen from Figure 13.37. The higher the carrier frequency the less the CQI decreases while the downlink data volume per cell increases. Also the more the bandwidth available the more downlink data volume can be delivered with certain CQI value and therefore the slope is much improved for high band 20 MHz bandwidth compared to 10 MHz bandwidth cells (also note that the transmission power per sub-carrier is the same regardless of the carrier bandwidth).

13.4.2 Physical Downlink Control Channel

PDCCH is used to deliver all the resource allocation grants (number of allocated PRBs, MCS, used transmission mode, etc.) for PDSCH and PUSCH; therefore, the capacity of PDCCH defines the amount of UEs that can be scheduled in single TTI. The PDCCH contains CCEs and each CCE is formed from nine resource-element groups and the total size is 36 quadrature phase shift keying (QPSK) symbols, that is, 72 bits. The resource allocation grants, called DCI messages, are then mapped onto one or several CCEs depending on the coding rate required for successful reception by the UE. In case low coding rate is needed then several CCEs might be needed and hence aggregation of CCEs is needed. The DCIs are fixed in terms of size and purpose according to [2] (uplink resource allocation grant, downlink resource allocation grant for dual stream and so on) and the level of aggregation is fixed to steps of 1, 2, 4 or 8 CCEs per DCI. This is done in order to reduce the UE complexity to decode the DCIs without any prior knowledge of what to expect, that is, UE is required to perform blind decoding of several different DCIs with different aggregation level per each TTI. PDCCH CCEs are split into two groups of CCEs:

- UE-specific search space CCEs are used for DCIs that are UE specific and therefore meant for only single and known UE. The aggregation level can be 1, 2, 4 or 8 depending on the requirement per UE.
- Common search space CCEs are used for common DCIs that are targeted to be received by all the UEs under the cell, for example, for paging message, for system information or for message 2 resource allocation grant delivery. The aggregation level can be either 4 or 8.

According to [7] the CQI reports from the UEs can be used to decide the needed CCEs, that is, coding rate, for successful resource allocation grant delivery per UE. Therefore, the worse the UE-reported CQI is the more CCEs, that is, lower coding rate, are needed from PDCCH and the lower amount of UEs can be scheduled per single TTI. This applies for the UE-specific search space resource allocation grants but as the common search space resource allocation grants are targeted for all the UEs under the cell the coding rates are typically fixed to 4 or 8.

Figure 13.38 shows the impact of average CQI on average aggregation level and the worse the CQI the higher the aggregation level. Note that Figure 13.38 contains DCIs for both UE-specific and common search space and the common search space resource allocation grants are all using fixed aggregation level of four. The higher the traffic the more the UEs needed to be scheduled per TTI and the more CCEs are therefore needed but the higher the traffic the higher the inter-cell interference and therefore per UE the required amount of CCEs increases. Therefore, in terms of PDCCH usage the control of inter-cell interference increase due to the traffic increase is crucial. The low-frequency band cells are having higher average aggregation

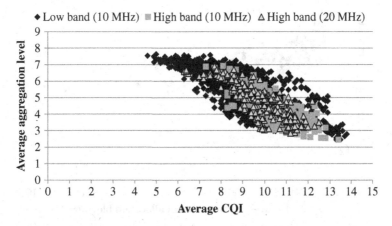

Figure 13.38 Average aggregation level as a function of average CQI, reported by the UEs, per cell using 15-minute statistics samples

level due to the lower average CQI compared to the high-frequency band cells. The carrier bandwidth does not seem to have strong impact on average aggregation level or CQI.

As explained in [7] due to the UE decoding cost reduction, the blind decoding attempts a UE shall monitor on every TTI is reduced by hashing function. Hashing function defines the starting positions of each aggregation level (1, 2, 4 or 8) based on the UE identification e.g. cell radio network temporary identifier (C-RNTI) for UE-specific search space and system information RNTI (SI-RNTI) for common search space. Based on hashing function there are six candidate positions defined for aggregation level 1 and aggregation level 2 and two candidate positions are defined for aggregation level 4 and aggregation level 8. This means that, for example, there can be maximum two UEs scheduled per TTI in case both of those UEs require aggregation level 8 for their resource allocation grants. Also due to the hashing algorithm the candidate positions of several UEs can be overlapping and therefore blocking of CCE allocations (using the required aggregation level) can occur even if all the CCEs are not used. The CCE allocation blocking increases rapidly during mass events due to the increased traffic and especially due to the CQI decrease and thus more aggregation level 8 is required as shown in Figure 13.39 below.

The high aggregation level can be decreased by increasing the power of those CCEs used for UE requiring high aggregation level (doubling the aggregation level can be replaced by using the same aggregation level but 3 dB higher power) and therefore the CCE blocking, especially due to hashing algorithm, can be reduced. This type of mechanism requires advanced algorithms in the eNodeB so that the power of CCEs used for the UEs requiring lower aggregation level than the allocated one can be re-allocated to UEs requiring higher aggregation level than the one available.

If the downlink inter-cell interference can be improved the spectral efficiency will be improved as well as the PDCCH capacity in terms of number of UEs that can be scheduled per TTI. Figure 13.40 shows the average aggregation level as a function of CQI for the highest

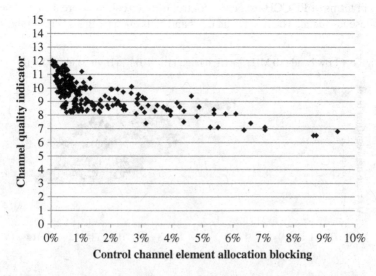

Figure 13.39 Average channel quality indicator per cell per 15 minutes as a function of control channel element allocation blocking

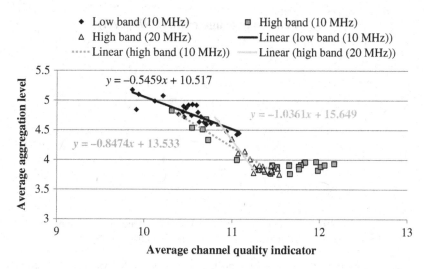

Figure 13.40 Average aggregation level as a function of average channel quality indicator for three different bands using 15-minute statistics

loaded cells per different frequency bands and bandwidths. As it can be seen the high carrier frequency band has lower average aggregation level compared to the low carrier frequency band allocations.

As can be seen from Figure 13.40 the average aggregation level of high band 20 MHz bandwidth allocation increases much faster compared to the high band 10 MHz bandwidth allocation. This is due to the fact that 20 MHz bandwidth has double capacity for PDCCH compared to 10 MHz bandwidth and this double capacity means that necessity for lowering the aggregation level by power increase is much lower hence the average aggregation level can be increased for more UEs compared to the 10 MHx bandwidth.

13.5   Signalling Load and Number of Connected Users Dimensioning

The signalling load and number of connected users per eNodeB and cell during the mass event has to be carefully dimensioned.

Each eNodeB has limited signalling load capacity given, for example, in terms of call setups per second or control plane actions per second. In case the signalling load capacity of eNodeB is exceeded then depending on the eNodeB overload control efficiency either the call setup success rate and dropped call rate degrade or eNodeB collapses out of service. Latter scenario in case of no proper overload control needs to be avoided by proper dimensioning.

Each cell or eNodeB also has upper limit of number of supported simultaneously RRC-connected state users for which PUCCH capacity is allocated (CQI and scheduling request transmissions) and synchronization with eNodeB maintained. Exceeding this capacity limit would cause call setup rejections at RRC connection setup phase. Any rejected RRC connection setup leads to re-attempt by the terminal which in turn causes increase in random access procedures and uplink interference levels.

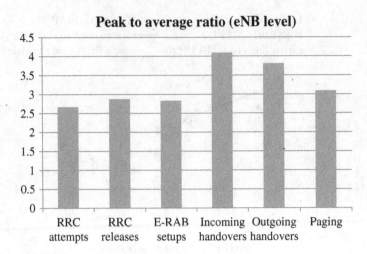

Figure 13.41 Peak to average ratios for signalling actions per eNodeB, average is given based on 15-minute samples per eNodeB

13.5.1 Signalling Load

The eNodeB limits for signalling load values are given as call setups per second or transactions per second and therefore the peak signaling load value per second during the event should be evaluated. Typically the statistics from eNodeBs are collected in 15-minute intervals containing sums of counters over the period. Then calculating any average per second values based on those 15-minute sums does not reflect the real peak value. Therefore, peak to average values are needed to multiply the average values calculated from 15-minute average to get the peak per second value. Figure 13.41 shows one example of the peak to average values for some signalling messaging averaged over the highest loaded eNodeBs. Peak value is the highest value per second during the measurement period and average per second is calculated based on 15-minute measurement period eNodeB level counters.

For mass events the signalling load can be estimated per cell-based average number of RRC-connected users as shown in Figure 13.42. The signalling attempts per RRC-connected user decreases while the number of RRC-connected users increases. At low (~100–200) RRC-connected users per cell the average RRC attempts per RRC-connected user is between 42 and 48 but at high RRC-connected users (~300) the average RRC attempts per RRC-connected user is around 42. By combining the graphs in Figures 13.41 and 13.42 it is then possible to calculate the amount of signalling messages per second per eNodeB limit. The number of RRC-connected users can also be translated into number of subscribers as explained in Section 13.5.2.

13.5.2 RRC-Connected Users

The peak number of RRC-connected users per cell needs to be properly set for mass event to avoid unnecessary RRC setup attempts rejections by eNodeB admission control. This means that the parameters limiting the number of RRC-connected users and corresponding PUCCH

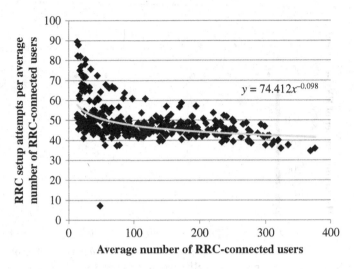

Figure 13.42 RRC connection setup attempts per average number of RRC-connected users versus average number of RRC-connected users per cell per 15 minutes

capacity allocation parameter values should be set large enough to avoid any avalanche of RRC connection requests and RRC connection rejects. Figure 13.43 shows the average number of RRC-connected users versus maximum number of RRC-connected users per cell per 15 minutes in mass event area. As it can be seen there are two types of cells in the area: cells for which the max number of RRC-connected users is 700 and cells for which the max number of RRC-connected users is 580. When the maximum supported RRC-connected users limit

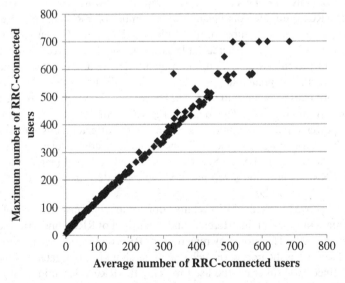

Figure 13.43 Average number of RRC-connected users versus maximum number of RRC-connected users per cell per 15 minute

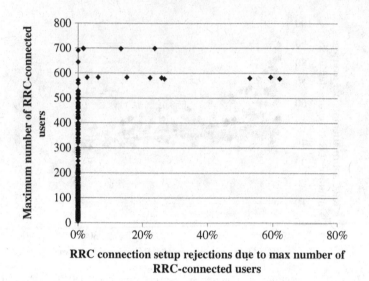

RRC connection setup rejections due to max number of RRC-connected users

Figure 13.44 RRC connection setup rejections due to max number of RRC-connected users reached the maximum limit set for the cell versus maximum number of RRC-connected users

per cell is reached the admission control of the eNodeB starts to reject the RRC connection setups. The rejection rate reaches extremely high values easily as shown in Figure 13.44. These admission control rejections will increase the total signalling load due to the terminals retrying until connection is established, that is, capacity granted by the eNodeB. Example of this is shown in Figure 13.45 which shows the number of preambles needed per RRC setup attempt for cells and times (15-minute samples) for which there are admission control rejections for RRC setup attempts because max number of RRC-connected users limit was reached. The increase in the number of preambles needed per RRC connection attempt is due to the interference which causes increased UE transmission power requirement and therefore higher number of preamble power ramp-up rounds (increased interference causes message 3 reception problems). The preamble power increase for UEs under the cells rejecting the RRC connection setups causes increase in inter-cell interference and therefore reduction in PUSCH SINR for the surrounding cells. This degraded PUSCH SINR in turn causes increase in UE transmission power for the preamble process under the affected cells which in turn causes problems to the surrounding cells due to the increased interference level. Therefore, at the end there is overall decrease in PUSCH SINR across several cells and increase in the number of preambles per RRC connection setup attempt and rapid decrease in random access success rate as shown in Figure 13.20.

As explained above the rejected RRC connection setups can cause increase of signalling and therefore increase of inter-cell interference and the impact of RRC connection setup rejections spreads to several cells. Therefore any rejections of RRC connection setups should be avoided and in case of rejections the impact of such rejections should be mitigated as much as possible. The 3GPP-defined t302 timer can be used to delay the RRC connection setup repetitions in case of admission control rejection [8]. However, t302 timer only delays the RRC connection setup re-attempts by t302 time given in RRC connection reject message by eNodeB. Increasing

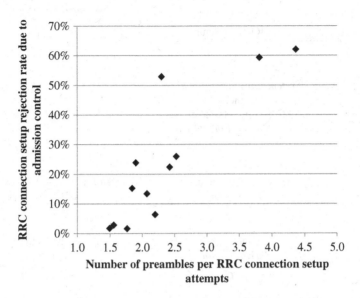

Figure 13.45 RRC connection setup rejection rate versus number of required preambles per RRC connection setup attempt – only cells and samples for which there are admission control rejections are included

the value for t302 to 16 seconds, which is the maximum possible [8], is beneficial during the mass events. Alternatively 3GPP-defined [8] service-specific access class barring can be used to bar the cell for given time. The UE uses the service-specific access class barring algorithm, shown in Figure 13.46 for mobile-originated data call type as an example, to evaluate whether the cell should be considered to be barred or not based on [8]. First the UE will evaluate if the received system information block (SIB) number 2 contains access class barring parameters for the initiated call setup cause or not. Then in case the access class barring parameters are included and the UE has all the valid access classes within the 11–15 range, as stored in the USIM, set to non-zero values, the UE considers the cell to be barred for the specific call setup cause. Next the UE will draw the random number, *rand* in Figure 13.46, uniformly distributed within the range $0 \leq rand < 1$ and in case the *rand* value is higher than or equal to the access class barring factor – parameter (ac-BarringFactor, delivered to the UE in SIB2) value – then the cell is considered to be barred. The barring time, Tbarring, is defined according to [8] and given in Eq. 13.1:

$$(0.7 + 0.6^* rand)^* ac - BarringTime \qquad (13.1)$$

With the service-specific access class barring it is possible to define barring probability depending on the call type, for example, mobile-originated data call can have different probability for cell barring than mobile-terminated data call. By enabling the service-specific access class barring it is possible to limit the RRC connection setup rejections due to the RRC-connected users reaching maximum capacity and therefore avoid the PUSCH SINR collapse and huge signalling load.

For dimensioning purposes the signalling amount per subscriber needs to be estimated and therefore the number of subscribers needs to be estimated. This can be done based on mobility

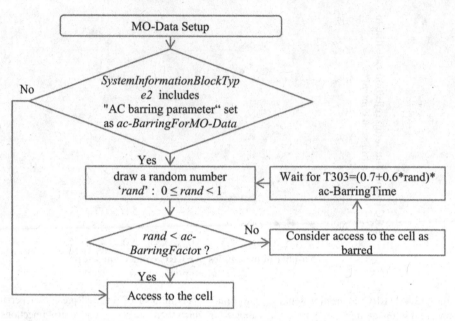

Figure 13.46 Service-specific access class barring for mobile-originated data call attempt according to Reference 8

management entity (MME) statistics by calculating the ratio between registered users, which is the number of subscribers, and number of connected users, which is the number of evolved packet system connection management connected users. Figure 13.47 shows one example result of ratio of RRC-connected users and MME-registered users which can then be used to estimate subscriber amount based on the RRC-connected users. It should be noted that the inactivity timer setting, that is, how long the RRC-connected users is kept connected without any uplink or downlink traffic, plays a significant role in the ratio of connected users and registered users.

Figure 13.42 can then be updated based on Figure 13.47 as shown in Figure 13.48 which shows that with high number of subscribers per cell around 3.4 RRC setups are generated per subscriber per 15 minutes. Then all types of signalling events per subscriber can be derived and used together with peak to average ratio calculations based on Figure 13.41 to estimate signalling load per eNodeB for mass event.

13.6 Load Balancing

Load balancing functionality can be activated between the different frequency band cells (or also between same carrier band cells) to balance the load. Idle mode load balancing algorithm typically directs certain amount of RRC-connected users for which the RRC connection release message is about to be sent (and RRC connection released), from one carrier frequency to another. Similar load balancing functionality can be done during RRC-connected state using measurements and handovers. But as in case of special event analysed here each carrier

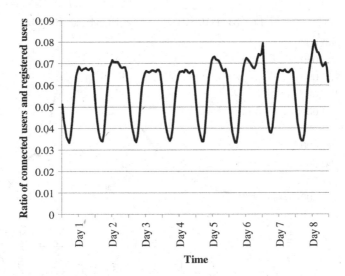

Figure 13.47 Hourly ratio between RRC-connected users and mobility management entity (MME)-registered users based on MME statistics over 8 days (RRC connection inactivity timer = 10 seconds)

frequency has full coverage and users are mainly outdoors there is no need to perform any target frequency (inter-frequency) measurements and UEs can be moved from lower frequency band to higher or vice versa blindly. Figure 13.49 shows how call setup attempts can be balanced between frequency bands using idle mode load balancing. In Figure 13.49 the idle mode load balancing is enabled at sample 27 and disabled at sample 40.

As can be seen from Figure 13.49 the high band cells collect call setups faster than the low band cells. This is because in idle mode the high band is prioritized so that all UEs that support high band and low band are camping on the high band cells.

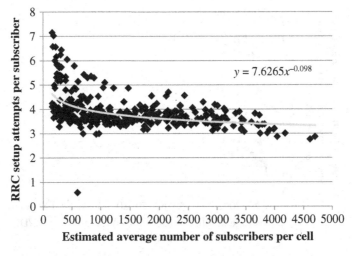

Figure 13.48 RRC setup attempts per subscriber versus estimated number of subscribers per cell per 15 minutes

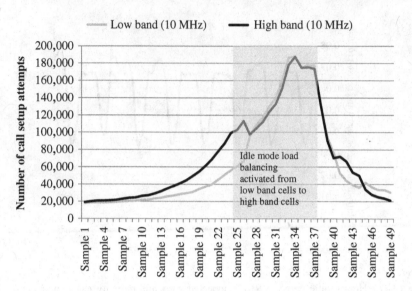

Figure 13.49 Load balancing activated to equalize the number of call attempts for both carrier frequencies. Idle mode load balancing activated for a cluster of cells and 15-minute statistics samples used

13.7   Capacity Bottleneck Analysis

During the mass events the PRB utilization typically reaches 100% (all PRBs from all available, e.g. 50 for 10 MHz, LTE are used) especially for uplink but also possibly for downlink as shown in Figure 13.50. The circled area 1 contains the samples for which the uplink data volume is fast increasing as indicated in Figure 13.2b. The average number of users with data

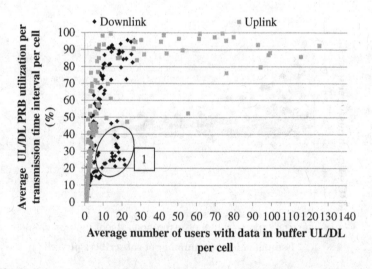

Figure 13.50 Uplink and downlink physical resource block utilization versus average number of users with data in buffer per cell and per 15 minutes

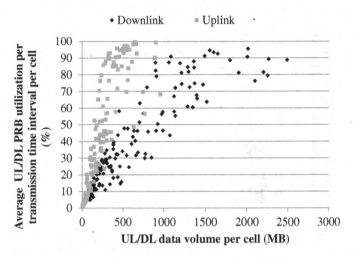

Figure 13.51 Uplink and downlink physical resource block utilization versus uplink and downlink data volume in megabytes per cell and per 15 minutes

in buffer in uplink reaches much higher values than in downlink due to the fact that uplink spectral efficiency is much lower than downlink (no MIMO nor 64QAM). Also the uplink SINR decreases rapidly as the number of users with data in buffer in uplink increases as shown in Figure 13.56. The CQI remains almost constant regardless of the number of users with data in buffer in downlink as shown in Figure 13.57. Therefore, it can be said that this network is uplink limited during the special event.

As the number of users grows, and therefore the amount of users with data in buffer increases as well as the data volume increases as shown in Figure 13.51, the scheduling delay (the time instance between user's data packet arrival to the eNodeB and the packet successfully transmitted to the user) increases as shown in the Figure 13.52. This increased scheduling delay causes end-user-experienced throughput decrease as more and more TTIs are required between single UE scheduling instances.

The results in Figure 13.51 can be used to derive spectral efficiency results. In downlink during maximum PRB utilization, around 2500 MB per 15 minutes can be delivered which is 22.2 Mbps on average. Considering 10 MHz bandwidth cells the spectral efficiency becomes 2.22 bps/Hz/cell which is at the upper end of the 3GPP-simulated results [4]. In uplink correspondingly 1000 MB per 15 minutes is received which indicates around 0.9 bps/Hz/cell which is also at the upper end of the 3GPP-simulated results when considering 2rx receiver [4].

The scheduling delay shown in Figure 13.52 can also be increased due to the PDCCH blocking which delivers the uplink and downlink scheduling grants. Therefore, the CCE blocking shown in Figure 13.39 can cause increase in scheduling delay as shown in Figure 13.53. This is due to the fact that in case required aggregation level cannot be used for scheduling certain UE and transmission power increase cannot be done to boost the available aggregation level decoding performance to the level of required aggregation level, the UE cannot be scheduled in corresponding TTI and the data remains in the eNodeB buffer (and scheduling delay increases). Both high PRB allocation and CCE blocking cause increase in scheduling delay

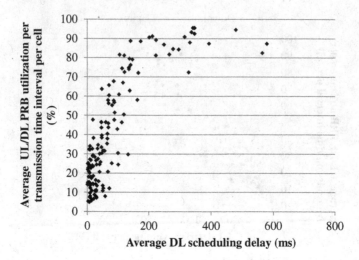

Figure 13.52 Uplink and downlink physical resource block utilization versus average downlink packet scheduling delay per cell and per 15 minutes

and increase in number of users with data in buffer in uplink and downlink and therefore lower end-user-experienced throughput as shown in Figures 13.54 and 13.55, respectively.

As shown in Sections 13.3 and 13.4, uplink and downlink quality can also limit the network performance and capacity. Each new user that has some data to be transmitted causes small reduction in uplink SINR and downlink CQI as shown in Figures 13.56 and 13.57, respectively.

The example cells are clearly uplink limited due to the PUSCH SINR collapse when the number of users with data in buffer in uplink increases (as indicated in Figure 13.56) and

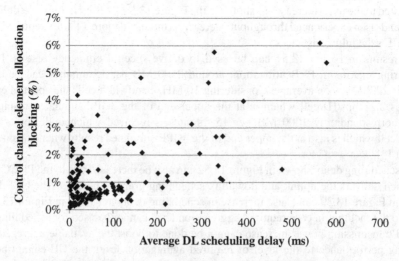

Figure 13.53 Control channel element allocation blocking versus average downlink scheduling delay per cell using 15-minute statistics

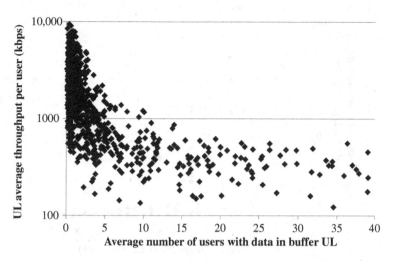

Figure 13.54 Uplink throughput per user as a function of number of users with data in uplink buffer per cell using 15-minute statistics

no impact to average CQI value when number of users with data in buffer in downlink increases.

When the number of users with data in buffer increases also the end-user-experienced uplink and downlink throughputs rapidly decrease as shown in Figures 13.54 and 13.55, respectively. The number of users with data in buffer (uplink and downlink) increases as a function of number of RRC-connected users as shown in Figure 13.58 for one mass event.

The capacity limits for each cell can be calculated based on the number of RRC-connected users required to be supported and predictions can be made for any coming mass event as

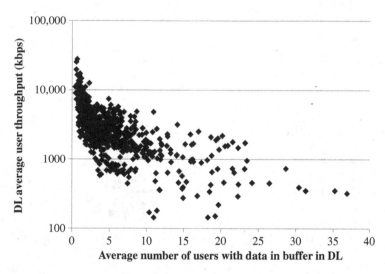

Figure 13.55 Downlink throughput per user as a function of number of users with data in downlink buffer per cell using 15-minute statistics

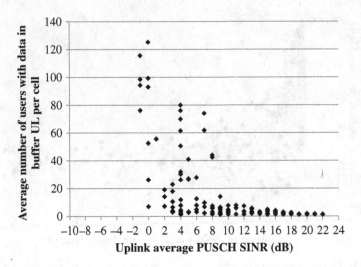

Figure 13.56 Average number of users with data in buffer in uplink versus uplink average PUSCH signal to interference ratio per cell per 15 minutes

long as the increase in RRC-connected users is known per cell. Example of the increase of the cell-level average and maximum RRC-connected users between mass event and normal day is shown in Figure 13.59. As can be seen the average (averaged over the highest loaded cells) increase of average RRC-connected users per cell is up to 15 and up to 11 for maximum RRC-connected users for the event day compared to the normal day.

When the maximum and average RRC-connected users are estimated per cell then all the other limitations can be estimated and possible bottleneck on cell level analysed.

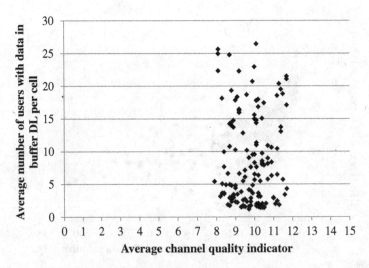

Figure 13.57 Average number of users with data in buffer in downlink versus average channel quality indictor per cell per 15 minutes

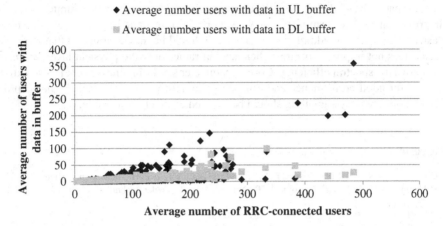

Figure 13.58   Average number of users with data in buffer as a function of average number of RRC-connected users per cell based on 15-minute statistics

13.8   Summary

Each cell can have different upper limits in terms of capacity. Some cells are uplink or downlink PRB utilization limited and therefore uplink or downlink end-user throughput limited. In case conservative-enough uplink power control is not used the cells typically become uplink interference limited. Also in case of poor physical layer performance the cells can become

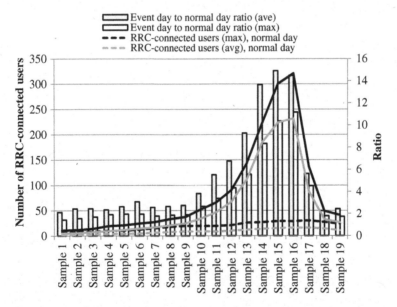

Figure 13.59   Average and maximum RRC-connected users during event day and normal day with the increase ratio between normal and event day. Hourly statistics is used for the highest loaded cells within the mass event cluster

PDCCH capacity limited. Typically the cells become end-user throughput limited and that can be improved only by performing physical layer optimization or adding more cells or increasing the carrier bandwidth or adding more carriers. It should be, however, noted that adding extra capacity does not improve spectral efficiency but rather only the physical layer optimization can improve the spectral efficiency. Correct parameter set during mass event is also extremely important for good performance and especially the uplink power control should be properly set and all unnecessary signalling should be avoided as much as possible.

References

[1] C. U. Castellanos, D. L. Villa, C. Rosa, K. L. Pedersen, F. D. Calabrese, P.-H. Michaelsen, J. Michel, 'Performance of Uplink Fractional Power Control in UTRAN LTE', IEEE Vehicular Technology Conference (VTC), Spring 2008.
[2] 3GPP TS 36.213, 'Physical Layer Procedures', v.10.9.0, 2013.
[3] 3GPP TSG RAN R1–072261, 'LTE Performance Evaluation – Uplink Summary', May 2007.
[4] H. Holma and A.Toskala. *LTE for UMTS – OFDMA and SC-FDMA Based Radio Access*, John Wiley & Sons (2009).
[5] 3GPP TR 37.868, 'RAN Improvements for Machine-Type Communications', v.1.0.0, October 2011.
[6] 3GPP TSG RAN R1–072444, 'Summary of Downlink Performance Evaluation', May 2007.
[7] F. Capozzi, D. Laselva, F. Frederiksen, J. Wigard, I. Z. Kovacs, and P. E. Mogensen, 'UTRAN LTE Downlink System Performance Under Realistic Control Channel Constraints', IEEE Vehicular Technology Conference Fall (VTC 2009-Fall), 2009.
[8] 3GPP TS 36.331, 'Radio Resource Control (RRC); Protocol Specification', v.10.11.0, 2013.

14

VoLTE Optimization

Riku Luostari, Jari Salo, Jussi Reunanen and Harri Holma

14.1 Introduction

LTE radio is designed for the packet connections without any support for circuit switched (CS) connections. Therefore, other voice solutions for voice service are required instead of the traditional CS voice. The voice service is important from the operator revenue point of view even if the share of voice traffic from the data volume point is very low. Every single LTE-capable smartphone needs to have a solution for offering high-quality voice connection. The different voice options including CS fallback and voice over LTE (VoLTE) are presented and the optimization is discussed in this chapter. Section 14.2 illustrates the timing and the main differences of the voice options. Section 14.3 presents CS fallback and Section 14.4 VoLTE solution. The handover from VoLTE to CS voice is called single radio voice call continuity (SRVCC) and is discussed in Section 14.5. The chapter is summarized in Section 14.6.

14.2 Voice Options for LTE Smartphones

The first LTE smartphones used dual radio solution where the terminal included two parallel radios: one for CDMA voice and another for LTE data. It looks like two separate terminals

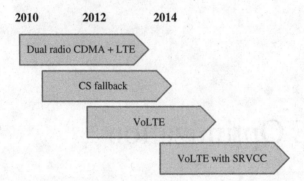

Figure 14.1   Availability of different LTE voice solutions

from the network point of view. Such a solution is simple from the network point of view since no network interworking is required. The dual radio has the customer benefit that simultaneous CS voice and LTE data are available. The drawback of the dual radio is the complexity from the device point of view since more radio components are required increasing the cost, size and power consumption. The dual radio approach has been used only by the CDMA operators. Some CDMA/LTE devices use just single radio which is simpler for the device implementation but does not support simultaneous voice and data.

All GSM/WCDMA operators started voice with CS fallback solution where the connection is moved from LTE to GSM/WCDMA when the voice call is started. CS fallback is also used by some CDMA operators. CS fallback solution is simple for the devices since only single radio is running at a time. CS fallback requires some network interworking in the radio and in the core network. CS fallback has an impact on the call setup time since UE must switch from LTE to 3G during the call setup.

Wide-scale commercial VoLTE started in South Korea during 2012. Since the coverage in South Korea was practically 100%, it was possible to offer VoLTE without interworking to CS domain. VoLTE deployments became more widely spread during 2014 together with SRVCC functionality. The availability of the different voice options are illustrated in Figure 14.1 and the benchmarking is shown in Table 14.1.

14.3   Circuit Switched Fallback

14.3.1   Basic Concepts

Until native LTE voice service becomes globally supported, circuit switched fall back (CSFB) will be required as an option for LTE subscribers without PS voice-capable terminals and for roamers who do not have IMS VoLTE subscription provisioned in the HSS of the home PLMN. During attach or tracking area update a voice-capable UE signals its voice preference to MME as one of the following [1]:

- CS voice only
- Prefer CS voice with IMS PS voice as secondary
- Prefer IMS PS voice with CS voice as secondary
- IMS PS voice only

Table 14.1 Benchmarking of different voice options

	SV-LTE	CDMA/LTE	CS fallback	VoLTE
Voice bearer	CDMA CS voice	CDMA CS voice	WCDMA, GSM or CDMA CS voice	LTE VoIP
Network requirements	Simple, no interworking requirements	Minor interworking requirements	Interworking in radio and core network	IMS + VoIP support in radio + SRVCC
Terminal requirements	Complex with dual radios	Simple with single radio	Simple with single radio	VoLTE support required but simple with single radio
LTE coverage requirement	No need for full LTE coverage	No need for full LTE coverage	No need for full LTE coverage	Good LTE coverage preferred to avoid many SRVCCs
Simultaneous voice + data	CS voice + LTE data	No	CS voice + HSPA data	LTE VoIP + LTE data
Voice setup time	Equal to CDMA CS voice	Equal to CDMA CS voice	CS voice setup time + 1–2 seconds	Faster than CS voice

For the second and third cases the response from MME determines whether the UE will establish mobile-originating call using PS or CS voice. For mobile-terminating calls the CS/PS domain selection is made by IMS and HSS. In the first three cases the UE tries to attach to CS voice services. The attach request is relayed by MME to MSC Server over the SGs interface. In the attach request MME includes the Location Area Code for the UE, based on an operator-configurable LAC-TAC mapping table. From this point on, the MSS knows the location of the UE at MME level and can send paging to the correct MME over the SGs interface. In this section, the emphasis is on the performance optimization aspects of CSFB; the interested reader can find further SGs protocol details and background information in [2] and [3].

Figure 14.2 is a simplified signalling diagram of a mobile-originating CSFB call in the case where the LAC of the target RAT is different from LTE LAC. UE initiates the call by sending Extended Service Request to MME (RRC signalling not shown). The transition to target RAT can be made via RRC release with redirect, via PS handover or (in case of 2G target RAT) inter-RAT cell change order. The decision of target RAT and how to move the UE there depend on eNodeB configuration and UE radio capability. Once the UE is in target RAT, it initiates Location Update procedure since in this example it is assumed that the LAC is different from the one it received from MME during attach or TAU. Depending on CS core configuration the MSS may or may not be the same as the one serving the LTE LAC. If MSS is changed, this will increase the Location Update time, and hence also delay the call setup in target RAT. Moreover, depending on the MSS configuration authentication, IMEI checking and TMSI reallocation may be done at this point. After Location Update is accepted, the UE is able to start the actual CS call setup by sending CM Service Request as it would do in a normal mobile-originating call in the target RAT.

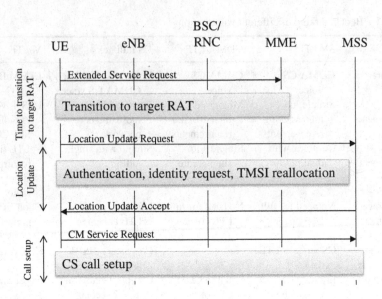

Figure 14.2 Mobile-originating CSFB call signalling where LAC in target RAT is different from LTE LAC, and both LACs served by the same MSS

After the CS call is finished, the UE should return back to LTE. This is possible by either UE-controlled normal inter-RAT cell reselection or redirect/handover based on a vendor-proprietary trigger.

The main CSFB-specific optimization tasks to be discussed in this section are minimization of call setup time and accelerating the return to LTE. From the example in Figure 14.2 it can be seen that the CSFB call setup time is essentially the CS call setup in target RAT plus Location Update time, the time to transfer the UE from LTE to the target RAT and connection setup time in LTE (if UE is idle). On the other hand, call setup failure rate is the same as CS call setup failure rate in the target RAT plus the probability that Extended Service Request (connection setup) or the transition to target RAT fails. Neither the CSFB call setup time nor the CSFB call setup success probability can be better than the corresponding performance in the target RAT. However, the gap can be minimized by proper design choices and optimization.

14.3.2   CSFB Call Setup Time, Transition to Target RAT

RRC Release with Redirect

There are two basic methods for directing the UE to the target system in CSFB call setup: RRC release with redirection and PS handover. These methods can be either blind or measurement based, depending on the vendor implementation.

The blind redirect to the target RAT can be considered the baseline method. It is easy to configure and requires very little planning effort, since only the carrier frequency (or BCCH frequencies in case of 2G) of the target RAT needs to be known. With blind redirect the transition time from LTE to 3G for a non-stationary UE is around 2 seconds, based on field measurements. On the other hand, after initial access to target RAT the UE may cache[1] the target RAT system information, and in subsequent CSFB calls a stationary UE does not have to re-read the whole system information from air interface. Reading pre-stored SIBs from UE memory reduces the time to access a 3G target cell to about half (less than 1.5 seconds typical). This makes the simple CSFB with blind redirect an attractive initial solution in terms of deployment complexity versus call setup time performance.

The time to read system information in target RAT can be removed by providing the required system information in RRC release. If the RRC release is preceded by target RAT measurements then it is possible to provide the exact target cell system information to the UE. This completely eliminates the time to read system information, and in case of 3G the UE is able to access the target RAT in less than a second (around 0.6 seconds typical). On the other hand, the optional measurements and RIM signalling to fetch the system information from target RAT BSC/RNC prior to RRC Release consumes some of the call setup time gain. Blind redirect with system information is also possible, as 3GPP Release 9 specification allows sending system information of up to 32 2G cells and up to 16 3G cells. The system information could even be regularly updated via RIM signalling and cached at the source eNodeB. From a purely call setup time perspective blind redirect with system information is an attractive solution, as it avoids the delay of configuring and performing target RAT measurements. The main risk is if the provided information is misaligned or outdated, in which case the UE is forced to make the call setup as if there were no system information available. Another drawback is the increased signalling bandwidth consumption in source cell, due to large size of system information that is sent in the RRC Release message.

Deferred Measurement Control Reading (DMCR), specified in 3GPP Release 7, is a non-CSFB-specific method to accelerate access to a 3G cell. DMCR allows the UE to skip reading SIB11/11bis/12 system information which may be distributed to several segments, depending on the size of the neighbour list. Field tests indicate that DMCR reduces the SIB acquisition time from 2 seconds to about 1 second. The information whether the 3G cell supports DMCR is broadcast in SIB3, which the UE will have to acquire before deciding if SIB11/12 reading can be skipped. Therefore, SIB3 scheduling has impact on the benefit of DMCR.[2] UE informs the RNC that it has skipped reading the neighbour SIBs in RRC Connection Setup Complete uplink message. As mentioned earlier, the UE may have pre-stored the system information and read it from its cache when later accessing the cell. In this case, the DMCR will not bring any benefit.

From deployment point of view, a simple while efficient initial CSFB solution is to combine basic blind redirect (without system information) with DMCR in 3G. As there are no measurements made in the source LTE cell, redirect can be triggered immediately. In the 3G target cell the UE will either read SIBs from its memory cache or use DMCR, depending on the availability of the target cell information in the UE-internal SIB database. As DMCR is UE-based feature, there is very little planning and configuration to do in the 3G side, while in the LTE side there is no need to plan and optimize neighbour lists.

---

[1] Caching rules and implementation is not specified and is UE dependent.

[2] Obviously, the 3G SIB scheduling should be optimized by default even without DMCR and CSFB as it has major impact on cell access time and UE battery consumption [4].

CSFB with PS Handover

CSFB with PS handover can be made with or without target RAT measurements. The measurement quantity is either received level (RSCP for 3G and RSSI for 2G) or signal quality (Ec/N0 for 3G). The measurement-based handover is especially beneficial in interference prone areas such as high-rise buildings because the UE can be handed over to the least interfered layer. Therefore, one important use case for PS handover is the capability to use Ec/N0 measurements to select the target layer.

CSFB PS handover may also improve the call setup time relative to redirect. If the handover is preceded by measurements however, part of this speed-up is consumed by measurement configuration in source LTE cell, target RAT measurements and handover preparation. For 3G target RAT, these components amount roughly to the following:

- Time to first measurement report from Extended Service Request depends on the length of the 3G neighbour list and time to trigger. The range is from 0.1 seconds to more than 1 second.
- Handover preparation to target RAT is typically about 0.3 seconds.
- From Handover Command in LTE to Handover Complete in target 3G cell is about 0.3 seconds.

The total time to access 3G is typically around 1 second; obviously this assumes optimized neighbour list and target RAT with sufficient radio quality. This is faster than blind redirect with DMCR or redirect with pre-stored SIBs, but slower than blind redirect with system information. On the other hand, the resources in target system have already been reserved during preparation phase and UE can send the initial NAS message very quickly after the handover is completed. With 13.6 kbps signalling bearer this saves another 0.3 seconds relative to redirect which has to establish signalling connection prior to sending first NAS message. Another less obvious benefit from CSFB with PS handover is that the system is able to collect handover counters at LTE-3G cell pair level which provides some visibility to the CSFB call setup performance at cell level.

Call setup time with PS handover can be optimized by shortening the time to trigger to first measurement report and by optimization of target RAT neighbour lists. The latter, while usually amounting to tedious optimization, is important as missing neighbours will result in delayed or absent measurement report. Thus, non-optimized neighbour lists may result in degraded call setup time. Some guard timer should be implemented in the source LTE cell to resort to blind handover or redirect in case of missing measurement report.

Comparison of Call Setup Times with Different Methods

Figures 14.3 and 14.4 summarize the transition time to target RAT with different methods. With redirect, most of the time is spent on system information acquisition in the target system. With system information provided in RRC release this time can be efficiently minimized; less than 1 second delay from Extended Service Request to initial NAS message is possible for both 2G and 3G. For CSFB to 2G the redirect with system information provides large savings in call setup time which would otherwise be more than 2 seconds longer. On the other hand, if the provided system information is misaligned, the benefit is lost. Depending on the

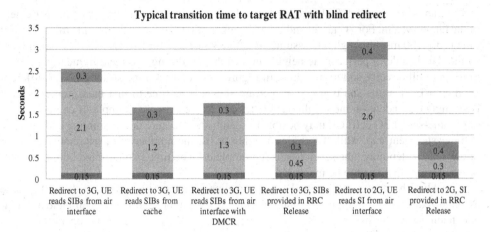

Figure 14.3 Typical transition time from LTE to target RAT with redirect without target RAT measurements (ESR = Extended Service Request)

implementation, there may also be additional delay if up-to-date system information needs to be retrieved over RIM signalling from BSC or RNC in the target system.

The transition from LTE to 2G can also be either blind or preceded by measurements. The 2G measurements require both RSSI measurement and BSIC decoding and therefore are slow compared to 3G measurements; with several BCCH frequencies in the measurement list the typical time from measurement command to first measurement report is between 1 and 2 seconds, which can be considered too long for the CSFB call setup.

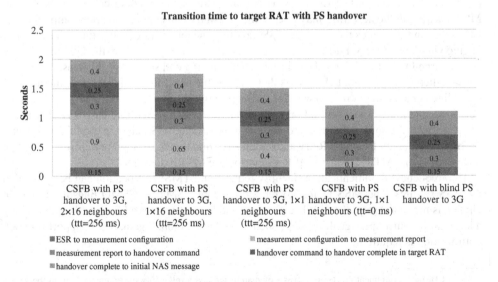

Figure 14.4 Transition time from LTE to 3G with PS handover. The notation 2x16 means that neighbour list has a total of 32 neighbours on two 3G frequencies, 16 neighbours per frequency

The transition time with PS handover depends on the time to obtain first valid measurement of the target system. For 3G target and time to trigger of 0 ms the best achievable measurement time is 50–100 ms, although this assumes that the 3G neighbour list has only few neighbours so that UE is able to perform the neighbour search step during one measurement gap. Time to trigger will directly add to this time. In Figure 14.4 measured transition time for different neighbour list lengths and times to trigger are shown. The time to first measurement report is strongly dependent on the neighbour list length. In the extreme example of one neighbour (e.g. same-sector 3G cell) UE may be able to search and measure the scrambling code in only one measurement gap, while for a long neighbour list the number of measurement gaps for initial search step may take several gaps to complete.[3] The time is multiplied by the number of frequencies that need to be measured.

With blind handover, one second transition time is achievable which is about the same as blind redirect with system information. Obviously vendor- and network implementation-dependent variations in either direction are possible, especially in handover preparation which could be faster than the 0.3 seconds shown in the figure.

Location Update
If the target RAT Location Area Code differs from the one UE obtained in LTE, the call setup begins with a Location Update procedure. The time to complete the procedure depends on core network configuration. Namely, the delay is increased in the following cases:

- if the new LA is served by a different MSS and the MSS needs to be changed
- if authentication, identity request or TMSI reallocation is made as part of the Location Update

Changing MSS is especially problematic for mobile-terminating call since the new MSS (to which the UE sends the LU request) is not aware of the ongoing call setup in the old MSS. As a result the new MSS may release the UE signalling connection after completing the LU, which in turn forces the gateway MSS handling the terminating call setup to initiate new paging via the new MSS. Furthermore, if the new paging is sent too late, the UE may already have returned back to LTE, resulting in failed terminating call setup. For this reason, UE sets the so-called CSMT flag in LU Request that informs the new MSS to delay releasing the NAS signalling connection in order to wait for the incoming call setup from the gateway MSS. The new MSS and the UE have to support the CSMT flag which is 3GPP release 9 NAS feature. There are two variants in the standard on how to handle the MSS change in MTC call setup: Mobile Terminating Roaming Retry and Mobile Terminating Roaming Forwarding. The main difference is that MTRR requires HLR support while MTRF does not.

The frequency of authentication, identity request and TMSI reallocation can be controlled by MSS settings. The MSS is able to recognize from CSMT or CSMO flag that Location Update is the initial step of a CSFB call setup, and it can accept the LU directly without any additional intermediate signalling steps such as authentication. Again, MSS support for such optimization is required.

---

[3] The WCDMA measurement procedure consists of search step and level/quality measurement step. In the search step the UE makes code acquisition for the strongest cells of the provided neighbour list. This is followed by the actual measurement level/quality measurement for the strongest cells found in the search step.

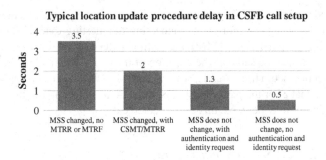

Figure 14.5 Location Update delay in CSFB to 3G call setup

From radio access point it is advisable to use fast signalling channel to minimize message transfer delay in the air interface. For 2G this means early allocation of FACH for signalling. For 3G signalling bearers are recommended to be mapped to HSPA or DCH 13.6 kbps.

Depending on if MSS change is required, authentication configuration, MSS support of CSMx flag, the Location Update signalling may take from around a couple of hundred milliseconds to 4 seconds or even longer (e.g. MTC where MSS changes, CSMT flag not supported and re-paging is required).

The most efficient way to reduce the time spent on call setup signalling is to align Location Area Codes in LTE and the target RAT. In this case the UE will not need to make LU and will directly proceed to service request or paging response. A common encountered concern is that skipping LU in the beginning of the call setup might result in paging degradation for subsequent terminating calls. However, [5] requires that CSFB-capable UE will make the LU after the call has been released even when the LACs are the same.[4]

Figure 14.5 gives some idea of the Location Update delay for different CS core network configurations. The values should be considered typical ones rather than definitive; both radio and core network parameter configuration as well as radio conditions will cause variations to the procedure duration. The values shown assume 3G and 13.6 kbps signalling radio bearer. For 2G access, the delay will be larger.

It is worth noting that by appropriate CS core network configuration it is possible to make considerable savings in CSFB call setup time. Specifically, coordinated Location Area planning for LTE and the target system is especially important. Further savings are possible by reducing the frequency of authentication and IMEI checking (identity request) procedures.

Call Setup Signalling
The third phase of the CSFB call setup is the call setup signalling. The duration of this phase is basically the same for ordinary CS call in the target system. Optimization is possible by using fast signalling connection. Also if the UE falls back to the same MSS, reducing the frequency of authentication and IMEI checking will reduce the setup duration by around half a second.[5]

---

[4] In normal inter-RAT reselection from LTE to another RAT a CSFB-capable UE will make LU even if the LACs are the same. The 3GPP 24.008 option to postpone LU after the release of the CS call is to optimize the CSFB call setup time.

[5] Authentication and identity request may have been done already earlier during Location Update.

As mentioned in [2], for mobile-terminating call, the call setup time perceived by the calling party can be made to appear shorter by connecting the alerting tone as soon as the UE has responded to CS paging in LTE and MME has acknowledged the paging to MSS. The perceived time savings are almost the same as the duration of the end-to-end call setup less LTE paging and RRC connection setup time, that is, several seconds. MSS needs to support the capability. The drawback is that after the paging response in LTE the call setup might fail at a later stage and the calling party may perceive this as a rejected call.

14.3.3 CSFB Call Setup Success Rate

Average CSFB call setup success rate in case of redirect cannot be better than the average call setup rate in the target radio access. Therefore, increasing the CSFB call setup success rate boils down to improving the connection setup success rate in LTE and the target RAT; this can typically be improved only with basic physical RF and parameter optimization and possibly load balancing in the target RAT. In the case of CSFB PS handover it is the handover success rate that typically dominates the CSFB call setup success rate. Cell-level optimization of neighbour lists and target carrier (e.g. non-interfered target carrier) plays a significant role. It should be noted, however, that even if the PS HO fails, the call setup will be resumed by the UE at the expense of increased call setup delay and hence handover success rate is not equal to the call setup success rate.

A specific use case is where the target layer has bad radio conditions, due to lack of coverage or interference. A typical example is a high-rise building without indoor cell where a blind redirect to a polluted 3G layer can result in call drop, degraded voice quality and possibly a follow-up handover to a cleaner 3G layer or 2G. For 3G target, utilizing Ec/N0 measurement quantity and fine-tuned neighbour list with two or more target layers may help in this case. Besides cumbersome neighbour optimization, the obvious drawback is the increased call setup time due to UE measurements in LTE before handover can be triggered.

14.3.4 Return to LTE after CSFB Call

After the CS call finishes, it is beneficial for the LTE-capable UE to return back to LTE, provided that sufficient coverage and quality are available. This can be accomplished via idle mode inter-RAT cell reselection, radio connection release with redirect or handover. The latter two can be either blind or based on measurements.

Throughout the section it is assumed that LTE has higher priority than the 2G or 3G RAT where the CS call is released.

Inter-RAT Cell Reselection from 3G to LTE

UE inter-RAT reselection measurement scheduling from 3G to a higher priority LTE layer depends on the DRX cycle and priority search thresholds Spriontysearch1 and Spriority-search2 which are broadcast in SIB19. In line with the design of the priority-based reselection, if the UE is considered to be in good RF conditions – 'good' defined by the priority search parameters – higher priority layer cell search can be performed at the interval of 60 seconds times the number of higher priority layers broadcast on SIB19. Otherwise the search

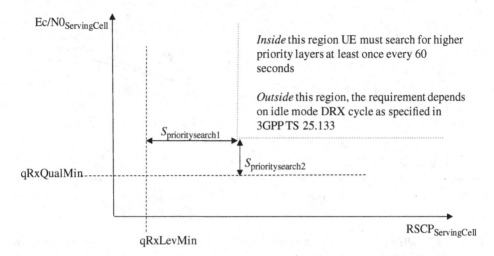

Figure 14.6 Dependence of 3G to LTE reselection measurements on priority search parameters

requirements are considerably stricter. This principle is illustrated in Figure 14.6 from where it is seen that if either Sprioritysearch1 or Sprioritysearch2, or both, are set to a high value the UE will search for LTE cells more frequently. The search requirement is per layer, so for two LTE layers the search requirement in good radio conditions would be $2 \times 60 = 120$ seconds.

When the received RSCP level and quality are outside the region shown in Figure 14.6, the UE cell search and measurement minimum requirements are given in Table 14.2. The DRX cycle length given in the table may be different in CELL_PCH and idle mode. The detection time in the second column is defined as the maximum time from cell search to cell reselection decision, that is, it includes cell measurements. The requirements apply only to LTE cells having RS and SCH subcarrier power higher than -124 to -121 dBm (depending on band) and subcarrier SNR of >-4 dB. The values are for one layer, and for multiple layers the requirements are multiplied by the number of layers.

Table 14.2 Minimum requirements for LTE measurements when UE is in idle mode or CELL_PCH in 3G

DRX cycle length (seconds)	$T_{detectE\text{-}UTRA}$ (seconds)	$T_{measureE\text{-}UTRA}$ (seconds) (number of DRX cycles)	$T_{evaluateEUTRA}$ (seconds) (number of DRX cycles)
0.08	30	2.56 (32)	7.68 (96)
0.16		2.56 (16)	7.68 (48)
0.32		5.12 (16)	15.36 (48)
0.64		5.12 (8)	15.36 (24)
1.28		6.4 (5)	19.2 (15)
2.56	60	7.68 (3)	23.04 (9)
5.12		10.24 (2)	30.72 (6)

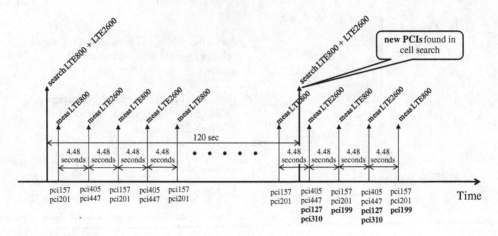

Figure 14.7 Idle mode and CELL_PCH search and measurement schedule for a commercial UE when accelerated high-priority search is not used. Two LTE layers broadcast in SIB19. New PCIs found in the second search step are shown in bold

The third column in the table is the minimum measurement interval of cells that have been found in the search step. The last column is the maximum cell measurement time before the cell reselection decision must be made if a suitable candidate exists. In other words it can be interpreted as the maximum allowed averaging time of measurements.

Figure 14.7 shows an example of a LTE search and measurement schedule of a commercial UE with two higher priority LTE layers (LTE800, LTE2600) when the UE is camping in one 3G cell without cell reselection triggered. In this case, Sprioritysearch1 = 0 and Sprioritysearch2 = 0 and consequently the search interval is 2 × 60 = 120 seconds; the UE under test searches both layers at one go once every 120 seconds which satisfies the 3GPP requirement. With three higher priority layers the corresponding search interval would be 180 seconds. The cell measurements are repeated every seven DRX cycles, alternating between LTE800 and LTE2600. In the example the DRX cycle is 0.64 seconds and hence the UE performs better than the 3GPP requirement of eight DRX cycles given in Table 14.2.

In the first search step of Figure 14.7 the UE detects PCIs 157 and 201 on LTE800 and PCIs 405 and 447 on LTE2600. These PCIs are then measured until the next search interval after 120 seconds, at which time two new PCIs (127, 310) are detected on LTE2600 and one new PCI (199) is detected on LTE800.

Figure 14.8 shows the measurement schedule of the same UE when Sprioritysearch1 = 62 dB, that is, the UE is using the accelerated measurement schedule of higher priority layers. In this case the cell measurement interval is the same as in the case with Sprioritysearch1 = 0 dB, but the search interval is reduced from 120 to 14.72 seconds (23 DRX cycles). This can considerably speed up the reselection from 3G to LTE.

Inter-RAT Cell Reselection from 2G to LTE

UE inter-RAT reselection measurement scheduling from 2G to a higher priority LTE layer is very loosely defined by 3GPP. In 2G, the priority-based cell reselection measurement is

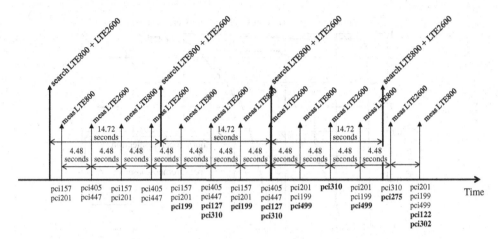

Figure 14.8 Idle mode and CELL_PCH search and measurement schedule for a commercial UE when accelerated high-priority search is used (Spriorirysearch1 = 62 dB). Two LTE layers broadcast in SIB19. PCIs detected in the search steps are shown in bold

not activated until the UE has received the priority-based reselection information over BCCH SI2quarter system information. This information may have to be split into multiple system information segments and hence there may be delay by several seconds until all segments are received by the UE and reselection to LTE is activated. Before the activation, pre-Rel8 cell ranking-based reselection is used. This often results in UE reselecting first from 2G to 3G instead of moving directly to LTE. Approaches to accelerate delivery of SI2quarter segments to UE are to use minimal number of LTE and 3G frequencies in SI2quarter (fitting to one segment) and deploy the so-called extended BCCH which effectively increases the scheduling rate of SI2quarter.

After priority-based cell reselection has been activated, the time to reselect to a cell in a higher priority LTE or UTRAN layer must be shorter than $70 + T_{reselection}$ seconds, where the minimum value of $T_{reselection}$ is 5 seconds. Every additional LTE or UTRAN higher priority layer adds another 70 seconds to the requirement. Therefore, the time requirements for 2G to LTE reselection are very loose. In practice, however, UEs tend to reselect to LTE faster than the 3GPP requirement after the priority-based reselection has been activated (all SI2quarter segments with received).

Inter-RAT Redirect and PS HO from 2G/3G to LTE

Similarly as from LTE to 3G, it is possible to use RRC release with redirect to quickly move UE from 3G to LTE or from 2G to LTE. In the radio connection release message the target RAT carrier frequency information is included. Depending on RNC and BSC implementation, LTE measurements, if supported by the UE, may be performed prior to RRC release. In normal radio conditions, the time from radio connection release in 3G to send Tracking Area Update in LTE is at most a few hundred milliseconds.

Also handover from 2G and 3G to LTE is possible, if supported by the system and UE. The triggering of the measurements and the handover are based on UE capability and

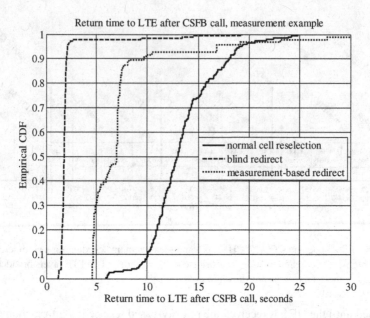

Figure 14.9 Example drive test measurement of return time from 3G to LTE after CSFB call finishes, with cell reselection

vendor-proprietary triggers, for example, serving cell received level and data buffer fill factor. In this case the return time depends on the time to measure LTE target cell.

Comparison of Return Time from 3G to LTE after a CSFB Call

Figure 14.9 shows an example of the measured empirical distribution of the return time from 3G to LTE, via cell reselection as well as blind and measurement-based redirect, obtained from drive test measurements. The return time is defined as time difference of disconnect message received from MSS in 3G to Tracking Area Update Complete message sent by UE in LTE. For the redirect methods the time to return depends heavily on the type of triggers and measurement thresholds implemented in RNC. Therefore, the results should be considered examples only.

For basic idle mode reselection in 90% of cases the UE returns to LTE in less than 18 seconds. Since the reselection is possible only in idle and CELL_PCH mode, UE data activity as well as DCH-to-FACH and FACH-to-PCH inactivity timers will have impact on the results. In the measured network slow high-priority search measurement schedule was used (Spriori-tysearch1 = 0 and Spriorisearch2 = 0).

For blind redirect back to LTE, the transition is the fastest method to return back to LTE, assuming that there is good enough LTE coverage. The average time is about 2 seconds, but it can be seen that in a few percent of cases the redirect is not triggered quickly, or the UE has problems accessing LTE or completing the TAU procedure.

In the measurement-based redirect the UE returns back to LTE in less than 7 seconds in half of the cases. In about 10% of the cases the return takes more than 10 seconds. Measurement

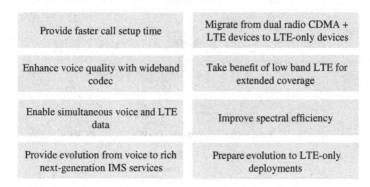

Provide faster call setup time	Migrate from dual radio CDMA + LTE devices to LTE-only devices
Enhance voice quality with wideband codec	Take benefit of low band LTE for extended coverage
Enable simultaneous voice and LTE data	Improve spectral efficiency
Provide evolution from voice to rich next-generation IMS services	Prepare evolution to LTE-only deployments

Figure 14.10   Motivations for VoLTE deployments

configuration and thresholds to trigger the redirect play a role in the return time. The criteria for triggering blind and measurement-based redirect are RNC vendor dependent and not specified by 3GPP.

14.4   Voice over LTE

There are a number of motivations for migrating from CS voice to VoLTE by the mobile operators. The main motivations are shown in Figure 14.10. VoLTE can enhance voice quality with wideband codec. VoLTE also provides faster call setup time and offers simultaneous LTE data together with the voice call. The data capability allows to introduce new IMS services. VoLTE can also make the device architecture simpler for CDMA operators when the dual radio implementation can be avoided. VoLTE also improves the spectral efficiency compared to CS voice. VoLTE can also enhance network coverage if LTE can use lower frequency with more favourable propagation.

Migration from CSFB to IMS-based VoIP solution, that is, VoLTE, is typically done in several steps starting from CSFB only, then combining of CSFB and VoLTE and finally full VoLTE where all voice calls (inter-operator, inter-network and roaming) are fully utilizing VoLTE. SRVCC enables handover from IMS-based VoIP to CS speech when the UE is running out of LTE coverage. Also reverse SRVCC (SRVCC based on release 11) enables handover from CS speech back to IMS-based VoIP. Simplified VoLTE system architecture based on [3] is presented in Figure 14.11.

14.4.1   Setup Success Rate and Drop Rate

The voice call performance in 2G and 3G networks has been extremely good and therefore the expectation for VoLTE performance in LTE networks is set very high. The share of VoLTE calls given as ratio of quality of service class identifier (QCI) one E-RAB setup attempts to all QCI E-RAB setup attempts, in Figure 14.12, shows that the amount of VoLTE calls is still very small for three example networks depending on the VoLTE deployment strategy. The number of VoLTE calls generated per single UE per day is just tens whereas the number of non-GBR E-RABs is several hundreds and therefore the VoLTE share from all E-RABs will be small even in case each UE would generate some VoLTE calls per day.

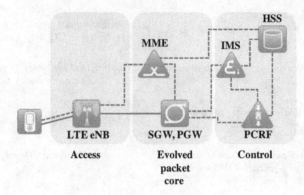

Figure 14.11   VoLTE system architecture

The share of QCI1 E-RAB setup attempts from all E-RAB setup attempts, containing QCI1 and non-guaranteed bitrate QCIs, is below 1% but rapidly increasing in case of example networks 1 and 3 which means quick transformation from CSFB to VoLTE. For example network 2 the main voice call method is still CSFB utilizing the existing 3G network as much as possible.

Setup Success Rate

The QCI1 E-RAB setup success rate shown in Figure 14.13 is the radio network statistics measurement of VoLTE setup success rate. It covers the E-RAB setup portion of the whole VoLTE setup signalling shown in Figure 14.14. The QCI5 E-RAB carries all the session initiation protocol (SIP) signalling and the QCI1 E-RAB then carries the actual voice user plane data. More details on the SIP signalling for VoLTE call setup can be found from [3]. The QCI5 E-RAB is setup at the time the terminal is attached to the network and SIP signalling registration to IMS is then done using QCI5 E-RAB.

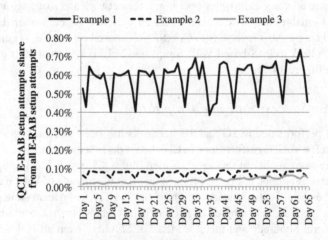

Figure 14.12   QCI1 E-RAB setup attempts share from all E-RAB setup attempts

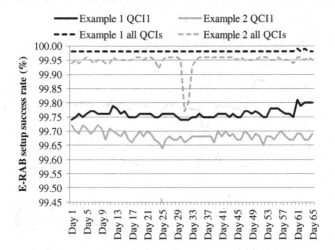

Figure 14.13   E-RAB setup success rate for QCI1 and all QCIs

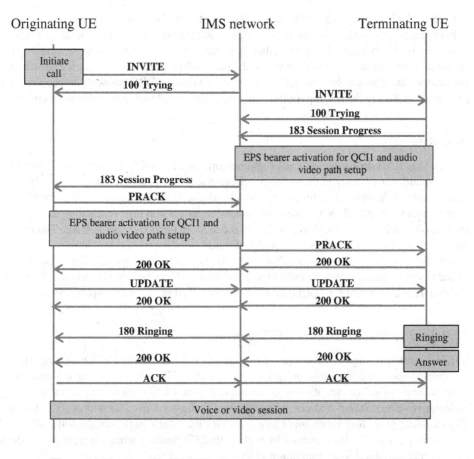

Figure 14.14   Simplified VoLTE call setup signalling. Based on GSMA IR.92

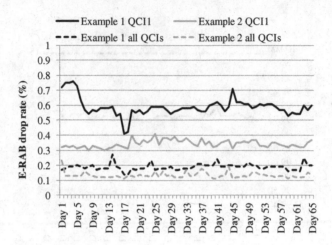

Figure 14.15   E-RAB drop rate for QCI1 and all QCIs

Figure 14.13 shows the E-RAB setup success rate for QCI1 and for all QCIs. The QCI1 E-RAB setup success rate is slightly worse than E-RAB setup success rate for all QCIs and this difference is mostly caused by the fact that there are so much less QCI1 setups compared to all QCIs that the weight of setups attempted under poor radio conditions is increased and therefore the success rate is worse. However, it should be noted that the RRC connection reconfiguration process (for QCI1 E-RAB setup) optimization was not completed in this example network yet.

Drop Rate

The QCI1 E-RAB drop rate is slightly higher compared to all QCIs as seen from Figure 14.15.

As explained in Chapter 12 the mobility performance in terms of late handovers and handover success rate in general has a high correlation with the E-RAB drop rate. This correlation is even stronger for QCI1 E-RAB due to the much longer session time (excluding any inactivity time) as shown in Figure 14.16 and therefore much more handovers per E-RAB. The more the handovers per E-RAB the higher the probability for mobility-related problems and mute calls (due to RRC re-establishment) and dropped calls. Therefore, the optimization actions related to the mobility performance (handover success rate and too late handovers) as discussed in Chapter 12 are essential for VoLTE end-user experienced performance optimization.

14.4.2   TTI Bundling and RLC Segmentation

VoLTE uplink coverage can be improved by using the feature called TTI bundling. The idea is to repeat the same transmission in four consecutive TTIs which enables to transmit four times more energy. The operation is illustrated in Figure 14.17. The voice packets from AMR codec arrive every 20 ms and can be transmitted in single 1-ms TTI without TTI bundling. TTI bundling gives four times more transmission time which improves the link budget. The average power over 20 ms increases by 6 dB with TTI bundling when the peak power during the TTI transmission hits its maximum value.

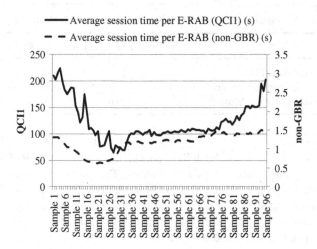

Figure 14.16 Average session duration per E-RAB for QCI1 and for non-GBR QCIs based on 15-minute samples over 24-hour period

The simulated benefit of TTI bundling in the uplink base station sensitivity is shown in Table 14.3. The gain of TTI bundling is approximately 4 dB. The simulation takes into account the maximum delay budget for VoLTE which is assumed to be 50 ms for one radio leg. Maximum six retransmissions are allowed without TTI bundling and three retransmissions with TTI bundling.

The measured benefit of TTI bundling is shown in Figure 14.18. The measurement is done in weak signal area where RSRP is below –120 dBm and UE transmits with full power. The uplink BLER without TTI bundling is very high above 70% which leads to bad voice quality. The uplink BLER drops to below 10% with TTI bundling providing a major improvement in the voice quality.

Another solution to improve the uplink coverage is segmentation of voice packets into smaller units. Figure 14.19 illustrates the segmentation of Radio Link Control (RLC) Service Data Unit (SDU) into four smaller Protocol Data Units (PDUs). Each smaller PDU is sent separately over the air interface. The packet segmentation enables to increase the uplink transmission time to boost the coverage. The concept is shown in Figure 14.20. The drawback of segmentation is the increase in the overhead. The voice packet of AMR 12.65 kbps including

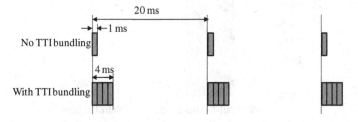

Figure 14.17 TTI bundling operation

Table 14.3 Simulated base station sensitivity for VoLTE with and without TTI bundling

	Without TTI bundling	With TTI bundling
Number of bundled TTIs	1	4
Max number of retransmissions	6	3
Base station sensitivity (dBm)	−120.6	−124.4

compressed IP header is 35 bytes. Each RLC packet needs additionally RLC header of 3 bytes and Cyclic Redundancy Check (CRC) of 3 bytes. The overhead without segmentation is (3+3)/35 = 17%. The overhead increases to (3+3)/(35/4) = 69% which will eat large part of the benefit from segmentation. Therefore, TTI bundling should preferably be used to improve the uplink coverage rather than packet segmentation and simultaneous use of TTI bundling and packet segmentation should be avoided.

14.4.3 Semi-persistent Scheduling

VoLTE packets can be transmitted with dynamic scheduling where every packet is scheduled separately. The dynamic scheduling has the benefit that it allows smart scheduling solutions in the frequency domain and it allows fast link adaptation based on the CQI feedback from UE. The challenge with dynamic scheduling is that PDCCH signalling capacity is required for the allocation of every voice packet arriving with 20-ms period. An alternative solution is semi-persistent scheduling (SPS) where eNodeB pre-assigns resources for each VoLTE connection so that no PDCCH signalling is needed. SPS is practical for VoLTE traffic since eNodeB knows that VoLTE packets will arrive every 20 ms. The limitation with SPS is that the resources in the frequency domain, the modulation and the coding schemes are fixed and cannot be modified based on CQI reporting. These two scheduling principles are shown in Figure 14.21.

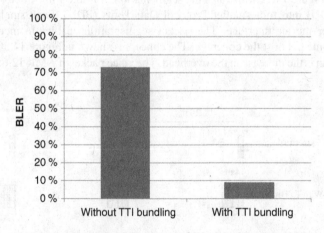

Figure 14.18 Uplink BLER in weak signal area with and without TTI bundling

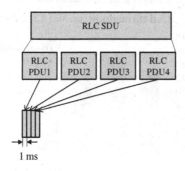

Figure 14.19 Packet segmentation principle

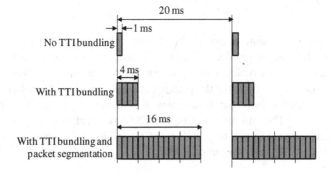

Figure 14.20 TTI bundling and packet segmentation operation

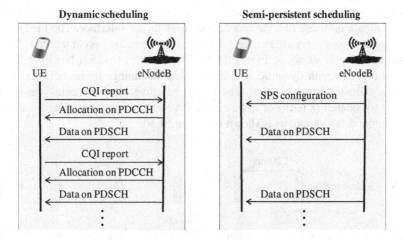

Figure 14.21 Dynamic and semi-persistent scheduling

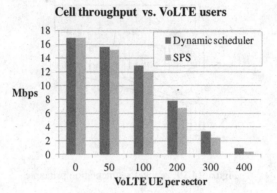

Figure 14.22 Cell throughput as a function of simultaneous VoLTE users with different scheduler solutions

It has been shown in simulations that SPS can enhance the VoLTE capacity by +5% compared to the dynamic scheduler in voice-only cases. A more relevant use case would be mixed voice and data since a typical LTE cell carries a lot of data on top of VoLTE traffic. Figure 14.22 shows the cell throughput for the packet data as a function of number of simultaneous VoLTE users. The results show that the dynamic scheduler can provide higher efficiency than SPS in the mixed case. The difference can be even 20% more cell throughput. We can conclude that the dynamic scheduling solution with fast CQI feedback is typically the most efficient solution in practical LTE deployment scenarios.

14.4.4 Packet Bundling

The PDCCH signalling overhead and Layer 1 and 2 overhead can be minimized by using packet bundling for VoLTE. The idea is to wait until there are two voice packets in the transmission buffer and then send those two packets at once over the air interface. The benefit is that the transmission becomes more efficient and the allocation signalling is cut into half. The concept of packet bundling is shown in Figure 14.23. The benefit of packet bundling is that it can be utilized together with dynamic scheduling while obtaining similar benefits as with SPS in terms of signalling minimization. The packet bundling increases delay variations which need to be considered in terms of number of allowed retransmissions. The impact of packet bundling to the delay variations is shown in Figure 14.30.

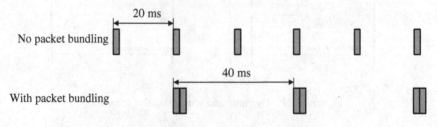

Figure 14.23 Packet bundling concept

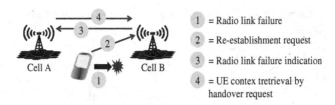

Figure 14.24   Re-establishment procedure

14.4.5   Re-establishment with Radio Preparations

The radio link may be dropped in some rare cases when UE is moving from one cell to another and the radio conditions change very rapidly. The connection can be recovered with the re-establishment procedure which is illustrated in Figure 14.24. UE moves from Cell A towards Cell B in this example but the handover fails and the radio link drops during the mobility. UE then sends Re-establishment request to Cell B after the radio link time-out has been identified. Cell B does not have any earlier information about the UE. Therefore, Cell B sends Radio link failure indication to Cell A to fetch the UE context from Cell A. Cell B can then re-establish the connection and the radio link failure does not lead to call drop. There is a short break of a few seconds in the user plane while the radio link failure timer runs and the connection is re-established. The break duration depends on the timer values. The major benefit of re-establishment is that the VoLTE call can be maintained and an annoying call drop can be avoided. This feature is also called as mobility robustness in 3GPP and it was included in Release 9.

14.4.6   Voice Quality on VoLTE

The voice quality depends heavily on the voice codec sampling rate and the resulting audio bandwidth. AMR Narrowband (NB) codec provides audio bandwidth of 80–3700 Hz while AMR Wideband (WB) extends the audio bandwidth to 50–7000 Hz. The CS connections can use either AMR-NB or AMR-WB while VoLTE in practice always uses AMR-WB. Also many OTT VoIP applications use similar wideband audio.

The voice quality using the legacy CS switched voice services was mostly impacted by layers 1–3 in radio interface while the voice quality of VoLTE depends on robustness of all layers. LTE provides dedicated bearers for voice with negotiated QoS settings which allow separate prioritization of VoLTE packets based on packet loss and delay targets to help various network elements in resource allocation decisions.

The current generation for MOS (mean opinion score) voice quality measurement methods is POLQA (perceptual objective listening quality assessment) which was adopted as ITU-T P863 [6]. It is particularly developed for HD Voice and is a successor of P862 PESQ which is not suitable for higher bandwidth voice codecs such as G722.2, which is also called WB-AMR. The POLQA is formed by OPTICOM, SwissQual and TNO, the group of companies who won the competitive standardization process run by the ITU-T, and KPN, who owns patents on the TNO technology [7].

Measuring MOS can be subjective to implementation of VoLTE in UE. Differences, such as the size of Jitter Buffer in the UE can lead to different MOS figures with same jitter profile

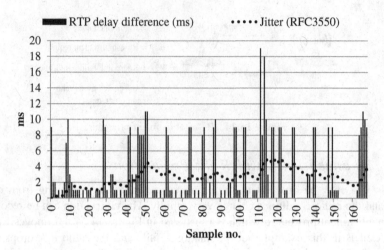

Figure 14.25 Jitter variation as a function of measured delay differences between consecutive RTP packets in relatively constant conditions; used AMR codec bitrate = 23.85 kbps

when measured using different UEs. This can make the comparison between networks difficult if not impossible as the results may not be comparable.

The factors mostly impacting the quality of the voice are the selection of the codec, used bandwidth, packet loss ratio, jitter and excessive delay of an individual RTP packet as well as the implementation of VoLTE in the UE. As VoLTE voice traffic is usually delivered using UM on RLC layer, only the optimization of the scheduler, delay budget algorithms and ensuring reliable delivery of the packets can provide a difference to the quality of the voice from the LTE network point of view.

Delay

The measure often used to indicate delay variations profile is jitter and it is specified by IETF in RFC3550 [8]

$$J(i)(i-1) + \frac{|D(i-1,i)| - J(i-1)}{16}$$

This works very well to indicate delay variations when they are maintained in relatively small range, example of which is shown in Figure 14.25. The picture shows the delay difference from the expected 20 ms inter-packet delay – ideal case here equals to 0 ms. The spikes at around 8 ms mark are caused by a single HARQ retransmission while ~16 ms cases are most likely caused by two consecutive HARQ retransmissions.

Typical drive test for MOS analysis contains series of, for example, 6-second-long audio samples where each sample gets analysed by the POLQA MOS algorithm. Analysed results are stored together with other measured statistical and radio parameters for further analysis.

During drive test data collection for MOS analysis, jitter is one of the commonly collected measurement values. As jitter averages the delay, it may not indicate long individual delays

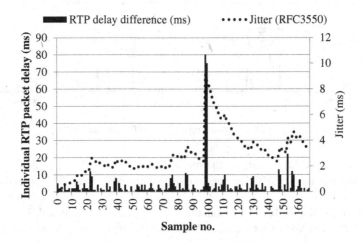

Figure 14.26 Jitter variation as a function of measured delay differences between RTP packets where one individual packet arrives 110 ms late; used AMR codec bitrate = 23.85 kbps

well. In the example shown in Figure 14.26, measurement collection was done over 170 RTP packets, jitter provided average of 2.9 ms while highest jitter value over the period is about 9 ms even though this measurement period contains a single peak delay value of 80 ms. POLQA MOS value in this example case was relatively poor with a value of 2.5. No RTP packets were lost indicating that even just one RTP packet having relatively long delay may have high impact on the overall MOS figure.

When the impact of peak delay within a MOS sample was analysed using very large number of 6-second-long MOS samples in varying radio conditions, a correlation between the peak delay and MOS value became evident as can be seen in Figure 14.27. The larger the RTP

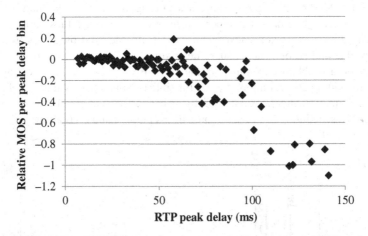

Figure 14.27 MOS versus peak RTP delay during repeated 6 seconds MOS sample. Average MOS per peak delay is calculated and compared to the average MOS within 20 ms RTP peak delay. Jitter buffer size in UE is unknown; used AMR codec bitrate = 23.85 kbps

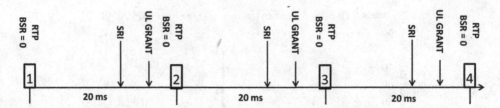

Figure 14.28   Twenty-millisecond RTP scheduling by UE

peak delay within the MOS sample, the more the average MOS per peak RTP delay value decreases from the average MOS (calculated as average MOS with RTP peak delay between 0 and 20 ms) value.

It is to be noted that different UE manufacturers have implemented different jitter buffer sizes in the UEs. The size of the buffer in UE will have impact on how well the UE is able to cope with increased jitter or delay of individual packets and therefore different UEs can have different MOS performance.

Packet Aggregation

The UE would typically attempt to schedule a transmission of RTP packets every 20 ms. A scheduling request would arrive to the eNB every 20-ms period as shown in Figure 14.28. eNB sends a Grant and UE will transmit the RTP packet with headers on PUSCH.

However, instead of transmitting one packet every 20-ms period the UE may aggregate the packets on higher layers and schedule aggregated transmission of two RTP packets simultaneously every 40 ms as shown in Figure 14.29. The eNodeB may then allow 40 ms scheduling or force 20 ms scheduling. Also it is possible that the eNB ignores the SRI and schedules the packets only every 40-ms period – that way eNB can ultimately control whether RTP packet aggregation is used or not.

As a result of RTP packet aggregation, shown in Figure 14.29, packet number 1 is already delayed by 20 ms, while packet number 2 arrives on time, but 20 ms early in comparison to previously delivered packet. Packet number 3 is also delayed by 20 ms from previously delivered packet and so on. The standard deviation of delays increases to 20 ms. That may not directly be a problem depending on the Jitter Buffer size in the receiving UE, but is unnecessarily adding 20 ms on top of other possible reasons for why individual packet is not delivered in a timely manner. Example of this behaviour is shown in Figure 14.30 where the UE changes the scheduling between 20- and 40-ms periods and eNB allows it.

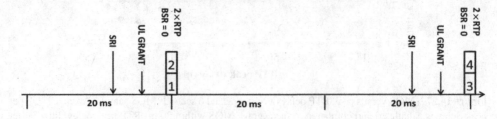

Figure 14.29   Forty-millisecond RTP scheduling by UE

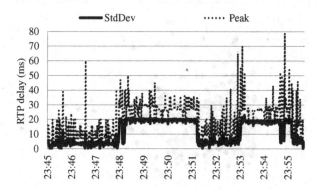

Figure 14.30 RTP delay variations caused by UE aggregating RTP packets over 10-minute period. Step changes are caused by UE switching between 20 and 40 ms scheduling; used AMR codec bitrate = 23.85 kbps

Also eNodeB may decide to aggregate RTP packets in downlink direction resulting in similar impact. While there is certainly a gain in reduced PDCCH usage, it adds to the delay variations and jitter.

As there seem to be many variables resulting in variations to MOS figures, such as jitter buffer size in UE, use of sound equalizers to compensate the quality of the tiny speaker in UE and even volume setting in some cases, it is impossible to compare networks by using absolute MOS values. However, it is evident that controlling the delay is an important factor to maintaining good voice quality.

As smaller peak delays (less than 60 ms) in this particular case, shown in Figure 14.30, seemed to have little or no impact on POLQA MOS, it is not surprising that HARQ retransmissions correlate poorly with MOS. The impact of HARQ BLER variations was verified in live network with moderate non-GBR loading and relatively low GBR loading to have no correlation with average POLQA MOS as shown in Figures 14.31 and 14.32.

Packet Loss

During the testing it was also found that RTP Packet Loss Ratio seemed to have no or very small impact on POLQA MOS as is shown in Figure 14.33.

MOS is impacted by both packet loss and delay variations, but it seems to be a bit unclear whether dropping a packet after certain delay is a better option than putting a lot of emphasis delivering it too late.

Coverage Impact on Voice Quality

The previously shown measurement results in this chapter concentrated on voice quality within reasonable coverage where UL BLER can be maintained at low level. When the pathloss increases there comes a point where BLER cannot be maintained any longer. Therefore, the increased number of retransmissions leads to increased delays and jitter and the probability of dropping RTP packets increase.

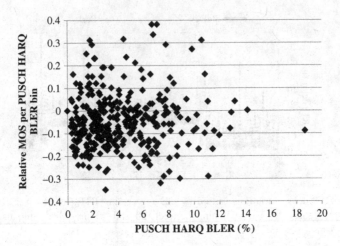

Figure 14.31 Relative (to PUSCH HARQ BLER = 0% MOS) POLQA MOS versus. PUSCH HARQ BLER. Used AMR codec bitrate = 23.85 kbps and target block error ratio of transport blocks for first transmission = 10%

An example case of laboratory measurements shown in Figure 14.34 without TTI bundling shows that UL BLER starts increasing rapidly at RSRP of −114 dBm. Used fading channel here is enhanced pedestrian A (EPA) with 3 km/h UE velocity and with 3 dB reference signal boosting. Used AMR codec bitrate = 12.65 kbps and target block error ratio of transport blocks for first transmission = 10%; 2 × 15 W was used over 10 MHz band.

The MOS value is maintained up to RSRP level of −114 dBm after which the increasing jitter and packet loss start to impact negatively on the MOS values as shown in Figure 14.35.

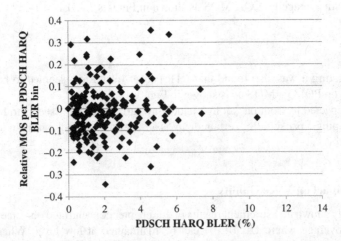

Figure 14.32 Relative (to PDSCH HARQ BLER = 0% MOS) POLQA MOS versus. PDSCH HARQ BLER. Used AMR codec bitrate = 23.85 kbps and target block error ratio of transport blocks for first transmission = 10%

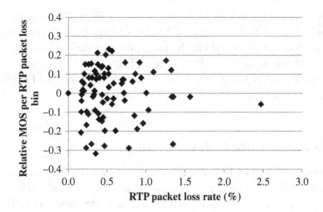

Figure 14.33 Relative (to RTP packet loss rate = 0%) POLQA MOS versus RTP packet loss ratio. Used AMR codec bitrate = 23.85 kbps and target block error ratio of transport blocks for first transmission = 10%

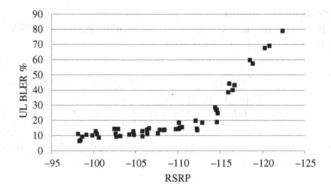

Figure 14.34 Laboratory measurements of RSRP and PUSCH BLER without TTI bundling

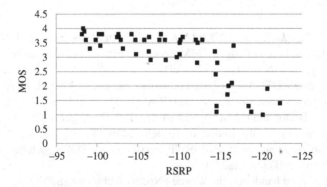

Figure 14.35 Laboratory measurements of RSRP and POLQA MOS without TTI bundling

As shown in Figures 14.34 and 14.35 the VoLTE call quality degrades rapidly in poor radio conditions. In order to maintain the VoLTE call quality in poor coverage conditions TTI bundling or RLC segmentation can be used (as explained in Section 14.4.2) and hence coverage can be extended. However, in case the UE is at the LTE coverage edge the VoLTE call needs to be handed over to the other radio access technology (RAT). This is explained in the next section.

14.5   Single Radio Voice Call Continuity

When the terminal runs out of LTE coverage area and maintaining an ongoing VoLTE call becomes difficult, EPS VoLTE bearer connection can be handed over to CS connection in 3G or 2G network. The procedure that provides the means to the handover is SRVCC.

14.5.1   Signalling Flows

UE and MME Capability

Attach and Tracking Area Update

The SRVCC UE informs the network about its capability of performing SRVCC in the Attach Request message and in Tracking Area Update:

* The UE includes the SRVCC capability indication in 'MS Network Capability' IE where SRVCC to GERAN/UTRAN capability is set.
* If the SRVCC UE supports GERAN access, it includes the GERAN MS Classmark 3
* MS Classmark 2 is included if GERAN or UTRAN access or both are supported.
* The supported Codecs IE is in the Attach Request message and in the non-periodic Tracking Area Update messages.

Context Setup

When a new context is set up in eNodeB, the eNodeB gets informed about UE and MME SRVCC capabilities. The MME informs the eNodeB by including 'SRVCC operation possible' indication in the S1 AP Initial Context Setup Request.

Intra-EUTRAN Handover and EUTRAN to UTRAN Iu Mode Inter-RAT Handover

In case of handovers, both the UE and MME SRVCC capabilities need to be transferred to the target MME, target SGSN and target eNodeB.

In case of an intra-EUTRAN S1-based handover and EUTRAN to UTRAN Iu mode inter-RAT handover procedures, *the source MME* sends MS Classmark 2 and 3, STN-SR, C-MSISDN, ICS Indicator and the Supported Codec IE to the target MME/SGSN.

'SRVCC operation possible' indication is included by the *target MME* in the S1-AP Handover Request message and in case of IRAT HO, the target SGSN includes the indication in the RANAP Common ID message.

In case of X2-based handover, the source eNodeB includes a 'SRVCC operation possible' indication in the X2-AP Handover Request message to the target eNodeB.

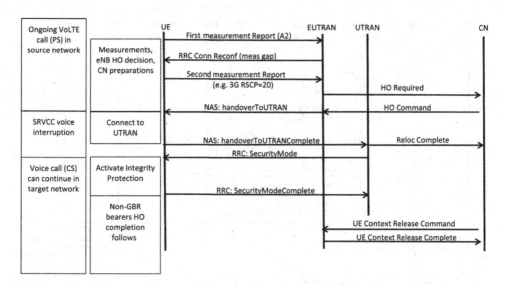

Figure 14.36 QCI1 HO signalling flow from EUTRAN/UTRAN point view, excluding the non-GBR bearers' HO completion part (shown in Figure 14.37)

SRVCC

In this chapter the signalling flow of inter-system EUTRAN to GERAN or UTRAN handover for both voice and data bearers is discussed. Figure 14.36 shows the signalling between the UE and EUTRAN and after the handover command between the UE and UTRAN.

As the voice bearer in EUTRAN is packet switched, it requires a conversion into CS domain and involves rather large number of network elements. The complete signalling flow in Figure 14.37 shows the signalling between core elements and radio access networks.

Measurements and Decision to Initiate SRVCC

Once the coverage is running thin and VoLTE call can no longer be maintained on LTE a relocation/handover can be performed to 3G or 2G network. This procedure is triggered by the UE sending measurement reports at threshold configured in measurement configuration. In one of the measurement reports, the UE reports, for example, the RSRQ or RSRP of the serving LTE cell followed by the RSCP or ECNO of the underlying 3G cell.

Based on the measurement reports the eNodeB makes the decision about starting the SRVCC procedure. It sends HO-required message to the MME and indicates there whether the target cell is both CS and PS capable or CS only capable and whether the request is for voice bearer only or for both voice and packet data bearers.

Splitting the Voice and Data Bearers

The minimum number of bearers required for a UE that has VoLTE call ongoing in LTE network is three bearers. They are QCI1 for VoLTE call, QCI5 for IMS signalling and, for example, QCI9 for the default bearer.

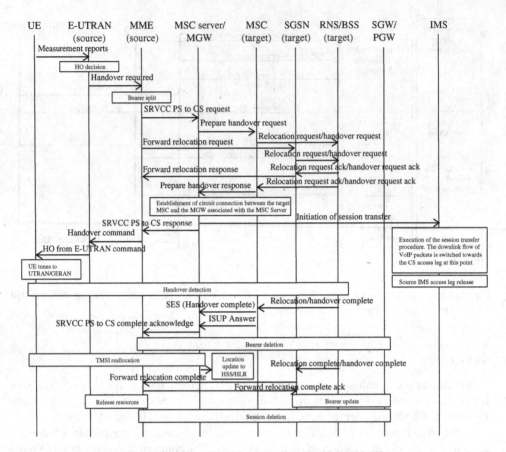

Figure 14.37   SRVCC from EUTRAN with PSHO [9]

Once the MME receives the HO required message from eNodeB it performs PS bearer splitting and separates the QCI1 voice bearer from the other data bearers and initiates their relocation towards MSC server and SGSN. The MME sends a single received transparent container to both the target CS domain and the target PS domain. So the CS and PS relocation procedures are split and are performed in parallel but separately in different CS core network and PS core network. In case the target system is incapable of both CS and PS, only the voice bearer is transferred and PS bearers are suspended.

Voice Bearer Transfer Preparations in Core Network

The MME initiates the PS to CS HO procedure for QCI1 bearer by sending a SRVCC PS to CS Request to the MSC server which then sends Prepare Handover Request message to the target MSC. The target MSC sends Relocation Request/Handover Request message to the target 3G or 2G radio network system which acknowledges the prepared CS relocation/handover.

Target MSC then sends Prepare handover response message to the MSC server and the circuit connection gets established between the MGW associated with MSC server and target MSC.

As the Session Transfer procedure is executed, the remote end is updated with the SDP of the CS access leg, downlink flow of VoLTE packets is switched towards the CS access leg and source IMS access leg is released. The MSC server sends to MME a SRVCC PS to CS response

PS Bearer Transfer Preparations in Core Network

In parallel to the voice bearer transfer preparations, the MME initiates the relocation of the PS bearers by sending Forward Relocation Request message with information of all bearers except for the voice bearer to the target SGSN over either S4 or Gn/Gp as in any inter-system PS handover.

Target SGSN sends Relocation Request/Handover Request message to the UTRAN or GERAN network in order to request for resources for the PS allocation. The target GERAN or UTRAN radio network system acknowledges the prepared PS relocation/handover by sending the Reloc Req Ack/Ho Req Ack message to the target SGSN which then sends a Forward Reloc Response message to the source MME.

Handover from EUTRAN

Once the above voice bearer and data bearers relocation/handover preparations are completed and the target 3G or 2G radio network system has received both the CS and PS relocation/handover requests, the appropriate CS and PS resources can be allocated. The source MME synchronizes the relocations and sends Handover from EUTRAN Command message to the UE with information of all the bearers (e.g. QCI1, QCI5 and QCI9) that are transferred to GERAN or UTRAN and then releases the LTE radio bearers.

Establishing Connection with Target Radio Network and CS Domain and Releasing VoLTE Resources from EUTRAN

IRAT handover takes place once UE has received the Mobility from EUTRAN Command in which a NAS message Handover to UTRAN (or GERAN) is included. The handover message also includes configuration for ciphering in the target network. Once UE has received the handover command, it will not acknowledge the reception of the message to the LTE network, but will tune to the target network frequency immediately and synchronizes with it.

Let us assume here that the target network is UTRAN. Once the UE has synchronized with the target network it will send a Handover Complete message to UTRAN. This message is already ciphered but not yet integrity protected, but contains the information for enabling integrity protection.

After UTRAN receives the Handover to UTRAN Complete message from the UE it sends Security Mode Command to the UE in order to activate integrity protection for CS domain and it also sends Relocation Complete/Handover Complete message to the target MSC, which then sends SES to the MSC server. The speech circuit gets connected in MSC Server.

After UE has processed the integrity protection information it should now be able to receive and transmit voice. UE responds with security mode complete message; the new ciphering key and Hyper Frame Number are used to calculate the RRC sequence number.

MSC server sends SRVCC PS to CS Complete Notification message to the source MME which acknowledges with SRVCC PS to CS Complete Acknowledge message. The MME then deactivates the voice bearer with Delete Bearer Command message to S-GW/P-GW and sets PS-to-CS handover indicator.

Establishing Connection with PS Domain and Releasing the Resources from EUTRAN

At this stage the UTRAN sends Mobility Information message to provide CN system information of the PS domain. Typically also a Routing Area Update is performed and if the HLR is to be updated because the IMSI is authenticated but unknown in the VLR, the MSC Server performs a TMSI reallocation towards the UE after which the MSC Server performs a MAP Update Location to the HSS/HLR.

In case the security context was not transferred, UTRAN performs Authentication and Ciphering procedures in order to reinitialize the keys on the PS domain and starts ciphering of the PS RABs by sending the Security Mode Command. The UE responds with Security mode complete message. The RRC sequence number is calculated using HFN and ciphering key. Routing area update accept and complete messages are exchanged. PSHO from UE point of view is now complete.

Simultaneously the target UTRAN or GERAN sends Relocation Complete/Handover Complete message to target SGSN which sends a Forward Relocation Complete message to the source MME which acknowledges the information by sending a Forward Relocation Complete Acknowledge message to the target SGSN. Target SGSN updates the bearer with S-GW/P-GW/GGSN; the MME sends Delete Session Request to the S-GW and a Release Resources message to the Source eNodeB. The Source eNodeB releases its resources related to the UE.

14.5.2 Performance

SRVCC – Voice Interruption

LTE network stops delivering voice packets to the UE after Handover to UTRAN command is sent; UE immediately tunes on UTRAN target frequency and establishes connection. Once Security Mode message is delivered to the UE from UTRAN and UE has processed it, the UE can start decoding the voice information delivered to it.

Lab testing was done in RF chamber and the results are shown in Figure 14.38. SRVCC handovers were made from EUTRAN to UTRAN while target network RF conditions were adjusted. The target network RSCP from last measurement report to EUTRAN was recorded and voice interruption time was measured. The number of samples is rather low, but it is

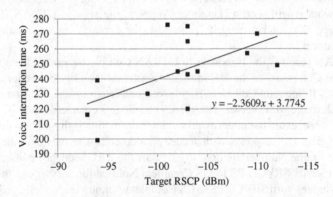

Figure 14.38 Voice interruption time in milliseconds as a function of target network (WCDMA) RSCP value. Used AMR codec is WB-AMR 12.65 kbps in LTE and NB-AMR 12.2 kbps in WCDMA

already shown that interruption times vary between about 200 and 280 ms which is below the 3GPP target of 300 ms. It is also evident that the average voice interruption time increases as the target network RSCP gets worse.

There is room for optimization, but knowing that generally only small percentage of the calls experience one SRVCC and that 300 ms interruption is hardly noticeable, the performance is good for the purpose.

SRVCC Success

The purpose of SRVCC is to handover the VoLTE call to the target system, for example, WCDMA, before the LTE coverage runs out. Therefore, in order to guarantee sufficiently high VoLTE call quality the SRVCC must have very high success rate. The source (LTE) cell must have good enough coverage to deliver the necessary signalling between the UE and the eNodeB for measurements as well as for the handover execution command itself.

The field and laboratory measurements of SRVCC handover to WCDMA have indicated that the MOS (POLQA) decreases rapidly when the LTE coverage runs out as well as the HARQ BLER increases rapidly indicating delays in successful PUSCH packet delivery from the UE. The laboratory measurements in Figures 14.39, 14.40 and 14.41 are done with enhanced pedestrian A fading channel 3 km/h UE velocity Doppler shift. The reference signal transmission power from eNodeB was boosted by 3 dB, eNodeB transmission power for 2 × 2 MIMO was set to 15 W+15 W and 10 MHz bandwidth was used. The used measurement tool averaged the UE-measured RSRP and PUSCH HARQ BLER over 1 second and the voice samples for the used 12.65 kbps wideband AMR codec for POLQA MOS evaluation were

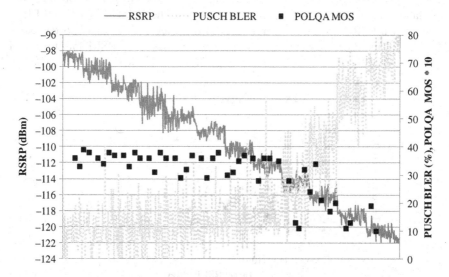

Figure 14.39 Laboratory measurements of RSRP, PUSCH BLER and POLQA MOS without TTI bundling. Used fading channel is enhanced pedestrian A (EPA) with 3 km/h UE velocity and 3 dB reference signal boosting. Used AMR codec bitrate = 12.65 kbps and target block error ratio of transport blocks for first transmission = 10%

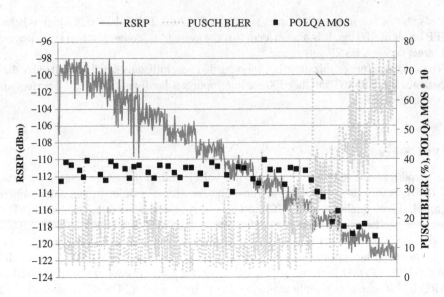

Figure 14.40 Laboratory measurements of RSRP, PUSCH BLER and POLQA MOS with TTI bundling. Used fading channel is enhanced pedestrian A (EPA) with 3 km/h UE velocity and 3 dB reference signal boosting. Used AMR codec bitrate = 12.65 kbps and target block error ratio of transport blocks for first transmission = 10%

taken on 12 seconds intervals. Figure 14.39 shows the results without TTI bundling (TTI bundling explained in Section 14.4.2) and Figure 14.40 shows the results with TTI bundling.

The laboratory testing in Figure 14.39 shows how the PUSCH HARQ BLER rapidly increases when RSRP decreases to around −114 dBm level. POLQA MOS decreases after PUSCH BLER increase when the path loss is further increased and RSRP falls to around −116 dBm level (−119dBm without 3 dB RS boosting). The 70% PUSCH BLER is reached at about

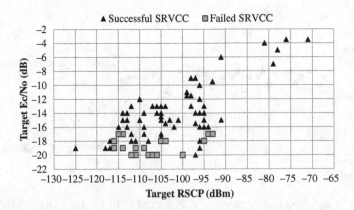

Figure 14.41 Successful and failed SRVCC attempts as a function of WCDMA CPICH Ec/No and CPICH RSCP. Used fading channel is enhanced pedestrian A (EPA) with 3 km/h UE velocity. Used AMR codec bitrate = 12.65 kbps in LTE and 5.9 kbps in WCDMA

RSRP $\sim$−120 dBm ($\sim$−123 dBm without RS boosting) which is similar result as indicated in the measurement results in Section 14.4.2.

As can be seen from Figure 14.40 the TTI bundling greatly helps to maintain the voice quality (POLQA MOS) and PUSCH HARQ BLER until about −118 dBm (−121 dBm without 3 dB RS boosting) RSRP level, that is, performance gain from TTI bundling is around 2–3 dB. Also Figure 14.40 indicates that the SRVCC should be triggered latest at around −118 dBm (−121 dBm without 3 dB RS boosting) RSRP level in order to maintain the voice quality. The target system measurements should be started couple of decibels earlier (from RSRP point of view) at around −115 dBm (−118 dBm without 3 dB RS boosting) RSRP level in order to guarantee fast delivery of measurements (PUSCH HARQ BLER still <30%).

The field and laboratory measurements of SRVCC handover to WCDMA have indicated that SRVCC failure rate increases rapidly once the target WCDMA common pilot channel (CPICH) Ec/N0 falls below −16 dB with varying target CPICH RSCP conditions. The results of the testing as shown in Figure 14.41 suggest that it is more beneficial to configure CPICH Ec/N0 as SRVCC criteria to evaluate target radio conditions instead of CPICH RSCP. The used AMR codec in LTE was wideband 12.65 kbps and narrowband 5.9 kbps in WCDMA.

Optimization

All handovers pose an additional risk to an ongoing call; the more components and procedures are required in the process, the higher the risk of dropping an ongoing call gets. SRVCC involves inter-frequency and inter-system handovers as well as a PS to CS domain switch, making it a complex procedure and risk of dropping a call is high hence it is beneficial to keep the SRVCC probability low. First target for SRVCC optimization is to reduce the number of SRVCCs by providing adequate LTE coverage, by implementing features extending the coverage (e.g. TTI bundling) and by not setting the triggering threshold for SRVCC too early.

However, when LTE coverage is getting thin, at certain point the risk of dropping the call due to the poor coverage starts to increase and there will be a point where the risk of dropping a call during SRVCC is lower than the risk of dropping the call by maintaining it on LTE. Therefore, the second target for SRVCC should be to find the point where the probability of dropping the call during SRVCC gets lower than the probability of dropping the call by maintaining it on poor radio environment as illustrated in Figure 14.42.

For successful SRVCC the quality of coverage in the target network plays an important role. The problem is that often the LTE networks utilize and reuse the existing 2G/3G sites and quite possibly the coverage of the SRVCC target system is also poor when the LTE coverage is getting poor. Third target for optimization is to find the best target network, for example, prefer UMTS on 900 MHz over UMTS on 2100 MHz as SRVCC target or combination of both depending on coverage.

Before 3GPP release 11, in case the SRVCC handover has gone that far that the SRVCC target system MSS has already sent PS-to-CS response and received 200 OK to session transfer INVITE, the SRVCC cannot be cancelled and call continued in LTE when UE returns back to EPC. This is due to MME sending PS-to-CS cancel notification to SRVCC MSS and SRVCC MSS will release towards ATCF/SCC AS with SIP BYE the session transfer leg (access transfer leg) with incorrect cause code and call is released. This cancellation procedure is improved in 3GPP release 11 as specified in section 12.3.3.4 in [10] and once implemented the SRVCC can be cancelled and call can return back to LTE. This 3GPP procedure enhancement

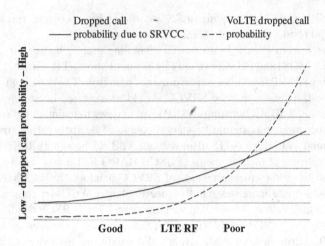

Figure 14.42 Probability of dropping a call due to LTE RF Quality and due to SRVCC as a function of RF Quality in LTE

can improve the SRVCC success rates and ease the optimization of SRVCC measurements and triggering thresholds in LTE.

Another SRVCC procedure improvement introduced in 3GPP release 10 is the alert phase SRVCC which means that it is possible to trigger the SRVCC to target system during the VoLTE call setup phase. The alert phase SRVCC can be triggered during the VoLTE setup signalling phase, see Figure 14.14. This feature once implemented in terminal and in IMS improves the voice call setup success rate from end-user point of view and can also ease the SRVCC triggering thresholds optimization.

Figure 14.43 shows the probability of SRVCC in an example network. The SRVCC probability was reduced from 7% to below 3% by network optimization. More than 1 million SRVCC attempts are included in this graph.

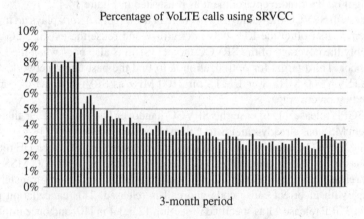

Figure 14.43 Percentage of VoLTE calls using SRVCC

14.6 Summary

In LTE the voice calls can be done either by CS fall back, that is, moving the voice call, before the actual setup, to target system, for example, WCDMA, and complete the signalling in the target system or by VoLTE, that is, running the voice over the LTE radio. CSFB requires no new core network element, that is, IMS is not needed, but voice call setup delay can be quite significant causing degraded end-user experience. VoLTE requires new core network, that is, IMS and new interoperability in terms of SRVCC handover between the LTE packet switched voice and target system CS voice. SRVCC handover triggering points need to be carefully designed in order to maintain the high voice quality, low handover delay and high success rates.

References

[1] 3GPP TS 23.221, 'Technical Specification Group Services and System Aspects; Architectural Requirements', v.9.5.0, December 2012.
[2] 3GPP TS 23.272, 'Technical Specification Group Services and System Aspects; Circuit Switched (CS) Fallback in Evolved Packet System (EPS); Stage 2', v.9.14.0.
[3] M. Poikselkä, H. Holma, J. Hongisto, J. Kallio, and A. Toskala. *Voice over LTE (VoLTE)*, Wiley (2012).
[4] A. Catovic, M. Narang, A. Taha, 'Impact of SIB Scheduling on the Standby Battery Life of Mobile Devices in UMTS', Mobile and Wireless Communications Summit, 2007, 16th IST, IEEE, 2007.
[5] 3GPP TS 24.008 Mobile radio interface Layer 3 specification; Core network protocols; Stage 3V8.6.0, June 2009.
[6] ITU Recommendation P.863, September 2014, Retrieved from http://www.itu.int/rec/T-REC-P.863 (accessed February 2015)
[7] POLQA, September 2014, Retrieved from www.polqa.info/ (accessed February 2015)
[8] The Internet Engineering Task Force (IETF®), RFC3550, May 2003, Retrieved from www.ietf.org (accessed February 2015)
[9] 3GPP TS 23.216, 'Single Radio Voice Call Continuity (SRVCC); Stage 2', v.10.6.0, June 2013.
[10] 3GPP TS 24.237, 'Technical Specification Group Core Network and Terminals; IP Multimedia (IM) Core Network (CN) Subsystem IP Multimedia Subsystem (IMS) Service Continuity; Stage 3', v.11.13.0, December 2014.

15

Inter-layer Mobility Optimization

Jari Salo and Jussi Reunanen

15.1 Introduction

In most networks LTE will be deployed on multiple frequency bands and carrier frequencies. For this reason the design and optimization of mobility parameters between LTE layers and different radio access technologies (RATs) easily develops into an unmanageable tangle of cell change rules controlled by a large number of interdependent parameters. In order to easily maintain and optimize complex multi-layer networks it is therefore beneficial to have some kind of systematic – and preferably simple – approach to the management of mobility parameters. The purpose of this chapter is to summarize some potentially useful concepts and information for handling mobility in multi-layer LTE networks.

The mobility design problem can be split into two regimes: coverage and capacity-limited cases. During initial LTE network roll-out the main emphasis is on guaranteeing continuous service coverage for the subscribers. In this phase the average network utilization is low, and the mobility thresholds can be set assuming that service level is limited by radio conditions rather than network capacity.

At a later stage, when parts of the network become heavily utilized, it is the resource congestion that dominates the service-level degradation. This degradation can to certain extent

LTE Small Cell Optimization: 3GPP Evolution to Release 13, First Edition.
Edited by Harri Holma, Antti Toskala and Jussi Reunanen.

be mitigated by balancing resource usage between cells of the network, necessitating steering of traffic to layers with free capacity. Without carrier aggregation the steering can be realized either by tuning of static mobility thresholds or dynamically by employing some form of load-adaptive mobility rules. While carrier aggregation is likely to be the long-term solution for inter-layer load balancing, single-carrier mobility will be needed for some time before multi-band carrier aggregation becomes widely available in networks and devices; the pace of evolution will obviously depend on the specific market.

The focus in this chapter is on inter-layer mobility of a classical cellular network without small cells. Mobility in heterogeneous networks is beyond the scope of this chapter.

Regarding terminology, the term 'layer' is defined by the combination of RAT and carrier frequency. For example, LTE2600, LTE2100, UMTS2100 and GSM900 are all considered as different layers.

15.2 Inter-layer Idle Mode Mobility and Measurements

The purpose of this section is to elaborate on the rather rich details of the idle mode cell selection/reselection and measurement process in LTE. In contrast to standard references the focus of this section is on the details of the UE measurement process, with some examples of how commercial LTE UEs implement the sometimes mystifying 3GPP idle mode measurement requirements.

15.2.1 Initial Cell Selection and Minimum Criteria for UE to Camp on a Cell

When the UE is switched on for the first time, it performs the public land mobile network (PLMN) and cell selection procedure, which involves sweeping UE-supported frequency bands for received total power and searching the strongest carrier frequencies for cells in non-forbidden PLMNs. The selection of PLMN can be automatic or manual. At subsequent PLMN selections, pre-stored frequency and RAT information can be used to accelerate the procedure. The order in which to search different RATs is not specified by 3GPP and is UE implementation specific, unless specific RAT priorities are encoded in universal subscriber identity module (USIM) for the PLMN [1].

The UE also enters the PLMN and cell selection procedure in order to recover from loss of coverage. Loss of coverage is defined as a two-step procedure. In the first step UE searches for cells on all known frequencies received in system information. This first search step is triggered if the serving cell does not fulfil the cell camping criteria for N_{serv} idle mode discontinuous reception (DRX) cycles, where N_{serv} is defined in Table 15.1. As the second step, if the search of known frequencies and RATs is still unsuccessful after 10 seconds, the UE enters PLMN selection procedure [2]. In scenarios with high downlink interference (e.g. high buildings without indoor solution) the cell camping may also fail because broadcast control channel (BCCH) decoding of system information fails repeatedly, leading to PLMN selection. This can be to some extent compensated by using lower channel coding rate for BCCH.

In order for UE to camp on a cell the level and quality criteria must be fulfilled. The cell camping criteria (the so-called S criterion) in 3GPP Release 9 and later are based on reference

Table 15.1 Maximum number of DRX cycles (N_{serv}) before triggering PLMN selection

DRX cycle length (s)	N_{serv} (number of DRX cycles)
0.32	4
0.64	4
1.28	2
2.56	2

signal received power (RSRP) and (optionally) on reference signal received quality (RSRQ). The simplest form of the cell camping criteria read as

$$\text{measured RSRP} - \text{qRxLevMin} > 0$$

and

$$\text{measured RSRQ} - \text{qRxQualMin} > 0$$

where qRxLevMin and qRxQualMin are broadcast in system information block one (SIB1). If the optional qRxQualMin is not broadcast the UE applies the default value of negative infinity meaning that all cells satisfy the quality criterion. The cell camping criteria in Release 9 are based on both level *and* quality while cell reselection to a cell with different absolute priority is based on either level *or* quality; this is to be discussed shortly. The above basic form assumes that the cell range should not be limited because of UE transmit power limitation. If such uplink cell range limitation is desired (e.g. UEs having different maximum transmit powers in the network) UE maximum transmit power can be optionally broadcast in SIB1. In this case the level criterion is modified as

$$\text{measured RSRP} - \text{qRxLevMin} - P_{compensation} > 0$$

where UE autonomously computes the positive number $P_{compensation}$ from maximum allowed UE transmit power in the cell and the UE power capability. For example, if the maximum UE transmit power in the cell is broadcast as 33 dBm and the UE maximum power capability is 23 dBm, the UE would apply a 10 dB penalty term and the cell radius would effectively shrink by 10 dB for this UE. Improper selection of UE maximum transmit power can result in ping-pong cell reselection; the simplest solution is to omit the parameter on BCCH in which case the UE automatically sets the power compensation term to 0 dB.

As will be discussed next, 3GPP Release 8 and beyond specify absolute priority-based cell reselection. However, before proceeding it is worth remarking that absolute priorities provided to the UE in BCCH system information or in radio resource control (RRC) release message are not used in the initial cell selection process.

15.2.2   Summary of Cell Reselection Rules

Cell Reselection to Equal Priority Layer

Reselection to cells with the same priority as the serving cell is based on best RSRP; however, the candidate target cell must be better by an offset which is the sum of hysteresis, target cell-specific offset and target frequency-specific offset. Only the cell reselection hysteresis is mandatorily broadcast, while the latter two parameters have a default value of 0 dB if not broadcast. Quality-based reselection to an equal priority layer or cell is not supported.

Cell Reselection to Higher Priority Layer

From 3GPP Release 8 and onwards idle mobility towards non-equal priority layer is based on the concept of absolute priorities. The difference to pre-Release 8 mobility is that reselection to a cell with higher priority is not based on choosing the cell with the best signal level. Instead the UE reselects to the cell with the highest priority as long as the pilot signal level or quality from the candidate is above a threshold broadcast in the serving cell. In 3GPP Release 9 inter-frequency cell reselection based on quality criterion (RSRQ) was added. Therefore, in plain text the 3GPP Release 9 reselection criterion towards a higher priority layer reads

measured RSRP from a higher priority cell > min RSRP high + level threshold high

or

measured RSRQ from a higher priority cell > min RSRQ high + quality threshold high

where the thresholds on the right-hand side of the inequality are broadcast in the serving cell. The quality-based criterion is used if the 'quality threshold high' parameter is broadcast in SIB3. Otherwise (and in case of Release 8 UE) level-based criterion is used.

Cell Reselection to Lower Priority Layer

Reselection to a lower priority cell requires that the serving cell level or quality is below a minimum requirement and the corresponding target cell-level measurement quantity is better than a threshold. In plain text both of the following conditions must be fulfilled if RSRP is used as measurement quantity:

RSRP serving cell < min RSRP serving + level threshold serving low

and

RSRP lower priority cell > min RSRP low + level threshold neighbour low

If RSRQ is used as measurement quantity, the criteria are

RSRQ serving cell < min RSRQ serving + quality threshold serving low

Table 15.2 Summary of Rel8 and Rel9 reselection criteria for LTE, UTRA FDD and GERAN

Target cell layer (source is LTE)	Level-based criterion	Quality-based criterion (Rel9), optional
Equal priority LTE	$RSRP_n > RSRP_s + Qhyst + TgtCellOffset_{s,n} + FreqOffset_n$	n/a
Higher priority LTE	$RSRP_n > qRxLevMin_s + threshHigh,P$	$RSRQ_n > qRxQualMin + threshHigh,Q$
Lower priority LTE	$RSRP_s < qRxLevMin_s + threshServLow,P$	$RSRQ_s < qRxQualMin_s + threshServLow,Q$
	$RSRP_n > qRxLevMin_n + threshLow,P$	$RSRQ_n > qRxQualMin_n + threshLow,Q$
Higher priority UTRA	$RSCP_n > qRxLevMin_n + threshHigh,P$	$Ec/No > qRxQualMin_n + threshHigh,Q$
	$Ec/N0 > qRxQualMin_n$	
Lower priority UTRA	$RSRP_s < qRxLevMin_s + threshServLow,P$	$Ec/No > qRxQualMin_n + threshLow,Q$
	$Ec/N0 > qRxQualMin_n$	
Higher priority GERAN	$RSRP_s < qRxLevMin_s + threshServHigh,P$	n/a
Lower priority GERAN	$RSRP_s < qRxLevMin_s + threshServLow,P$	n/a

and

RSRQ lower priority cell > min RSRQ low + quality threshold neighbour low

Additionally for UTRA FDD target the received Ec/N0 must always be higher than the minimum threshold (this does not apply to UTRA TDD). It is not allowed to have the same priority for different RATs [3]. However, different frequency layers within a RAT can have the same priority.

The quality-based cell reselection to UTRA TDD, GSM EDGE radio access network (GERAN) and CDMA2000 is not supported.

Table 15.2 summarizes the reselection criteria for LTE, UTRA FDD and GERAN. In the table the minimum level (qRxLevMin) is based on either RSRP, RSCP or RSSI for LTE, UTRA FDD and GERAN, respectively. Similarly, qRxQualMin means either RSRQ or Ec/N0, depending on the measured system.

The underlying idea of the absolute priority-based cell reselection is that the UE cannot camp on a layer that does not fulfil a minimum level requirement. Moreover, from all the layers satisfying the minimum level requirement the UE camps on the one with the highest priority. This makes planning of mobility thresholds intuitive and simple, which is especially useful since there may be several different RATs, carrier frequencies and bandwidths in the network. A mandatory criterion for exiting to a lower priority layer (level/quality threshold low) becomes hence the minimum required performance on the given layer, rather than the relative level/quality with respect to other layers, as with pre-Rel8 cell reselection.

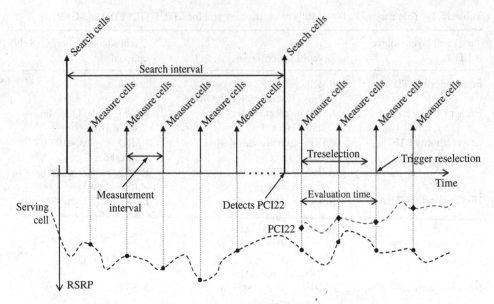

Figure 15.1   Illustration of cell search, measurement and reselection process

Besides broadcasting the serving and neighbour cell priorities, it is also possible to assign dedicated priorities for a UE in RRC connection release. Such mechanism is useful for load balancing purposes, for example. UE uses the dedicated layer priorities in cell reselection until either a new RRC connection is set up or the LTE timer T320 expires. The dedicated priorities can also include neighbour layers that are not broadcast on BCCH. When and what priorities can be assigned to a UE are implementation dependent. An example will be shown in a later section.

15.2.3   Idle Mode Measurements

The following topics, illustrated in Figure 15.1, are of some interest when troubleshooting and optimizing idle mode mobility:

- How often does UE measure the level and quality of detected neighbour cells?
- What is the delay to make a cell reselection decision?
- How often does UE search for new neighbour cells to be measured?

In the beginning of the example in Figure 15.1 the UE is measuring only the serving cell since no other cells have been found by the cell search step. For the intra-frequency case the measurement interval is the same as the serving cell idle mode DRX cycle, to maximize UE sleep time. The search step is repeated periodically and if a new cell is found it is added to the list of measured cells (PCI22 in the example). Even though PCI22 is stronger than the serving cell, cell reselection cannot be triggered immediately. The reselection timer is started at the first instant when the neighbour cell ranks better than the serving cell. As long as the timer value is less than the broadcast value Treselection, reselection must not be triggered. The ranking criterion and the timer value are typically evaluated once every DRX cycle. It should be noted that the value of the Treselection timer is an integer number of seconds (0–7 seconds) while the idle mode DRX period in seconds is a non-integer. Therefore, reselection may not be triggered exactly at the instant of timer expiry but at the next DRX cycle after the timer expiry.

The time from first measurement to reselection decision is called the evaluation time. The sum of search interval and evaluation time is called detection time. Search is initiated at cell reselection after reading the new neighbour information from BCCH. The UE idle mode measurement requirements are specified in terms of detection time, measurement interval and evaluation time; the search interval is not directly specified except for the higher layer search in good radio conditions (i.e. at most 60-second interval per higher priority layer). Nonetheless, it is convenient to think of the reselection process executed by the UE as a continuous cycle of search, measurement and evaluation steps.

How Often Does UE Measure Neighbour Cells?

Once the search procedure has produced a list of neighbour cells, the UE starts the collection of level and quality measurements of these cells. The measurement interval depends on the idle mode DRX cycle, the RAT being measured and the number of inter-frequency layers within the RAT. A summary of minimum requirements is given in Table 15.3.

From the table it can be seen that the serving cell must be measured at least once every DRX cycle, while neighbour cell measurements have more relaxed requirements. Specifically, other RATs need to be measured less frequently than LTE layers. The time requirements in the two right-most columns in Table 15.3 for non-serving cells are per layer. In other words, if there are two layers to measure the interval between layers measurement would be twice the value given in the table.

Example: Assuming that DRX cycle is 0.64 seconds and that there are two LTE inter-frequency layers and one 3G layer to measure, UE would measure one LTE inter-frequency layer once every 1.28 seconds and the 3G layer once every 5.12 seconds. Since there are two different LTE frequency layers, the time between measurements of cells on one frequency would be 2.56 seconds. These are the minimum requirements that every compliant UE in the market should satisfy.

Table 15.3 Time period per layer to measure cells that have already been found in the search procedure, minimum requirement from 3GPP TS 36.133

LTE idle mode DRX cycle length, seconds	Measurement interval for LTE serving cell, seconds (DRX cycles)	Measurement interval for LTE intra-/inter-frequency neighbour cells, seconds (DRX cycles)	Measurement interval for GERAN and UTRA FDD cells, seconds (DRX cycles)
0.32	0.32 (1)	1.28 (4)	5.12 (16)
0.64	0.64 (1)	1.28 (2)	5.12 (8)
1.28	1.28 (1)	1.28 (1)	6.4 (5)
2.56	2.56 (1)	2.56 (1)	7.68 (3)

Table 15.4 Maximum time to determine if a measured cell satisfies cell reselection criterion

LTE idle mode DRX cycle length, seconds	Evaluation time for LTE intra-/inter-frequency neighbour cells, seconds (DRX cycles)	Evaluation time for UTRA FDD cells, seconds (DRX cycles)
0.32	5.12 (16)	15.36 (48)
0.64	5.12 (8)	15.36 (24)
1.28	6.4 (5)	19.2 (15)
2.56	7.68 (3)	23.04 (9)

What Is the Delay to Make a Cell Reselection Decision Once a Cell Becomes Suitable?

If a non-serving cell in the measurement list becomes eligible for reselection the UE will have to be able to make the cell reselection decision within a time window defined in 3GPP [2, 3]. The reselection decision should not be made too quickly but on the other hand it should not be delayed excessively either. The UE is not allowed to make cell reselection unless the candidate cell has fulfilled the reselection criteria for at least $T_{reselection}$ seconds, where $T_{reselection}$ is a serving cell broadcast parameter that may be different for every RAT. It is also required that the minimum time between cell reselections is defined as 1 second even if $T_{reselection} = 0$, and hence the 3GPP standard prohibits making two reselections within a window of less than 1 second. On the other hand, assuming $T_{reselection} = 0$, the maximum delay to trigger a cell reselection to a suitable cell must not be more than the value given in Table 15.4 multiplied by the number of layers.

Comparing Tables 15.3 and 15.4 it can be deduced that for intra-LTE reselection the worst-case decision delay is five measurement intervals while for UTRA FDD measurements it is three measurement intervals.

What Is the Total Time to Trigger Reselection to a Cell That Satisfies Cell Reselection Criteria?

The UE can only measure RSRP and RSRQ of cells whose physical cell identity (PCI) and frame timing relative to serving cell are known. This is since the location of the reference signals[1] in time and frequency depends on the subframe boundaries, PCI and the number of antennas.[2] For this reason, as already discussed, the idle mode measurement procedure performed by the UE can be divided into two steps: cell search and cell measurement. Only those cells found in the search phase are subjected to the actual periodic level/quality measurements. If the cell search procedure was deactivated there would be no neighbour cells to measure, and consequently only the serving cell level/quality would be monitored. To control

---

[1] RSRP and RSRQ are measured only for the OFDM symbol where the reference signals of the first two antennas are transmitted, that is, the first and the sixth OFDM symbol of the 1 ms subframe.

[2] The number of antennas of neighbours is broadcast on the serving cell BCCH; possible values are one or two antennas. Otherwise, determining the number of antennas would require decoding PBCH of the neighbour.

how often UE searches for new intra-/inter-layer neighbour cells the so-called sIntraSearch and sNonIntraSearch parameters are used.

Intra-layer cell search may be deactivated if the serving cell level is better than a broadcasted threshold. Optionally for Release 9-compliant UE and network, if both RSRP and RSRQ are better than the broadcasted threshold, the UE is allowed to stop intra-frequency cell search. The rationale is that if the UE is camping in good radio conditions – 'good' here being defined by the sIntraSearch parameters – there is no need for neighbour cell search, which potentially results in some battery savings. Even so, using a low search threshold to start intra-frequency cell search is not recommended as this may result in the UE being interfered by a much stronger intra-frequency cell without being able to reselect to it. As an example, if sIntraSearch = –70 dBm and the UE could move to another sector of the same site while still camping on the much weaker sector. This results in low SINR leading to potential connection setup and paging reception problems. For this reason intra-frequency search should be kept active even at high serving cell signal levels.

Inter-layer cell search for higher priority layers (obtained on BCCH or from dedicated signalling) is always performed regardless of the value of sNonIntraSearch. Non-intra-layer search parameters merely define how often the UE has to search for cells on the higher priority layers. If both serving cell RSRP or RSRQ (optional) are above corresponding sNonIntraSearch level and quality thresholds then the UE has to search for higher priority layer cells at least once every 60 seconds, cyclically across the layers. Therefore, for example, if there are two higher priority layers in the list each layer would be searched at least every 120 seconds, and possibly more often depending on UE implementation. This is illustrated in Figure 15.2. The sNonIntraSearch level and quality thresholds essentially determine when the UE is in good radio conditions and is therefore allowed to schedule cell search less often. In power consumption tests with commercial UEs in the market (at the time of writing) the battery savings from disabling non-intra-layer search have been found negligible, and hence the benefit from using low sNonIntraSearch thresholds is debatable.

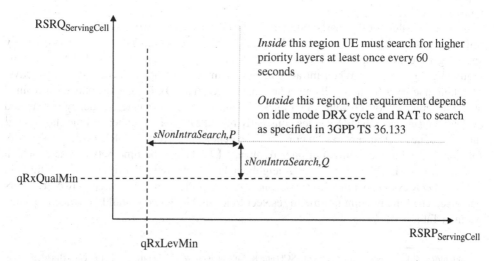

Figure 15.2 3GPP search requirement for higher priority cells when UE is camping in LTE

Table 15.5 Detection time requirement per layer to search, measure and trigger reselection to a suitable cell [2]

LTE idle mode DRX cycle length, seconds	Detection time, LTE intra-/inter-frequency, seconds (DRX cycles)	Detection time for one UTRA FDD frequency layer, seconds
0.32	11.52 (36)	
0.64	17.92 (28)	30
1.28	32 (25)	
2.56	58.88 (23)	60

Lower priority layers are searched only if the serving cell level and quality are outside the region (defined by sNonIntraSearch,P/Q) shown in Figure 15.2, or in other words, when the UE is considered to be in unacceptable radio conditions.

In addition to periodic search within the current cell, the cell search is also restarted after successful reselection and reading of the system information in the new cell after reselection. Also in the case that the serving cell does not fulfil the cell camping criteria, discussed in the beginning of this chapter, the UE will start neighbour cell search and measurement procedure for all layers broadcast on BCCH, regardless of their priorities and the values of sIntraSearch and sNonIntraSearch.

If either serving cell RSRP or RSRQ (optional) is below the corresponding sNonIntraSearch level and quality threshold then the UE is considered to be in bad RF conditions and has to search for higher priority cells more often. The search interval in this case is not specified exactly in [2]. Instead the so-called maximum allowed 'detection time' is given. The detection time is the maximum time allowed to identify, measure and evaluate the cell reselection criteria for a candidate.[3] A summary of detection times is given in Table 15.5 for one layer. The table applies to lower priority cells and higher priority cells when UE is considered to be in bad radio conditions.

A nominal value for the search interval can be deduced as the difference between detection time and the evaluation time, although it should be remembered that UE does not strictly speaking have to comply with such nominal value as long as the overall detection time requirement is satisfied. When intra-frequency or inter-frequency LTE cell search is active, searching one layer for new cells must not take longer than 20 DRX cycles, which is obtained as the difference between values in Tables 15.4 and 15.5. The serving frequency is searched (at least) once every 20 DRX cycles constantly regardless of the number of other layers. For inter-frequency LTE layers one layer must be searched once every 20 DRX cycles, and hence, for example, with two configured inter-frequency LTE layers the time between layer search can be up to 40 DRX cycles. For 3G neighbour layers, the detection time is not defined in terms of DRX cycles, but in terms of seconds. Again, if there are, for example, two 3G layers being searched the maximum time to reselect to a suitable 3G cell would be twice the value given in Table 15.5.

---

[3] A suitable candidate is essentially a cell with SCH and RS subcarrier power higher than -124 to -121 dBm (depending on band) and average subcarrier SNR >-4 dB [2]. Such cell is called 'detectable' in the 3GPP specification.

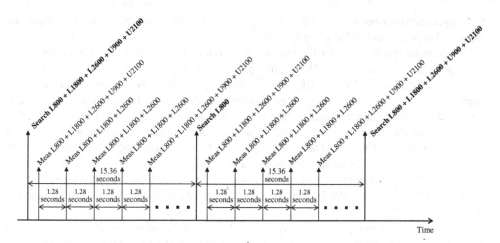

Figure 15.3 Example idle mode search/measurement schedule of a commercial UE. Serving cell is LTE800

The actual search procedure may be implemented in several steps over the search period or carried out as one-shot search through all PCIs or scrambling codes at once. As usual with 3GPP, only the minimum requirements are specified while the details are left to the UE implementation.

It should be noted that in case GERAN cells are part of neighbour information, there is no need to search for GERAN cells since the BCCH frequency list is provided in LTE system information, or in RRC release as part of dedicated priorities, and the measurement is based on received signal strength indication (RSSI) on the given GERAN carrier frequency (i.e., there is no need to acquire radio frame timing prior to RSSI measurement). However, GERAN cells are identified by the absolute radio frequency channel number – base station identity code (ARFCN-BSIC) parameter pair, and for this reason the UE is required to decode BSIC of up to four strongest neighbour cells once every 30 seconds; this procedure still requires acquiring timing of the GERAN synchronization channel.

Figure 15.3 illustrates the search and measurement implementation of a commercial LTE smartphone. In the example the following configuration is used:

- Serving cell is LTE800 (priority 5).
- Neighbour LTE layers are LTE1800 (priority 7) and LTE2600 (priority 6).
- Neighbour UTRA layers are UMTS900 (priority 3) and UMTS2100 (priority 4).
- The received level and quality from serving cell is less than intra- and non-intra search thresholds. In other words, the UE will search all layers using expedited search schedule for the higher priority layers and will also search all lower priority layers.
- Idle mode DRX cycle is 1.28 seconds.

It can be seen that in this case the UE searches for new intra-frequency cells once every 15.36 seconds (12 DRX cycles). Inter-layer cell search is done every 30.72 seconds (24 DRX cycles). The cells found in these searches are then added to the list of cells to measure. All

LTE layers are measured once every DRX cycle. UTRA FDD layers on the other hand are measured at the interval of four DRX cycles.

The nominal requirement for the inter-layer search of LTE cells is 20 DRX cycles. Since there are two LTE layers, the UE should search each layer at least once every 40 DRX cycles. It can be seen that the UE in this case searches both layers once every 24 DRX cycles, and hence performs better than required by the specification. Assuming a single layer the LTE inter-frequency cell detection time requirement for the 1.28-second DRX cycle is 32 seconds from Table 15.5 and since there are two such layers the detection time requirement in this case is 64 seconds. Since the UE searches both inter-frequency layers at a period of 30.72 seconds this leaves ample time to make measurements and evaluate the cell reselection criteria. The same applies for UTRA FDD layers for which the detection time requirement is 30 seconds multiplied by the number of layers (60 seconds in this case).

So far only the steps leading to cell reselection decision in the source cell have been discussed. Successful camping in the new cell still requires reading all SIBs (unless pre-stored in UE memory), and possibly a successful Tracking Area Update if the new cell has a different tracking area code. In case of inter-RAT cell reselection, a short message service/circuit switched fall back (SMS/CSFB) capable UE will always initiate a Location Area Update procedure in the 2G/3G target cell, regardless of the location area code of the new cell. These procedures may fail which, in turn, causes the UE to initiate new cell search and reselection attempt to another cell. For example, if reading of SIB1 and SIB2 of the target LTE cell fails due to downlink interference the UE may internally bar the cell for a certain period [4]. This has been noticed to cause problems in high-interference areas, such as high buildings without indoor base stations.

15.3 Inter-layer Connected Mode Measurements

In this section inter-layer measurements in connected mode are briefly summarized. An exposition of basic concepts can be found in [5] and [6]. Throughout this section it is assumed that the UE needs measurement gaps to perform inter-frequency or inter-RAT measurements. A UE supporting carrier aggregation may be able to do inter-frequency measurements without gaps.

The inter-layer connected mode measurement and handover process consists of the following phases:

1. *Measurement gap activation:* If the UE does not support simultaneous reception on multiple carrier frequencies, measurement gaps must be configured before inter-layer measurements can be performed. The eNodeB configures the frequencies and RATs to be measured either at RRC connection setup or at the time of measurement gap activation. In case of GSM and UTRA FDD, a list of scrambling codes and BCCH carriers is also specified at this stage.[4] For GSM/UTRA RATs, the UE is only required to measure the scrambling codes and BCCH frequencies provided in the neighbour list. The measurement configuration

---

[4] For LTE, there is no need to specify PCIs unless specific cell-specific measurement offsets are to be used.

also lists the measurement quantity, the events that trigger a measurement report, filtering parameters and time to trigger the measurement report.

2. *Inter-layer measurement:* UE measures the configured frequencies and RATs during measurement gaps that are repeated at either 40-ms or 80-ms interval. The measurement gap length is 6 ms.[5] During a measurement gap the UE cannot be scheduled in uplink or downlink and therefore measurement gaps induce throughput loss for the UE; there is no loss of cell throughput, however. The measurement phase consists of cell search and cell measurement phase, similar to the idle mode measurement process. The measurement results are averaged prior to evaluation of an event trigger. If connected mode DRX (C-DRX) is active the UE is allowed to use relaxed measurement schedule.

3. *Measurement report:* If one or more cells satisfy a measurement event the UE sends a measurement report. Unless reporting amount is restricted by measurement configuration, the UE will keep sending measurement reports for all cells that trigger a configured event until measurement gaps are deactivated, UE is moved to idle or it receives a handover command. If C-DRX is active, the UE is allowed to wait until next C-DRX active time before sending the measurement report.

4. *Handover command:* After receiving measurement report, eNodeB makes target cell selection based on the reported cells. If successful, it initiates handover preparation towards the target cell or system. If the preparation succeeds, the target eNodeB sends the handover command to the UE through the source cell (transparent to source eNodeB).

Figure 15.4 illustrates the cell search and measure procedure similar to what is used in some commercial modems at the time of writing. Connected mode DRX is not in use in the example. After measurement gap configuration has been completed the UE searches for cells in the target LTE layer; this can take several measurement gaps since all 504 PCIs need to be scanned. After the search step has been completed, the UE enters measurement mode where within one gap it measures up to four cells found in the search phase [2]. In the example, the second search discovers PCI 122 which is then added to measured cells. The time to trigger the measurement report depends on filter coefficient and time to trigger.

If there are several LTE frequency layers the search and measurement time is multiplied accordingly since in the example the UE can only tune its RF receiver to one frequency during a measurement gap.

The measurement procedure is similar if the target layer is UTRA. However, in that case the number of gaps required for cell search depends on the number of neighbour cells provided in the measurement configuration. It should be noted that scrambling codes that are not provided in the neighbour list need not be measured by the UE or, in other words, no detected set reporting requirements during handover measurements have been specified by 3GPP.[6]

---

[5] The measurement gap length of 6 ms is enough to be able to search for primary and secondary synchronization channels of inter-frequency LTE cells, even for the worst-case frame timing where the gap starts at the beginning of the second or seventh subframe of the measured cell. This is because PSS/SSS is transmitted at 5-ms interval at the end of the first and sixth subframes.

[6] Inter-RAT measurement to detect new neighbour cells for self-organizing network (SON) purposes can be triggered by eNodeB separately.

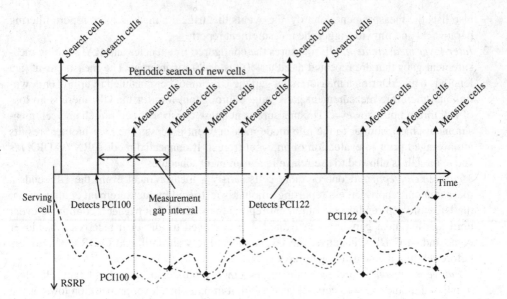

Figure 15.4 Inter-frequency measurement example in connected mode without C-DRX

For GSM target layer, there is no need to perform initial cell search since the GSM frame timing need not be known for power (RSSI) measurement. The UE is, however, required to decode the BSIC on the eight strongest measured BCCH frequencies.

If a cell is detectable and eligible to trigger a measurement report, the maximum time for UE to send the measurement report is called the cell identification time. The maximum allowed identification time depends on the C-DRX long cycle and the number of layers being measured. For example, the time from activating measurement gaps to the first measurement report of an inter-layer cell must not be longer than the value listed in Table 15.6, assuming

Table 15.6 Maximum allowed time to identify a detectable cell with measurement gaps, per layer (from 3GPP TS 36.133)

Long DRX cycle length, seconds	Max time to identify a detectable E-UTRA FDD inter-frequency cell		Max time to identify a detectable UTRA FDD cell		Max time to decode BSIC of a GERAN cell (no other layers measured)	
	40 ms gap	80 ms gap	40 ms gap	80 ms gap	40 ms gap	80 ms gap
No C-DRX	3.84 s	7.68 s	2.4 s	4.8 s	2.16 s	5.28 s
40 ms	3.84 s	7.68 s	2.4 s	4.8 s	2.16 s	5.28 s
64 ms	3.84 s	7.68 s	2.56 s	4.8 s		30 s
80 ms	3.84 s	7.68 s	3.2 s	4.8 s		30 s
128 ms	3.84 s	7.68 s	3.2 s	4.8 s		30 s
256 ms	5.12 s	7.68 s	3.2 s	5.12 s		30 s
320 ms	6.4 s	9.6 s	6.4 s	6.4 s		30 s
>320 ms	20 cycles	20 cycles	20 cycles	20 cycles		30 s

Table 15.7 Maximum measurement interval of inter-layer cells, per layer 3GPP TS 36.133

Long DRX cycle length, seconds	Max measurement period to measure E-UTRA FDD inter-frequency cell		Max measurement period to measure UTRA FDD cell		Max measurement period to measure GERAN BCCH frequency	
	40 ms gap	80 ms gap	40 ms gap	80 ms gap	40 ms gap	80 ms gap
No C-DRX		0.48 s	0.48 s*	0.8 s		0.48 s
40 ms		0.48 s	0.48 s*	0.8 s		0.48 s
64 ms		0.48 s	0.48 s	0.8 s		0.48 s
80 ms		0.48 s	0.48 s	0.8 s		0.48 s
128 ms		0.64 s	0.64 s	0.8 s		0.64 s
>128 ms			5 DRX long cycles			

*For one layer; for two or more layers the requirement is 0.4 seconds.

of course that the cell is eligible to trigger the measurement report in the first place. The requirement in the table assumes time to trigger of 0 ms as well no L3 filtering of the physical layer measurement results.

An E-UTRA cell is considered 'detectable' if the SCH and RS subcarrier power is above −125 to −122 dBm (depending on the frequency band) and the subcarrier to noise and interference power ratio for RS and SCH is above −6 dB. For UTRA FDD, in turn, the requirement is that UE must be able to detect a cell with CPICH Ec/N0 above −20 dB and SCH Ec/N0 above −17 dB. In 3GPP Release 9 requirements, stricter requirements to identify a detectable UTRA FDD cell apply if both CPICH and SCH Ec/N0 are above −15 dB.

The requirements given in Table 15.6 are for one layer. If there are two or more layers being measured the values should be multiplied by the number of layers. With measurement gap period of 40 ms one GERAN band counts as one layer, while with measurement gap period of 80 ms one GERAN band counts as one layer for up to 20 BCCH carriers and as two layers otherwise.

It can be seen that C-DRX slows down the time to first measurement report and hence handover decisions. Especially for long DRX cycles the time to send a measurement report from measurement gap activation may be quite long, for example, up to 20 × 0.64 = 12.8 seconds for 640-ms-long cycle. For two measured layers, the time would be doubled to 25.6 seconds, and so on. Obviously the UE vendor is free to implement faster search and measurement scheduling than required by 3GPP, but network parameters will in most cases need to be optimized for the worst-performing fraction of UE models in the network. For this reason, if allowed by the radio configuration parameters, C-DRX should be disabled for the duration of urgent inter-frequency measurements to reduce the probability of call drop due to measurement delay.

Table 15.7 summarizes the maximum allowed measurement period for cells that have already been found in cell search.[7] Again, it can be seen that C-DRX, if not deactivated, may lengthen

---

[7] As mentioned, no BSIC search is required to measure power on a BCCH frequency. However, after measuring RSSI on all the BCCH frequencies provided in measurement configuration, the UE is required to decode BSIC of the eight strongest BCCHs.

the measurement interval considerably. With C-DRX long cycle of 320 ms and two layers the measurement interval should be at most $2 \times 5 \times 0.32 = 3.2$ seconds. In comparison, intra-frequency measurement interval should be 200 ms or less when C-DRX is not in use, and five DRX long cycles otherwise.

The values in the table apply to 'raw' physical layer measurement before any L3 filtering. The actual event triggering delay depends on the L3 filtering coefficient and time to trigger for the event.

So far the discussion has revolved around physical layer measurement requirements. The raw measurement values provided by the UE physical layer are passed to RRC layer at the interval given in Table 15.7. RRC layer then processes the measurements in two steps:

- Filtering raw measurement results based on a recursive filter where the amount of averaging can be controlled by the filter coefficient parameter that can be optionally sent to the UE in measurement configuration
- For the filtered measurements, continuously evaluating if entry condition for any of the events configured to be reported is satisfied; if yes, then sending a measurement report to the eNodeB

The L3 filtering is defined for physical layer log-scale measurement samples M_n, $n = 0,1,2,\ldots$, as [4]

$$F_n = (1 - a)F_{n-1} + aM_n$$

where $a = 2^{(-fc/4)}$ is the filter coefficient. The value of fc is optionally signalled to the UE in measurement configuration, the default value being fc $= 4$ ($a = 0.5$). The filtered measurement quantity (e.g. RSRP, RSRQ, RSCP) F_n is used in the evaluation of event triggering. Since E-UTRA measurement bandwidth is at least 1.4 MHz, and typically much larger, there is usually no motivation to use long averaging since small-scale fading is already averaged by the large measurement bandwidth. This is in contrast to narrow-band RATs, such as GSM, where the amount of signal power fading is stronger and consequently either longer time averaging or larger handover margins (or both) are required.

Figure 15.5 illustrates the L3 filtering with two different coefficients, fc $= 4$ and fc $= 13$. Also the continuous wideband power and the raw measurement sampling points at 480-ms interval are shown; the sampling points correspond to L3 filter with fc $= 0$. The first measurement sample at time $= 0$ ms sets the initial filter value. The signal filtered with the coefficient value fc $= 13$ results in long bias since the initial value happened to be higher than the average over the observation interval. This could result in false triggering of a measurement event if the time to trigger is shorter than the filtering delay.

In general, controlling measurement event triggering delay is easier and more intuitive with time to trigger than with filter coefficient. Therefore, a simple approach is to set the filter coefficient to a fixed value and simply tune the time to trigger to control ping-pong. Time to trigger can also be defined per measurement event whereas filter coefficient is defined per measurement quantity (RSRP, RSRQ).

Table 15.8 summarizes the basic measurement events. In the table 'Ms' and 'Mn' indicate serving and neighbour cell measurements, respectively. The LTE measurement quantity can be

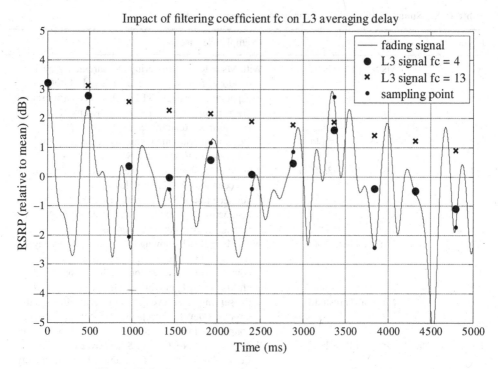

Figure 15.5   Example of L3 filtering; Doppler frequency is 3 Hz

RSRP or RSRQ while inter-RAT measurement quantity may be RSCP, Ec/N0 or GSM RSSI. All measurements and thresholds are in decibel-milliwatts or decibels.

In the table, the hysteresis parameter and other constants are subsumed in the threshold on the right-hand side of the inequality. Often there is no use case for adding superfluous additional constants (like hysteresis or frequency offset), as it typically only complicates the simple event triggering rule without adding benefit; indeed, all such extra modifiers can be lumped into one constant in order to keep the number of tunable network parameters as small as possible.

Measurement gaps can be activated with A1 and A2 events or by a service or load-based trigger such as CSFB call setup or cell congestion. Having measurement gaps active constantly is not in general recommended, since for 40-ms gaps, this results in a theoretical 6/40 = 15% loss of scheduled UE throughput. In practice the throughput loss is even bigger since the UE cannot be scheduled immediately before the gap in downlink and the UE cannot transmit immediately after the gap in uplink. Furthermore, transmission control protocol (TCP) throughput could suffer even more since delay peaks increase the burstiness of the TCP flow, possibly resulting in increased TCP retransmissions.

Most of the connected mode measurement events have their natural idle mode counter-part. For example, the idle mode 'dual' of the A1 event is the sIntraSearch when used for intra-frequency measurement activation/deactivation. In a similar fashion, the A3 event corresponds to cell reselection between equal priority layers with hysteresis, and A4 event is the connected

Table 15.8 Summary of events that can trigger a measurement report [4]

Event	Event trigger	Example use case
A1	Ms > a1Threshold	With Ms = RSRP, triggering measurement gaps for handover from coverage layer (e.g. LTE800) to capacity layer (e.g. LTE2600). Deactivating serving layer measurements, i.e., connected mode counter-part to sIntraSearch
A2	Ms < a2Threshold	Triggering measurement gaps, i.e. connected mode counter-part to sNonIntraSearch
A3	Ms < Mn + a3Offset	With Ms = RSRP, basic power budget handover to keep connected to the cell with least path loss, i.e. connected mode counter-part to idle mode reselection to a equal priority layer where A3 offset is equal to reselection hysteresis
A4	Mn > a4Threshold	With Ms = RSRP, moving to a higher priority layer if the level is good enough, i.e. connected mode counter-part to reselection to a higher priority layer
A5	Ms < a5Threshold1 Mn > a5Threshold2	With Ms = RSRP, moving to a lower priority layer if the serving layer level is below acceptable and the target cell level is acceptable, i.e. connected mode counter-part to reselection to a lower priority layer
B1 (inter-RAT)	Mn > b1Threshold	With Mn = Ec/N0, CSFB PS handover where Ec/N0 threshold is set to a value where call quality and call setup success rate can be expected to be acceptable. Handover to a higher priority 3G layer if RSCP or Ec/N0 is above a threshold
B2 (inter-RAT)	Ms < b2Threshold1 Mn > b2Threshold2	With Ms = RSRP and Mn = RSCP, coverage handover to a lower priority 3G layer if the serving cell level is below acceptable, but lower priority layer offers sufficient RSCP (coverage)

mode equivalent of moving to a higher priority layer whose level or quality is better than a minimum threshold. Such equivalence can be useful in unifying and simplifying the parameter design and optimization of inter-layer mobility thresholds, as will be discussed in the next section.

15.4 Inter-layer Mobility for Coverage-Limited Network

15.4.1 Basic Concepts

In a cellular network each UE should be served by the cell that offers the best service, at least if the mobility parameter design problem is approached purely from UE viewpoint. From network perspective, connecting each and every UE to the best server may not be the best strategy if the target is to maximize network spectral efficiency. The seemingly simple problem of mobility parameter design has received considerable attention in research community where

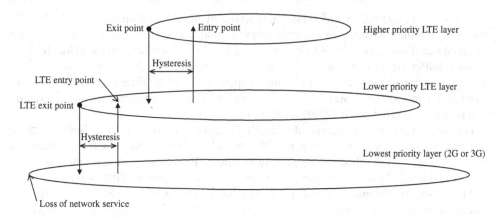

Figure 15.6 Illustration of inter-layer mobility exit and entry points for a three-layer network

it is often cast as a so-called 'user association' problem. The optimum solution to the user association problem is the assignment of each UE to a cell so that some optimality criterion such as network spectral efficiency is maximized. Various approaches to solve the problem for different optimality criteria can be found in literature, for example, [7, 8], and in many other references listed therein. Especially the introduction of small cells has boosted the efforts on the optimum user association research. However, most of the current research literature on the topic is not very useful in practical daily work of a radio planning engineer. This is partly because, at the time of writing, the algorithms introduced in literature have not been implemented in real-world products, but also because the basic operation of the algorithms relies on information that is not directly reported by UE to the network (e.g. downlink SINR).

In this section the focus is on inter-layer mobility in a classical macro-cell network. Specifically, small cells, carrier aggregation and coordinated multipoint are beyond the scope of discussion. In the long term, carrier aggregation and coordinated multipoint techniques are expected to drive a paradigm shift from classical 'slow' handover-based mobility to 'fast' millisecond scale mobility, although the transition period is likely to take some years.

To facilitate discussion it is useful to define some terminology for inter-layer mobility. Towards this purpose, Figure 15.6 illustrates a three-layer network with three different priorities. The highest priority layer could be 20 MHz LTE2600, while the second highest layer could be 10 MHz LTE1800, for example. The coverage layer – 2G or 3G – has the lowest priority and is to be used as the last resort when LTE service level is insufficient. In spirit of the priority-based layering, the idle and connected mode mobility thresholds are set so that the UE is moved to a higher priority layer whenever the service level offered by that layer is better than some minimum requirement; this is called the entry point of the layer. In reverse direction, the exit point for a layer should reflect a service level that is below an acceptable level. Obviously, some hysteresis is required in order to avoid unnecessary layer changes. The design problem is to choose initial entry and exit thresholds to satisfy the given service-level requirement, while the optimization problem is to tune those thresholds based on actual monitored performance.

To be able to discuss the mobility threshold design in a practically meaningful way, definition of service level is needed. For non-guaranteed bit rate data service a natural key performance indicator (KPI) to define service level is the downlink- or uplink-scheduled user throughput, which is the average rate at which the UE is served by the eNodeB during active TTIs. TTIs during which the transmit data buffer is empty should not be included into the calculation of scheduled throughput since TTI inactivity is caused by lack of data in the eNodeB or UE transmit buffer, for example, when downloading a web page the eNodeB may transmit only during a fraction of the available TTIs since it spends most of the time waiting for more data from the internet server. A useful user throughput KPI which is independent of traffic patterns and packet sizes is defined in 3GPP TS 36.314.

For Guaranteed Bit Rate service the mobility criteria could be slightly different since the UE only needs sufficient bit rate instead of the best achievable one. Nevertheless, choosing the cell that offers highest scheduled throughput minimizes PRB consumption for the guaranteed bit rate and hence the user throughput criterion could also be used for GBR services.

Many eNodeB implementations do not support handover triggering based on UE throughput, or any other service-level measure (such as low modulation and coding scheme, MCS), and triggering exit from the serving layer must in practice be based on some other measurement quantity. Potential measurement quantities include the following:

- *Signal to noise and interference ratio (SINR):* Throughput can be obtained from average SINR and the available bandwidth. Hence, a natural mobility trigger would be when SINR drops below a threshold level. Unfortunately, downlink SINR is not reported by the UE. In theory, the downlink SINR can be calculated from RSRQ if the RF utilization of the measured cell is known. For example, the serving cell should know its own PRB utilization, while for the calculation of a neighbour cell SINR the load could be obtained from the X2 Resource Status Reporting procedure [9], if supported by the system. Of course, the eNodeB is capable of measuring uplink SINR which could be used to trigger handover in case uplink quality drops. Even if the calculation of SINR is supported by the system for connected mode, SINR cannot be used as a trigger for idle mode, which makes controlling of 'ping-pong' difficult.
- *RSRQ:* This is a function of cell RSRP, own cell load, other cell interference and thermal noise power. In principle, downlink SINR can be estimated from RSRQ, which in turn allows approximate mapping of RSRQ to downlink throughput, if the cell load is known.
- *RSRP:* This is independent of own cell load, interference and thermal noise. RSRP is basically a path loss[8] measurement and as such insensitive to network load changes, unlike RSRQ which fluctuates based on own and other cell load. Thus, the main drawback of RSRP is that it cannot be used to estimate SINR, except in an isolated cell without other cell interference.

A systematic way to approach the inter-layer mobility design problem is to first choose a minimum acceptable uplink- and downlink-scheduled throughput for each layer and then map the chosen throughput target to SINR, RSRQ or RSRP thresholds. Often the decision of which measurement quantity to use in the mobility design is made considerably easier by the constraints of the eNodeB implementation which commonly supports only a subset of measurement triggers discussed thus far.

---

[8] In this section the term 'path loss' is used in the 3GPP sense to mean the difference between reference signal transmit power and RSRP, in other words antenna system gain is included. For 3G, the corresponding path loss measure is the difference between CPICH transmit power and RSCP.

15.4.2 Mapping Throughput Target to SINR, RSRQ and RSRP

The basic design question is to determine the minimum required service level, below which the UE should exit the layer by, for example, handover, redirect or idle mode reselection. In most business environments, the question has as much to do with operator marketing as it does with engineering. In this section it is assumed that the service level is based on minimum required UE throughput in uplink and downlink. Realistic cell edge throughput values in a 20 MHz 2×2 MIMO FDD system would be in the range of 4–10 Mbps for downlink and 0.5–3 Mbps uplink. As will be seen in the sequel, more often than not, it is the uplink that limits the service area.

As an intermediate remark, in field testing the throughput results have to be normalized per scheduled bandwidth (to remove impact of other serving cell users). Although popular in practice, speed test applications or FTP should not be used to assess radio interface throughput performance due to well-known limitations of the TCP protocol. The recommended way is to use user datagram protocol (UDP[9]), since it is less sensitive to transport network and protocol limitations.

In what follows mapping of throughput to SINR and RSRQ is discussed. Prediction of downlink throughput based on RSRP is also shown by means of a measurement example. Later these results are used to demonstrate the planning of mobility entry and exit points in the inter-layer scenario shown in Figure 15.6.

Mapping Throughput to SINR

Assuming that the minimum required throughput in an LTE layer is defined, the throughput target could be translated to minimum required SINR. This can be done by using link level performance curves such as the one shown in Figure 15.7. The figure shows uplink and downlink UDP throughput in log-scale for fading channel in low mobility (Doppler frequency is 5 Hz). The device under measurement is a commercially available LTE Cat3 modem connected to a commercially available base station via a channel fading simulator. The average SNR is determined by injecting artificially generated Gaussian noise to the receiver input.[10] The downlink transmission scheme is 2×2 3GPP TM3 with rank adaptation, which is sometimes also called dynamic open-loop MIMO. Uplink receiver is a MMSE equalizer with two-branch maximum ratio combining. In the uplink, power control has been disabled and transmission bandwidth is a constant 90 PRBs. One PDCCH symbol is used in the downlink.

It can be seen that at low SNR the performance curves are within 3 dB of each other, especially if the slightly larger L1/L2 overhead of the uplink is compensated. Downlink spectral efficiency (throughput per PRB) of the 2×2 system is slightly better due to four-path diversity. At high SNR the downlink is clearly better than uplink to dual-stream MIMO transmission with 64QAM.

---

[9] A common counter-argument among practitioners in the field is that TCP is the protocol used by end users. While this is surely correct, this section is about radio interface performance, not end-to-end performance. TCP throughput testing is suitable for end-to-end performance testing, but this is a separate topic beyond the scope here.

[10] In downlink, the SINR value measured internally by most commercial UEs (and reported to drive test tools) can be very unreliable, unless suitable calibration measures are used.

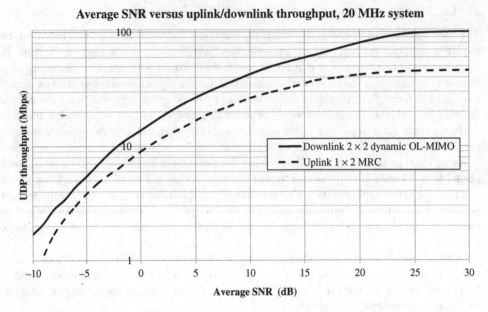

Figure 15.7 Example throughput measurement of a commercial Cat3 modem and base station at 20 MHz

For example, for the measured UE model, SNR > −2 dB is sufficient to reach 10 Mbps downlink throughput. For uplink, the evaluation of SNR target should take into account the number of PRBs allocated at a given SNR which in turn depends on the link adaptation implementation of the eNodeB. Assuming 90 PRBs, SNR = −5 dB achieves 4 Mbps uplink throughput. On the other hand, assuming that uplink link adaptation trades off the number of PRBs to higher power per PRB, say, by allocating 9 PRBs, this would provide 10 dB SNR gain thus shifting the SNR operating point to −5 + 10 = +5 dB. Since the uplink throughput at SNR = 5 dB is about 16 Mbps for 90 PRBs, then for 9 PRBs the uplink throughput would be around one-tenth of that, or 1.6 Mbps. Therefore, to evaluate uplink SNR requirement for a given throughput target it is necessary to know how the eNodeB link adaptation allocates uplink bandwidth; obviously, this is a vendor-specific eNodeB implementation issue. This noteworthy point will be further illustrated in the sequel where an example of RSRP versus throughput measurement is presented.

Figure 15.7 gives the full buffer throughput versus average SNR in low mobility. For a wideband system such as LTE, low SNRs can exhibit high throughput if the whole bandwidth is used for transmission. Even for the low spectral efficiency of 0.1 bits per second per Hz (corresponding to SNR = −9 dB in Figure 15.7), a 20 MHz system would still have a respectable 2 Mbps downlink data rate. In practice this is difficult to realize due to the control channel link budget limitations; in practice the link budget limitation often comes from the control channels and not PDSCH or PUSCH. As shown in Table 15.9 the control channel performance depends on UE speed, radio channel, channel coding parameters and receive antenna configuration. Based on the 3GPP requirements in [10], in high mobility with two transmit antennas PCFICH/PDCCH with 8 CCE aggregation starts limiting below about

Table 15.9 Example of uplink control channel SNR requirements from 3GPP

Control channel and number of receive antennas	SNR (dB)	Notes
PUCCH A/N 2Rx	−5	The ACK missed detection probability is the probability of not detecting an ACK when an ACK was sent. Target is 1%
PUCCH A/N 4Rx	−8.8	The ACK missed detection probability is the probability of not detecting an ACK when an ACK was sent. Target is 1%
PUCCH CQI 2Rx	−4	The CQI missed detection block error probability shall not exceed 1%
PRACH burst format 0 2Rx	−8	The probability of detection shall be equal to or exceed 99%
PRACH burst format 0 4Rx	−12.1	The probability of detection shall be equal to or exceed 99%
PUSCH msg3 80 bits 2Rx	−2.9	1 PRB, 3 retransmissions, residual BLER = 1% [3GPP R1–081856]

SNR = −3 dB (PDCCH misdetection probability >1%). For uplink with two receive antennas, a similar requirement states SNR of −3 to −4 dB for PUCCH CQI report and the initial RRC message in connection setup procedure (random access message 3 sent on PUSCH). Thus in the absence of better receiver performance test results, it is usually not advisable to use SNR targets approximately below −3 dB in either uplink or downlink. This also has some implications on handover offset planning since if intra-frequency A3 offset higher than 3 dB is used; this risks having SNR of below −3 dB in case one or more neighbour cells become fully loaded. For this reason, it is not in general recommended to use intra-frequency handover margins much higher than 3 dB.

Mapping Throughput to RSRQ

While SINR can in principle be estimated from reported RSRQ and own cell PRB utilization, most eNodeB implementations do not implement such option for downlink. Downlink SINR can be calculated from the relation RSRQ = RSRP/RSSI, where RSSI is the total received power normalized to 1 PRB bandwidth.[11] The goal is to define an RSRQ threshold that corresponds to downlink throughput that is below a target value. When downlink average subcarrier activity factor is γ the RSRQ can be written as

$$RSRQ = \frac{RSRP}{RSSI}$$

$$= \frac{RSRP}{12\gamma RSRP + 12I_{oth} + 12N}$$

---

[11] The number of PRBs in the 3GPP definition of RSSI in 3GPP TS 36.214 can be dropped to simplify notation. The RSSI in this section is normalized to one PRB bandwidth.

The average subcarrier activity factor γ is defined for orthogonal frequency division multiplexing (OFDM) symbols that carry RS of the first two transmit antennas and hence for a 1Tx cell the minimum value of γ is $2/12 = 1/6$ since there are always 2 RS symbols in the OFDM symbol even if the cell carries no PDSCH or PDCCH traffic. The maximum value is $\gamma = 1$ for the case when all PRBs in the 1Tx cell are utilized. The other cell interference (I_{oth}) and thermal noise power (N) have been normalized to subcarrier bandwidth. The equation can be related to SINR by rewriting it as

$$\frac{1}{\text{RSRQ}} = \frac{12\gamma\text{RSRP} + 12I_{oth} + 12N}{\text{RSRP}}$$

$$= 12\left(\gamma + \frac{I_{oth} + N}{\text{RSRP}}\right)$$

$$= 12\left(\gamma + \frac{1}{\text{SINR}_{sc}}\right)$$

where the subscript 'sc' emphasizes that the SINR is the average subcarrier SINR in the serving cell. RSRQ is hence related to SINR as

$$\text{RSRQ} = \frac{1}{12\left(\gamma + \frac{1}{\text{SINR}_{sc}}\right)}.$$

For example, if SINR is infinite then RSRQ is $1/12\gamma$ which for $\gamma = 2/12$ and $\gamma = 4/12$ give RSRQ $= -3$ dB and RSRQ $= -6$ dB, which are the well-known idle cell RSRQ values for 1Tx and 2Tx cells, respectively.[12] The case of 2Tx cell requires special attention since the resource elements transmitted from the second antenna should be included in the total power measured by RSSI. With the notation in use the simplest way to do this is to let the subcarrier activity factor take values larger than 1. For the fully utilized 2Tx cell there are 20 REs carrying power in the RS symbol at 1 PRB bandwidth. This is since the resource element carrying RS in one antenna is muted in the other antenna, in order to improve reference signal SNR (and hence the downlink channel estimation accuracy). In terms of average subcarrier activity factor the own cell contribution to RSSI of a fully loaded 2Tx cell can be modelled by simply letting $\gamma = 20/12 = 5/3$, while a 50% load would correspond to $\gamma = 1/2 \times 20/12 = 5/6$, and so on. The mapping between SINR and RSRQ is shown in Figure 15.8.

Using the above formulation it is possible to find a corresponding RSRQ value for a given SINR and own cell load. Assuming SINR $= -3$ dB and fully loaded 1Tx cell, $\gamma = 1$, RSRQ would turn out as $1/12/(1+2) = -15.5$ dB. As an application of this simple result, if an intra-frequency neighbour cell is 3 dB stronger (e.g. intra-frequency A3 RSRP offset is 3 dB), using an RSRQ trigger higher than -15.5 dB risks causing unnecessary triggering of layer change (for cell edge UEs) if both the serving and the neighbour cells become fully utilized. For a fully utilized 2Tx cell the corresponding value would be $1/12/(5/3+2) = -16.4$ dB. Thus the RSRQ–SINR conversion rules can also be used to lend some insight into how to set RSRQ-based event triggers.

---

[12] Throughout this section it is assumed that RS power boosting is not used.

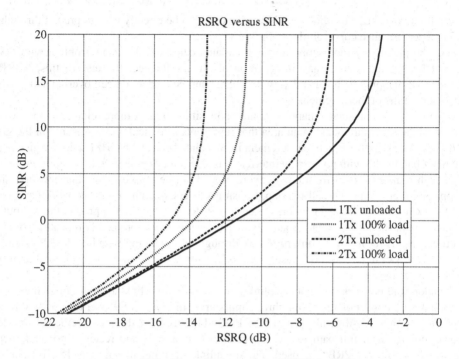

Figure 15.8 Theoretical mapping of RSRQ to SINR

In the absence of other cell interference SINR = 0 dB corresponds to RSRP of about −125 dBm, assuming UE noise figure of 7 dB. Triggering a layer change at such low received levels cannot be usually recommended since based on field experience uplink power starts to limit the performance with commercial handsets when RSRP is roughly below the rule of thumb value of −115 dBm, the actual value also depends on uplink noise rise, RS transmit power as well as UE and network capability (e.g. eNodeB receiver sensitivity and whether UE transmit antenna selection is used or not). At RSRP levels below −120 dBm the risk of an RRC connection setup failure or radio link failure tends to increase sharply. Furthermore, when RSRP is below −125 dBm the cell is not detectable in terms of neighbour cell search requirements as discussed in an earlier section.

Finally, with the mapping between SINR and RSRQ, downlink throughput as a function of RSRQ can be straightforwardly approximated. For example, for a 2Tx cell where average RSRQ = −16 dB and cell load is 100%, the SINR would be about −3 dB, which based on Figure 15.7 corresponds to about 8–9 Mbps downlink throughput. Obviously, uplink throughput cannot be obtained from RSRQ and hence RSRQ can be used as purely downlink-based mobility trigger.

Mapping Throughput to RSRP

If there is no other cell interference the downlink SNR is simply the difference between RSRP and UE thermal noise floor at one subcarrier bandwidth, which is −125 dBm assuming a UE noise figure of 7 dB. The obtained SNR value can then be directly used to predict throughput using a performance curve similar to Figure 15.7.

For the practically more important non-isolated cell case, the main problem with RSRP is that it does not contain any information about interference. Because of this, RSRP to throughput mapping is in general very unreliable and dependent on cell overlap and network load at the time of measurement.

One way to obtain such a mapping could be by simulation. A more reliable method would be to conduct measurements in an area that represents a 'typical' coverage area in the given network. An example of such measurement in an isolated cell at 2.6 GHz carrier frequency and 20 MHz bandwidth with no near-by interferers is shown in Figure 15.9. It is important to note that such a measurement from one network should not be reused in another network, since propagation environment, cell overlap, transmit powers, antenna types, network parameters and so on all differ between networks. Hence if RSRP is to be used to predict throughput the measurement should be redone separately in every network and preferably repeated from time to time. In this regard, downlink or uplink throughput modelling based on RSRP resembles path loss model tuning; in both cases the model is only usable in the network and area type where it was measured.

Vendor- and network-dependent uplink power control and link adaptation parameters can have a drastic impact on the uplink throughput curve. In Figure 15.9, average uplink throughput drops below 1 Mbps already at RSRP < –110 dBm. In the measurement, open-loop power control with full path loss compensation (alpha = 1) and received power target of $P_0 = -100$ dBm per PRB was used. The downlink RS transmit power is 15 dBm per subcarrier. It can be seen that the uplink throughput starts to drop at RSRP = –85 dBm which corresponds to UE-calculated path loss of 85+15 = 100 dB (which includes antenna gains). Plugging this path loss to the open-loop power control equation gives a UE transmit power of

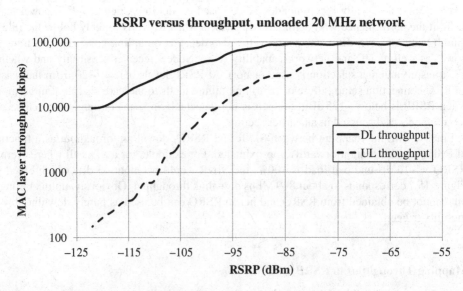

Figure 15.9 Example measurement of RSRP versus Cat3 UE throughput in an isolated cell in low mobility; system bandwidth is 20 MHz

$-100 + 100 = 0$ dBm per PRB. With 90 PRBs transmission bandwidth, the UE transmit power at this point is almost 20 dBm which is near the maximum possible with 16QAM modulation.[13] As the path loss increases further (RSRP decreases) the link adaptation in the measured network starts reducing PRBs to maintain the received power target per PRB in the uplink. Hence in the measured network the link adaptation sacrifices bandwidth to maintain received power target per PRB. At this point the uplink SINR is about 19 dB which is the difference between target received power per PRB ($P_0 = -100$ dBm) and the uplink thermal noise power per PRB, which is about -119 dBm assuming eNodeB noise figure of 2.5 dB. Comparing Figure 15.9 at RSRP $= -85$ dBm to Figure 15.7 at SNR $= 19$ dB, it can be seen that the throughput is about 40 Mbps in both cases and hence at this operating point the uplink performance curves still coincide.

Increasing path loss by 20 dB brings the RSRP down to about -105 dBm and SNR in Figure 15.7 to -1 dB. Now, comparing the performance in the two figures shows a 100% difference of about 4 Mbps in Figure 15.9 versus 8 Mbps in Figure 15.7. This difference is due to the different link adaptation configurations in the two cases. This illustrates that, for the uplink in particular, it can be dangerously misleading to blindly compare performance curves of different networks. Another point worth noting is the severe throughput limitation caused by the limited uplink transmission power. In the example of Figure 15.9 the maximum uplink transmit power is reached at RSRP $= -85$ dBm, below which the system peak throughput cannot be maintained anymore. Therefore, at RSRP of -115 dBm the uplink is 30 dB below the peak performance operation point. Uplink link adaptation attempts to compensate the missing power headroom of 30 dB either by reducing transmission bandwidth or by reducing MCS, or both. Regardless, the compensation works only up to a point and in many networks RSRP levels below -115 to -120 dBm start causing connectivity and throughput problems in the uplink. If downlink transmission power is very high, RS power boosting is used or uplink noise rise is high, the problems may start already at a higher RSRP.

The problem with directly measuring the dependence of RSRP and throughput is that it cannot predict the impact of network load on the throughput. If the network load changes the measurement would have to be repeated. If a high-quality scanner is available a more powerful result can be obtained by predicting SINR from RSRP measurement of strongest cells. This allows factoring network RF load into the SINR prediction, and furthermore gives a good indication of the impact of cell overlap on throughput performance. If accurate RSRP measurement of top-N strongest cells is available, the SINR at a measurement point can be calculated as

$$\text{SINR}_{\text{predicted}} = \frac{\text{RSRP}_0}{\sum_{n=1}^{N-1} \gamma_n \, \text{RSRP}_n + \text{thermal noise} + \text{EVM power}}$$

where $n = 0, \ldots, N{-}1$ are the measured RSRPs of the N cells, ordered from strongest to weakest. Subcarrier activity factors of the neighbours are denoted with γ_n .Thermal noise power depends on the UE noise figure which can be assumed to be in the range of 7–9 dB. The final term in the denominator is the error vector magnitude (EVM) power in the analogue parts of the eNodeB transmitter which limits the SINR to values of less than 30 dB even at very high

---

[13] The maximum UE power amplifier power of 23 dBm need not be supported with 16QAM modulation due to the required back-off to maintain PA linearity [10].

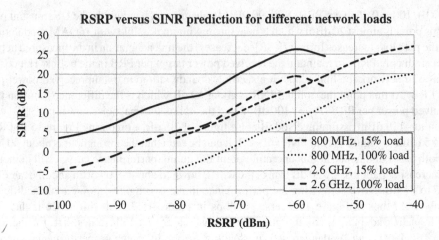

Figure 15.10 RSRP versus predicted downlink SINR for different downlink subcarrier utilization, based on RSRP drive test measurement

RSRP. Also other system imperfections such as subcarrier leakage due to carrier frequency offset could be included in the EVM term, but on the other hand these have impact mainly in the less interesting high-SINR regime, so accurate modelling of EVM is not typically required.

Figure 15.10 shows RSRP versus predicted SINR of a dual-carrier co-sited LTE800/ LTE2600 network for different network subcarrier activity factors; the subcarrier activity γ_n is assumed the same for all cells in the measured cluster. Up to 16 strongest RSRPs have been measured with a scanner with >20 dB dynamic range and an external antenna on the car roof. From the result it can be seen that especially at 800 MHz the cell overlap is alarmingly high, since the extrapolated SINR would cross 0 dB for RSRP below –85 dBm even for an unloaded network (15% subcarrier activity). As the network RF load increases the SINR and hence throughput in the measured network would decrease further. In the example, the SINR degradation is equal to the increase in subcarrier activity since all cells are assumed to have the same activity factor, and thermal noise impact is negligible. The measured network average inter-site distance (ISD) is only about 600 metres and hence there are no RSRP samples below –80 dBm at 800 MHz when the level is measured outdoors with external scanner antenna. At 2.6 GHz, the impact of interference from intra-site sectors is seen as a dip in the curve at RSRP above –60 dBm, indicating measurement samples near the site and between sectors.

For the purpose of designing mobility thresholds an RSRP–SINR curve similar to Figure 15.10 can be used for predicting SINR based on RSRP. As for a given network load, the SINR is determined by cell overlap; physical RF optimization is required to improve the fundamental RF performance of the network.

Dependency of Throughput on System Bandwidth and SINR

According to basic information theory, for a fixed SINR, throughput is linearly dependent on the system bandwidth. This should be taken into account in the planning of mobility thresholds, especially when using the A3 'power budget' event. Setting the A3 offset symmetrically for layers having different bandwidth may result in reduced throughput for the UE. The theoretical difference in SINR requirement for non-equal bandwidths could be obtained from the Shannon

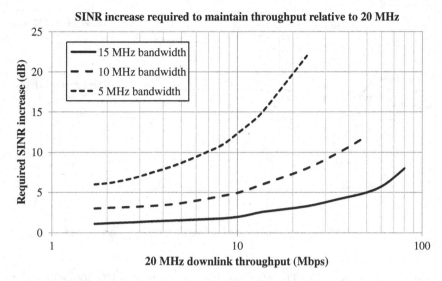

Figure 15.11   SINR increase required to maintain downlink throughput when reducing system bandwidth. Fading channel measurement result for 2×2 spatial multiplexing with a commercial UE

formula or from simulation or measurement results. The interesting quantity is the required increase in SINR to maintain the same throughput.

Figure 15.11 presents the additional SINR required when the source layer has 20 MHz system bandwidth and the target layer bandwidth is 15, 10 or 5 MHz. The curves are based on the measurement of a commercial UE under fading conditions. The delta SINR depends on the throughput; for low throughput the additional SINR required approaches 3 dB when the system bandwidth is halved while for high throughput the SINR difference is larger. As an example, for 10 Mbps throughput in 20 MHz system the SINR should be 5 dB higher for a 10 MHz system to reach the same 10 Mbps throughput. Similarly, 12 dB higher SINR is required for a 5 MHz system to reach 10 Mbps.

When tuning the A3 handover thresholds between layers having different bandwidths the throughput difference between the layers should be taken into account using design curves similar to that shown in Figure 15.11. To make the prediction more accurate the impact of different cell overlaps and loads on SINR characteristics between layers may be included using results such as shown in Figure 15.10.

15.4.3   Inter-layer Mobility Example #1 (Non-equal Priority Non-equal Bandwidth LTE Layers)

Equipped with the simple framework presented in the prequel, in this section the setting of mobility thresholds is illustrated by means of an example. The network used in the example is shown in Figure 15.12. The assumptions are

- Two LTE layers and one UMTS coverage layer, with priorities as shown in the figure.
- Minimum requirement for LTE uplink/downlink throughput at LTE1800 cell edge is 2 and 8 Mbps, respectively. LTE1800 is the higher priority layer with 20 MHz system bandwidth.

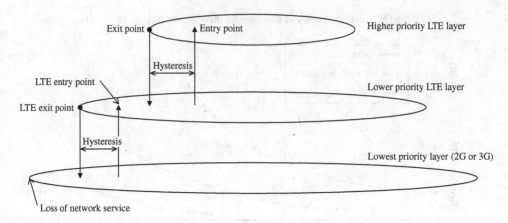

Figure 15.12 Network for the inter-layer mobility design example

- Minimum requirement for LTE uplink/downlink throughput at LTE800 cell edge is 1 and 4 Mbps, respectively. LTE800 is the lower priority LTE layer with 10 MHz system bandwidth.
- Downlink reference signal transmit power is 15 dBm per antenna branch on both layers.
- The thresholds are designed for the case when neighbour cells' average PDSCH PRB utilization is 20% in uplink and downlink.
- System supports RSRP- and RSRQ-based inter-frequency and inter-RAT measurement events in connected mode. In idle mode Release 9 RSRQ-based reselection is supported.

Based on the above minimal set of assumptions, the task is to set the layer exit and entry points for idle and connected mode mobility.

SINR Requirements for the Throughput Target

Based on the minimum LTE throughput requirement, the SINR requirements are obtained by using a figure similar to Figure 15.7. From purely PUSCH throughput point of view SNR = −7 dB would still satisfy the throughput requirement with 90 PRBs transmission bandwidth. On the other hand as already discussed it should be kept in mind that the values much below SNR = −3 dB should be avoided, unless eNodeB receiver performance curves indicate that the uplink control channels can be reliably detected at lower SINR. The downlink SINR requirement can be obtained similarly from Figure 15.7. The result is summarized in Table 15.10 for PUSCH and PDSCH.

Layer RSRP Exit Point for Thermal Noise Limited Case

It is instructive to first consider performance under thermal noise only, since this gives the best possible performance that cannot be exceeded in realistic interference conditions. If there is

Table 15.10 Nominal SINR requirements for the example based on downlink having two transmit antennas

Layer	Uplink PUSCH SINR requirement at 90% PRB utilization, 2Rx	Downlink PDSCH SINR requirement at 100% PRB utilization, 2Tx/2Rx
LTE1800 20 MHz	−7 dB	−3 dB
LTE800 10 MHz	−7 dB	−3 dB

Low UE mobility (Doppler frequency = 3 Hz).

no other cell interference, RSRP can be used as the mobility trigger since SNR can be derived with sufficient accuracy if thermal noise power in uplink and downlink is known.

For downlink the thermal noise level is about −125 dBm per subcarrier, depending somewhat on the UE noise figure (which in turn depends on the operating band). To satisfy SNR = −3 dB the RSRP should be higher than −128 dBm. The reference signal transmit power for one subcarrier is 15 dBm. This brings the maximum downlink path loss[14] to 125 + 15 = 140 dB. Uplink SNR needs to be checked to ensure robust control channel operation. At 23 dBm transmit power and 1 PRB bandwidth the received power is 23−140 = −117 dBm which is 2 dB above the eNodeB thermal noise floor (SNR = 2 dB). Thus for PUCCH the SNR target of −3 dB is still satisfied in the absence of uplink interference.

For the uplink data channel, power control will impact throughput at high RSRP. For low RSRP, it can be assumed that UE is already using its full transmission power of 23 dBm for any practical uplink power control scheme. For 90 PRBs bandwidth, the thermal noise power is roughly −100 dBm if MHA is used or the feeder loss is not excessive. The maximum path loss to satisfy PUSCH SNR = −7 dB under thermal noise is around 23 + 100 + 7 = 130 dB. This corresponds to RSRP of 15 − 130 = −115 dBm. Since the RSRP requirement for downlink throughput was determined to be −128 dBm, it can be seen that it is the uplink throughput requirement that limits coverage.

The path loss values calculated above can live a few decibels in either direction depending on various system parameters; the main purpose here is simply to get an idea on the uplink–downlink power imbalance. The result is that for the given throughput targets, path loss imbalance is 10 dB in favour of downlink. This implies that the mobility threshold design will most likely have to be based on uplink performance. Moreover, it can be already seen that RSRP should not be less than about −115 dBm since this is the RSRP level below which a noise-limited system cannot meet the uplink throughput requirement due to path loss. If the downlink transmission power was higher or RS power boosting was used, the imbalance would also increase correspondingly. With the given uplink throughput target PUCCH does not limit the coverage.

---

[14] Recall that for simplicity the term 'path loss' is in this chapter used in the 3GPP sense, that is, reference signal power per subcarrier minus RSRP. Thus it includes antenna system gain.

Layer RSRP Exit Point for Network with Interference

As was seen, in the noise-limited case RSRP can be used as the mobility trigger since it can be converted to SINR in a straightforward way. For the case with other cell interference a similar approach can be used for the uplink, if the average uplink noise rise is taken into account. For downlink, the situation is more complicated since the downlink noise rise depends on the location of the UE relative to the interfering eNodeBs; near the cell centre downlink noise rise is much lower than at cell edge. For this reason the mapping of RSRP to SINR will depend not only on network load, but on the RSRP level itself (user location in the cell). If a curve similar to Figure 15.10 is available, RSRP threshold configuration could in principle be based on that. However, the drawback of such an RSRP versus SINR mapping is that it presents the behaviour under a static network load whereas in reality the RF utilization of cells in the network fluctuates constantly.

RSRQ in turn can track the dynamic load fluctuations, but on the other hand it necessitates long time to trigger to avoid unnecessary cell changes. In some cases, such as high-rise buildings and cells overlooking a bay of water, the number of interfering cells seen by the receiver is so large that this fluctuation is averaged out and a stable interference power floor is effectively created. RSRQ could be an effective trigger in such cases, albeit there is a danger that the whole cell under interference could be emptied of UEs if the RSRQ trigger is set too high. The only solution to the root cause problem is physical RF optimization of the existing network or building local RF dominance by, for example, indoor solution or well-placed small cells.

To proceed with the example, the uplink layer change trigger will be based on RSRP and noise rise, while for the downlink RSRQ could also be used.

To add the impact of uplink noise rise to the uplink path loss calculations is readily done, since the noise rise is simply subtracted from the path loss target. The uplink noise rise in LTE can be much higher than in Rel99 WCDMA since there is no fast power control. Figure 15.13 shows an example simulation of uplink noise for full buffer data model where 20% of TTIs are fully utilized as a network average, corresponding to the boundary condition of 20% PRB utilization. In urban areas with small ISD the noise rise on 800 MHz can be quite high, since due to lower path loss UEs from several tiers away are still contributing to the uplink noise. Furthermore, in the example the non-optimal downtilt of 4 degrees is used in all cases; optimizing downtilts and uplink power control parameters would result in drastically reduced noise rise especially for the ISD of 500 metres scenario. Interference rejection receivers and intelligent interference-aware power control at eNodeB could also help.

For the purposes of this example uplink noise rise of 3 dB is assumed which is a compromise between different environments and also in line with the field experience from pre-launch optimized networks with similar parameters and low load. For urban area LTE800 with small ISD, the noise rise could be higher (if downtilts and power control are not optimized) but this is neglected for the time being and left for further optimization.

With the 3 dB uplink noise rise assumption, the RSRP trigger to exit to layer due to low uplink throughput changes from −115 to −112 dBm.

Finally, to determine the downlink exit point from RSRP requires a network-specific performance curve similar to what is shown in Figure 15.9. Skipping the details, it can be fairly easily deduced that the RSRP exit trigger based on downlink throughput target is lower than the uplink trigger, and hence it is the uplink performance that dictates the exit point.

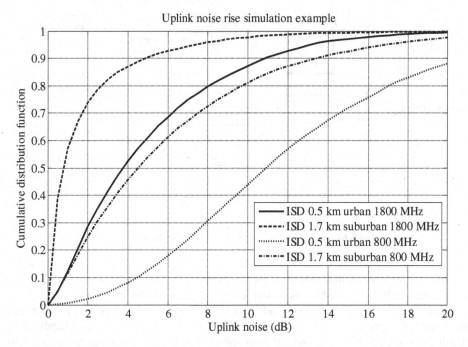

Figure 15.13 Uplink noise rise distribution example for full buffer data model with an average of 0.2 UEs scheduled in uplink per TTI. Open-loop power control with $P_0 = -100$ dBm and alpha = 1 is used. Downtilt is 4 degrees, antenna height is 30 metres. ISD = inter-site distance

Layer RSRP Entry Point for Network with Interference

At this point, the layer exit points have been defined based on the throughput target. The next step is to choose the layer entry points so that some hysteresis remains to avoid ping-pong layer changes.

For uplink the exit trigger is based on RSRP, so the entry trigger should also be based on the same measurement quantity since otherwise hysteresis is difficult to control. The main purpose of the hysteresis is to mitigate the effect of small-scale signal fading, and the hysteresis can be based on timer and level offset in both idle and connected modes. The recommended approach is to rather use small level offset and longer timer value, since this keeps stationary UEs connected to the most suitable layer; the design is optimized for low-mobility UEs since the assumed primary service is data.[15] Using too large level hysteresis can degrade throughput performance since even a 3 dB increase in SINR can improve the scheduled throughput by tens of percents if the layers have the same bandwidth. The main drawback from ping-pong cell change is the potential loss of paging in idle mode, and increased risk of dropping a call in connected mode. If ping-pong is too frequent, also an ongoing TCP download would suffer due to delay peaks and packet loss. A web browsing user is, however, unlikely to notice the effect unless the ping-pong is excessive.

---

[15] For native LTE voice service (VoLTE), a different mobility parameter design would be required.

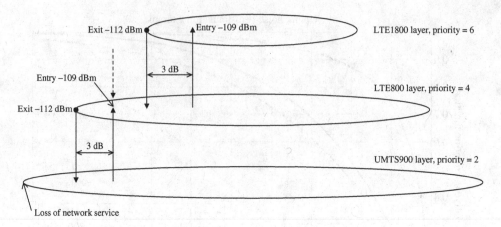

Figure 15.14   Layer exit and entry points for the network with low load

With this philosophy in mind, the RSRP hysteresis is set to a fairly low value of 3 dB and ping-pong will be controlled by increasing time hysteresis as needed.

Figure 15.14 summarizes the exit and entry points for the LTE layers. The actual measurement triggering events are selected according to the analogy discussed previously. In other words, A5 is used to move within LTE from high- to low-priority layer and A4 is used to move from low- to high-priority layer as these measurement events are the connected mode duals of the corresponding idle mode mobility rules.

Further refinements to the baseline case may be needed in high-interference areas where RSRP is strong but SINR is low. In this case the layer exit trigger could in addition be based on RSRQ to have a mobility trigger based on downlink throughput. Mapping the SNR target of −3 dB gives RSRQ threshold of −16.5 dB when serving cell is fully utilized. Assuming a fully utilized serving cell is logical since if the serving cell was not fully loaded the eNodeB could actually allocate more PRBs to the UE thus increasing its throughput.

The main problem with mixing RSRP- and RSRQ-based mobility triggers is that after a RSRQ-triggered handover the UE may swing back to the polluted high-priority layer if RSRP is used to control cell reselection, resulting in ping-pong. Hence in practice it is easiest to resort to a single trigger quantity, either RSRP or RSRQ. Essentially this means that the mobility should by default be based either on downlink throughput (RSRQ) or uplink throughput (RSRP combined with assumed uplink noise rise), unless some vendor-specific algorithms are available.

Threshold Settings

Tables 15.11, 15.12 and 15.13 provide the parameter settings for the example. The example design can be summarized as

- Layer exit point is RSRP < −112 dBm.
- Layer entry point is RSRP > −109 dBm.
- Hysteresis values are 3 dB with ping-pong controlled by timers.

Table 15.11 Idle mode mobility thresholds for LTE layers

Idle mode parameter	Value	Comment
qRxLevMin	−120 dBm	Both layers
threshServingLow,*P*	−112 dBm	LTE1800/LTE800 exit point in idle mode, broadcast on SIB3
threshX,low,*P*	−109 dBm	LTE800 entry point, broadcast on LTE1800 SIB5
threshX,high,*P*	−109 dBm	LTE1800 entry point, broadcast on LTE800 SIB5
Treselection	5 seconds	Long inter-layer reselection timer to avoid ping-pong due to 3 dB hysteresis, intra-layer reselection timer can be shorter

Table 15.12 Connected mode inter-layer mobility thresholds for the LTE1800 layer

Idle mode parameter	Value	Comment
A5Threshold1	−112 dBm	Serving cell RSRP must be below this value. The same value as threshServingLow,*P*
A5Threshold2	−109 dBm	LTE800 entry point, minimum required level from LTE800 to trigger handover. The same value as threshX,low,*P*
A2Threshold	−112 dBm	Serving cell RSRP to activate measurement gaps
A1Threshold	−109 dBm	Serving cell RSRP to deactivate measurement gaps

The values are based on uplink SINR target of −7 dB and 3 dB uplink noise rise which could be considered an optimistic assumption for a non-optimized network. A safety margin of 2–4 dB could be added.

The UMTS900 settings in exit and entry direction have been omitted for brevity, but they should be based on minimal service requirements to avoid UE ending out of network service completely.

The level thresholds summarized in Table 15.11 could be considered relatively high, but it should be remembered that the design was based on throughput targets of 8 and 2 Mbps for 20 MHz downlink and uplink, respectively. One could allow lower thresholds to be used if the throughput target was reduced or the thresholds were instead designed for voice service with certain voice quality target. Alternatively it can be imagined that a certain target probability of call drop is used as the design criterion which would again result in different threshold settings.

The connected mode parameters implement the same exit and entry points as in idle mode. The event reporting parameters (time to trigger, reporting interval, reporting amount, etc.) are omitted.

Table 15.13 Connected mode inter-layer mobility thresholds for the LTE800 layer

Idle mode parameter	Value	Comment
A4Threshold	−109 dBm	LTE1800 entry point. The same as threshX,high,*P*

Moving to a higher priority layer in connected mode corresponds to the A4 measurement event.

With the LTE800, the problem is when to trigger measurement gaps. Since the path loss difference to LTE1800 based on frequency difference is about 7 dB – neglecting modifiers from differences in penetration loss, antenna system gains and so on – there is a risk of having measurement gaps active for long periods of time without finding any suitable target cell on LTE1800. The A1 event could be used to trigger gaps if the level of the LTE800 serving cell is better than some threshold, indicating that the probability of finding a suitable target cell satisfying A4 is high. Utilizing 80-ms gaps minimizes throughput degradation at the expense of slowing down the measurement process. The measurement gap active time could also be limited by the system if the A4 event is not triggered within a certain time, to avoid having gaps active when there is no higher priority layer coverage available in the first place.

15.4.4 Inter-layer Mobility Example #2 (Equal Priority Equal Bandwidth LTE Layers)

The case where LTE carrier frequencies are assigned to high- and low-priority layers has one major shortcoming. While UE is always being served by the highest priority layer satisfying the minimum required RSRP/RSRQ, this does not guarantee that it receives the best available service level since there could be a strong good quality cell available on a lower priority layer. If both layers have the same system bandwidth and their path loss difference is not large (e.g. 1.8 GHz vs. 2.1 GHz), a more suitable approach may be to assign them the same priority, which leads to a different design of mobility parameters based on classical best server-based serving cell selection.

Figure 15.15 sets the background for this section. Both LTE layers have the same system bandwidth and the path loss difference between the 1.8 GHz and 2.6 GHz carrier frequencies is about 3 dB. In this case assigning different priorities to the layers may not produce any benefit but a simple equal priority layering scheme may be sufficient and be easier to operate and optimize.

In the baseline case with low network load, the mobility is based simply on the best server selection with some hysteresis, that is, A3 offset and idle mode hysteresis. There is no need to define any exit or entry point parameters for intra-LTE mobility. For moving to the UMTS

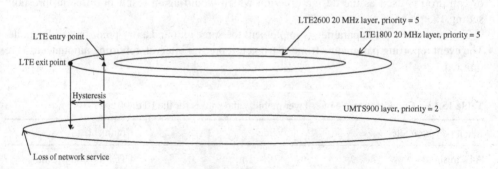

Figure 15.15   Network with two equal priority LTE layers

coverage layer an exit and an entry point are still needed since different RATs cannot have the same priority. This can be based on the minimum service-level requirement discussed earlier, for example, for 2 Mbps uplink throughput target RSRP should not be less than – 112 dBm (other assumptions as in the previous section). For intra-LTE mobility basically only hysteresis and timers need to be determined, which makes the design, configuration and optimization process delightfully simple. RSRQ as mobility trigger suffers from the earlier explained problem that it neglects the uplink direction (even more than RSRP does); RSRP is used as the default trigger in this example also and RSRQ is left for areas with high RSRP and high interference. Another problem in using RSRQ as connected mode handover trigger is that the idle mode reselection cannot be based on RSRQ between equal priority layers. This may result in difficulties controlling ping-pong.

As a rule of thumb the hysteresis between equal bandwidth layers should not be larger than 3 dB since otherwise throughput degradation can be considered excessive. Values of 1–3 dB are typical for a data-dominant network depending on UE mobility, with time hysteresis set accordingly to mitigate ping-pong. With voice service, retainability in high UE speeds becomes more important since a long time to trigger timer lengthens the time in bad voice quality conditions, or risks call drop if the signal strength drops rapidly. Therefore, a well-designed eNodeB implementation should support different sets of handover parameters for non-GBR and GBR services.

Once the level and time hysteresis have been decided, the parameter design is basically done. Tables 15.14 and 15.15 provide an example. A drawback of setting identical margins for both directions is that the traffic load of the layers tends to be uneven. As the network traffic increases this eventually causes the lower frequency layer to become congested first. This can be partly compensated with some form of load balancing, to be discussed in a later section.

In connected mode the measurement event corresponding to the equal priority hysteresis-based idle mode cell reselection is the A3 event.

Table 15.14   Idle mode mobility thresholds for LTE1800 and LTE2600

Idle mode parameter	Value	Comment
qRxLevMin	−120 dBm	Both layers
threshServingLow, P	−112 dBm	LTE1800/LTE800 exit point to UMTS900 in idle mode, broadcast on SIB3
threshX,low, P	−110 dBm	UMTS900 level entry point as RSCP, broadcast on SIB6. Value should guarantee network service with high probability. Should be set to a few decibels higher value than minimum RSCP threshold in UMTS900 (qRxLevMin)
qHyst	2 dB	The same hysteresis in both directions. Low hysteresis to keep UE camping in the cell with the near-optimal path loss. Network optimized for low mobility
Treselection	5 seconds	For intra- and inter-frequency mobility. Long reselection timer to avoid ping-pong due to 2 dB hysteresis

Table 15.15   Connected mode mobility thresholds for LTE1800 and LTE2600

Idle mode parameter	Value	Comment
a3Threshold	2 dB	To trigger intra- and inter-frequency handover, same as idle mode hysteresis. Parameters optimized for low mobility
a2Threshold	−100 dBm	Example value to trigger measurement gaps. Gaps should not be kept activated if inter-frequency handover is not triggered by A3 event after a guard time
Time to trigger	2048 ms	For A2 and A3 events, long timer to prevent ping-pong due to 2 dB offset

Again, the problem is when to trigger the measurement gaps. From purely mobility point of view, the best solution would be to have gaps active all the time, but this reduces the number of resource blocks that can be allocated and also causes delay peaks that increase burstiness of TCP packet flows, risking further throughput degradation. A possible strategy would be to periodically activate gaps for short period when the serving cell level is higher than a threshold (defined by A1 event) in order to probe if a better inter-frequency cell can be found. If the serving cell level is lower than a threshold (defined by A2) the gaps could be active continuously. Smartphones tend to go to idle mode frequently, if inactivity timer is not too long, and this allows changing layer via idle mode if needed. However, heavy data consumers which in practice generate most of the network load tend to stay in connected mode for extended periods of time. It is thus important to move such users to the best frequency layer by intelligently activating measurement gaps to allow inter-layer connected mode mobility. Already a 3 dB improvement in average SINR increases scheduled throughput by tens of percents or, for a fixed traffic demand, reduces PRB consumption by the same amount.

15.5   Inter-layer Mobility for Capacity-Limited Networks

In the case of non-congested network discussed in the previous section, the problem was approached by first defining a minimum acceptable UE throughput and then mapping the throughput target to layer entry and exit thresholds (RSRQ or RSRP). As the offered traffic in the network increases, at some point the PRB utilization hits maximum and multiple UEs have to share the radio resources in every TTI which in turn reduces the UE throughput and increases the packet scheduling delay. Ultimately with too many UEs sharing the radio resources this leads to a situation where the minimum acceptable UE throughput in LTE cannot be satisfied even when SINR or RSRQ is high, which was the design assumption for the unloaded network discussed in the previous section.

If all layers within a coverage area are congested, then there is not much that can be done besides increasing network capacity, for example, by deploying small cells in the centre of the traffic hot spots. Therefore, in such a case of both 'vertical and horizontal congestion' it is not possible to improve the situation by off-loading traffic to other cells or layers as all target cell candidates are also congested. On the other hand, if the congestion is vertical-only or horizontal-only, it may be possible to utilize the unused resources in other cells by mobility threshold tuning or congestion-triggered load balancing handover. Dealing with intra-layer (horizontal) congestion by means of off-loading traffic to another layer is the topic of this section.

The load balancing methods can be divided to two basic approaches:

- *Static load balancing via mobility threshold tuning.* Especially in the case where the load distribution between layers is uneven over a larger coverage area (instead of just few isolated sites or cells), load balancing can be achieved by changing the layer exit and entry points (unequal priority layers) or the cell change offset (equal priority layers). In idle mode, UE dedicated layer priorities can also be used. These methods are mostly based on 3GPP and can be used in a wide range of networks regardless of the eNodeB implementation.
- *Dynamic load balancing via eNodeB algorithms.* These methods include congestion-triggered load balancing handover and adaptive tuning of mobility thresholds based on cell load. The methods are vendor specific and hence not defined by 3GPP. The advantage of these methods is that they can dynamically adapt to the traffic fluctuation at cell level, unlike the static methods.

The static approach tunes the average load of layers over the coverage area while the dynamic methods can further fine-tune traffic load at cell level. A combination of both may be needed for satisfactory results since applying the dynamic methods alone can easily result in excessive ping-pong layer changes unless the average load of the layers within the coverage area is first tuned by the static mobility thresholds; cell-level local tuning is then handled by the dynamic load balancing.

To tune any system, some optimization criterion is needed. For load balancing purposes, several different criteria can be considered. The criteria can be applied to both static and dynamic load balancing. Example criteria include

- Balancing the number of RRC-connected UEs
- Balancing radio utilization (e.g. PRBs, PDCCH CCEs)
- Minimizing the number of UEs having throughput less than the minimum requirement
- Maximizing UE throughput

The choice of which criterion to use is often in practice constrained by the eNodeB implementation. Especially for dynamic load balancing the optimization criterion is typically hard-coded in the eNodeB implementation.

From end-user experience point possibly the most important performance criterion for non-GBR services is the UE-scheduled throughput. Figure 15.16 shows an example of the average UE throughput versus the average number of UEs sharing a TTI for a busy cell. Such performance curve can be used for capacity management and for load balancing.

15.5.1 Static Load Balancing via Mobility Thresholds

Load of the LTE layers can be to a certain extent balanced by tuning exit/entry points or cell change margin (idle mode hysteresis, A3 offset) for non-equal and equal priority layers, respectively.

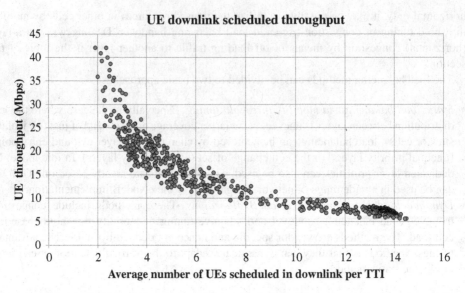

Figure 15.16 Example of UE downlink-scheduled throughput in a cell versus the number of UEs scheduled per TTI. Each dot corresponds to 1-hour average

Non-equal Priority Layers

When layers have different priority a natural way to tune the load distribution between the layers is to modify the layer exit and entry points. Since the difference between the two is the hysteresis – which can be kept constant – the number of variables to optimize consists only of a single parameter hence resulting in a simple and easily controllable process. This is illustrated in Figure 15.17 where the higher priority layer is made to absorb more traffic by lowering the layer exit point.

The amount of traffic that can be steered from one layer to another depends strongly on the spatial UE distribution. If UEs are concentrated near the cell centre, modification of layer exit/entry points is not likely to help much since mainly a few cell edge UEs are affected.

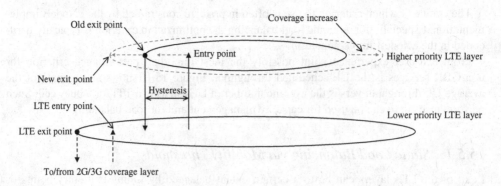

Figure 15.17 Principle of inter-layer load balancing via exit/entry point tuning for non-equal priority layers

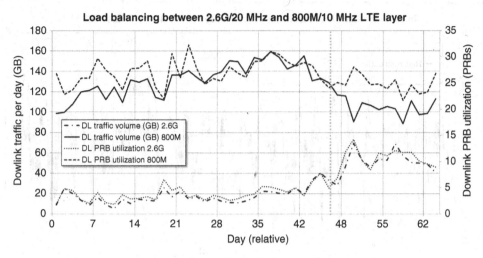

Figure 15.18   Example of intra-sector traffic balancing via static exit/entry mobility threshold tuning, from lower carrier frequency to higher

The principle shown in Figure 15.17 is illustrated by the example in Figure 15.18 where the high-priority layer is 20 MHz LTE2600 and the lower priority layer is 10 MHz LTE800. In this case the 2.6 GHz higher priority layer is almost unutilized. The LTE2600 layer RSRP exit and entry points are decreased by 8 dB on day 46 while keeping the hysteresis as a constant at 4 dB. This results in the LTE2600 layer absorbing more traffic. While the carried traffic and PRB utilization of the LTE2600 layer increase by a factor of almost three, the LTE2600 utilization at daily level still remains low relative to layer capacity. This demonstrates the expected outcome that the tuning of the static mobility thresholds has somewhat limited traffic steering capability which is strongly dependent on the UE distribution and the path loss difference between layers. In terms of service level the lower exit point also decreases the minimum UE throughput offered by the higher priority layer.

Figure 15.19 shows an example where the high-priority layer is 20 MHz LTE2600 and the lower priority layer is 10 MHz LTE800. In the example the LTE2600 layer is congested and the 800 MHz layer is underutilized in terms of served traffic volume. The LTE2600 layer exit point and LTE800 entry point are modified on day 17, in order to increase footprint of the 800 MHz layer. The change results in increased traffic on the LTE800 layer. However, closer analysis reveals that in this case most of the traffic volume increase is due to traffic obtained from other sites while the LTE2600 layer still remains highly loaded. This exemplifies another challenge with static load balancing via mobility threshold tuning: namely, that it can be difficult to force off-loading of the desired layer as the effectiveness of the method is strongly dependent on the UE distribution in the cell. In the example of Figure 15.19, the UEs congesting the LTE2600 cell are very near the site in good RF conditions and hence even several decibels shift in the exit point will not move the traffic away from the layer.

Besides static tuning of exit/entry triggers, basic 3GPP specification includes the option to send to the UE dedicated idle mode layer priorities in the RRC release. This can be used for load balancing purposes. For example, a certain percentage of UEs can be randomly assigned reversed layer priorities (relative to default layering) in order to make the UEs to reselect to the

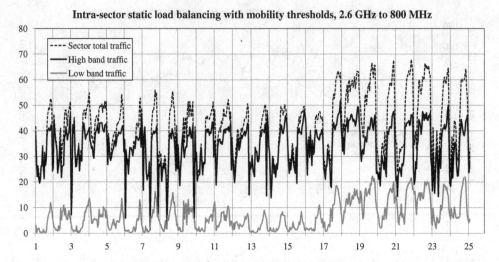

Figure 15.19 Example of intra-sector traffic balancing via static exit/entry mobility threshold tuning, from higher carrier frequency to lower

less congested layer. The main problem with this approach is that the dedicated priorities are erased when the UE establishes a new RRC connection. This can result in the UE being restored the default priorities from BCCH if the same dedicated priorities are not reassigned in a subsequent RRC release, thus resulting in ping-pong cell reselection back to the congested layer.

Equal Priority Layers

With equal priority layers traffic balancing can be done by suitably modifying the layer change offset, that is, A3 event offset and idle mode reselection hysteresis. The benefit compared to the non-equal priority case is that steering is not limited to cell edge UEs only, but is more uniform across the coverage area (assuming co-sited layers).

The optimization process in this case would be basically the same as with non-equal priority layers. In other words, A3 offset and idle mode hysteresis are gradually decreased on the layer to be off-loaded. Correspondingly the target layer cell change margin towards the congested layer may have to be increased to control ping-pong. The process can be continued until the desired average load distribution on the layers is achieved or some system limitation (e.g. drop call increase) is encountered.

Figure 15.20 illustrates the simple idea. With path loss exponent A and cell area reduction of $100(1-w)$ percent the path loss difference would be $5A \log 10(w)$. For example, with a path

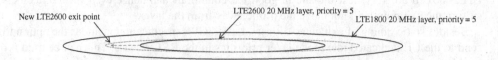

Figure 15.20 Example of static load balancing between equal priority layers

loss exponent of 4 and 30% reduction in cell area the theoretical change in A3 offset would be $5 \times 4 \times \log10(0.7) \approx 3$ dB. If the UEs are roughly uniformly distributed over the coverage area, a 3 dB reduction in the inter-frequency A3 RSRP offset and idle mode RSRP hysteresis would theoretically result in 30% reduction in the number of UEs served by the cell.

For instance, looking at the example shown in Figure 15.20 the coverage area of the LTE1800 layer is about 40% larger based on the path loss difference alone[16] (path loss exponent = 4). In order to balance the coverage areas of the two layers by the A3, handover margins should be made asymmetrical, and even negative. For example, if the A3 offset and idle mode hysteresis from 2.6G towards 1.8G are set to 3 dB and the reverse direction offset from 1.8G to 2.6G layer is set to −3 dB then the coverage areas would be approximately matched (assuming co-siting) but without any hysteresis. Adding 4 dB hysteresis to the theoretical setting would result in A3 offsets of 5 and −1 dB. Obviously practical systems rarely obey such simplified models due to various system link budget differences between the frequency bands, but such calculations can help in inferring a suitable starting point for further optimization.

15.5.2 Dynamic Load Balancing via eNodeB Algorithms

As mentioned, static load balancing via mobility threshold tuning is most usable for tuning average utilization between layers. As it often happens, congestion tends to be localized at cell or site level; one sector of a site can be congested while the other sectors have only moderate or little load. In this case static load balancing is difficult to apply and offers only limited counter-measures at cell level. For this reason cell-level dynamic load balancing algorithms are required. 3GPP has not defined any details, and the realization of such procedures is left to the eNodeB implementation. For this reason only general description can be given in this section.

Two typical template algorithms for congestion-triggered load balancing are

- Adaptive mobility threshold tuning
- Load balancing handover

The former method changes handover thresholds (e.g. A3 offset) adaptively based on some measure of cell load, therefore shrinking the cell size and off-loading cell edge UEs to another layer. Optionally also the idle mode thresholds broadcasted on BCCH may be changed.[17] The second method, in turn, directly triggers a handover for one or more candidate UEs in order to off-load the cell. While the former can be considered 'soft' load balancing with slow reaction time the latter one is able to handle fast congestion peaks and also off-load UEs that are not at the cell edge. Both methods can be used in parallel.

For the load balancing handover, the generic process could be as follows:

1. eNodeB constantly measures cell load based on some congestion metric, for example, PRB utilization, UE throughput, scheduling delay, transmit buffer occupancy, PDCCH blocking,

---

[16] Path loss difference based on carrier frequencies alone is around $20 \times \log10(2.6/1.8) = 3.2$ dB. Here all other factors are neglected for simplicity.

[17] UE can be informed of change in system information via the paging channel.

number of UEs scheduled per TTI or even S1 transport load [11]. The actual metric used, like the whole algorithm, is vendor specific. Some averaging would be typically applied to make the algorithm more robust.

2. If the congestion metric exceeds a pre-configured threshold, the cell is said to be congested.
3. *Candidate UE selection*: eNodeB should select one or more UEs as candidates for handover. The selection criteria could include UE buffer occupancy, UE PRB utilization, uplink/downlink quality, downlink level, uplink power headroom and so on. Also a UE making an RRC setup could be considered as a candidate.
4. Measurements with gaps are configured for the candidate UEs. The configured measurement event could be A3, A4 or A5, based on either RSRP, RSRQ, or both.
5. *Target cell selection*: Upon receiving a measurement report, eNodeB performs target cell selection and starts handover preparation towards the target eNodeB, possibly using X2 cause indicating that the handover attempt is due to source cell overload. The target cell selection can be blind or it could be based on X2 resource reporting procedure by which eNodeBs can exchange load information [9].
6. Upon successful handover preparation a handover command is sent to the UE.

The main optimization problem related to the above procedure is how to decide the threshold for high load in the source cell and if the target cell offers sufficient quality of service, that is, setting of the thresholds for the inter-frequency measurement events. This could also be automatically deduced by the eNodeB algorithm. For example, if the eNodeB has up-to-date information of the current PRB utilization of the candidate target cells this information could be used to automatically select the cell that offers the best throughput, at least approximately.

Figure 15.21 illustrates the load balancing inter-frequency handover from 2.6 GHz to 800 MHz. In this example the load measure is the number of UEs with unacknowledged data in downlink RLC transmit buffer. The high band (2.6 GHz) cell has an average of 14 UEs with an active data transmission ongoing during busy hour, while the 800 MHz same-sector cell has an average of 2 UEs. The load balancing inter-frequency handover is activated in the middle on day 9. As a result hourly average of the number of UEs loading the transmit buffer on the 2.6 GHz layer decreases by about 30% while the load on the 800 MHz layer is almost tripled using the same load measure. The amount of off-loading is controlled by the target number of UEs with data in transmit buffer which in the example was set to 10 on the congested 2.6 GHz layer; it can be seen that the target level is maintained approximately in the source cell. Setting a lower target would result in more aggressive off-loading at the expense of increased number of inter-frequency handovers. In the example the target cell must have RSRP higher than a pre-defined threshold to avoid handing UEs over to bad coverage. In general, the target cell should satisfy some level or quality criterion to custom-pick those UEs that can be handed over to a cell that does not degrade service level too much.

Since balancing of network resource usage by adaptive threshold tuning and congestion-triggered handovers will by definition modify the natural cell and layer boundaries the main problem with dynamic load balancing is the increased number of unnecessary layer changes with the increased probability of call drops and ping-pong. In a typical sequence of steps, the UE is first handed over to a less congested cell, and after returning to idle mode it reselects back to the congested cell, restarting the off-load cycle. In the worst case, if handover thresholds are not coordinated between layers, the UE may try to bounce back while still in connected mode

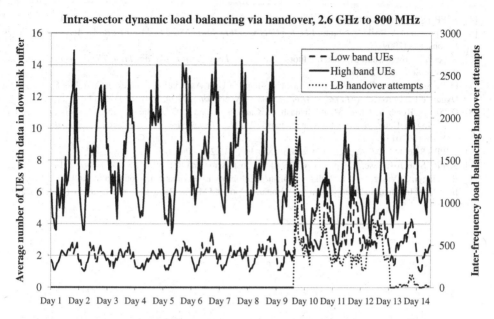

Figure 15.21 Dynamic load balancing via inter-frequency handover, from 2.6 GHz to 800 MHz

in the target cell. In other words, while load balancing between layers is achievable it may also result in undesirable side effects such as increased number of handovers and call drops. The conventional solutions of physical cell dominance optimization and addition of new cells should be considered as an alternative.

15.6 Summary

In this chapter idle and connected mode UE measurements were described. The classical form of inter-layer mobility applied to LTE is based on the best server-based rule where the cell change margin is implemented via hysteresis or A3 offset. From 3GPP Release 8 onwards it is possible in 3GPP RATs to assign different absolute priority to layers. In this non-equal priority case the layer entry and exit points need to be designed and optimized to satisfy some target service quality criteria. Usually the same exit and entry points are used in both idle and connected modes. In case of sporadic network congestion, it may be possible to balance traffic in the network by static or load-adaptive steering. To alleviate long-term intra- and cross-layer congestion, network capacity needs to be increased by physical optimization and addition of new cells.

References

[1] 3GPP TS 23.122, 'Non-Access-Stratum (NAS) functions Related to Mobile Station (MS) in Idle Mode', v.12.5.0, June 2014.
[2] 3GPP TS 36.133, 'Requirements for Support of Radio Resource Management', v.12.4.0, July 2014.
[3] 3GPP TS 36.304, 'Equipment (UE) Procedures in Idle Mode', v.10.8.0, March 2014.

[4] 3GPP TS 36.331, 'Radio Resource Control (RRC); Protocol Specification', v.10.13.0, June 2014.
[5] H. Holma and A. Toskala, editors. *LTE for UMTS – OFDMA and SC-FDMA Based Radio Access*, John Wiley & Sons, Ltd (2009).
[6] S. Sesia, I. Toufik, and M. Baker, editors. *The UMTS Long Term Evolution: From Theory to Practice*, John Wiley & Sons, Ltd (2011).
[7] B. Rengarajan and G. de Veciana, 'Practical Adaptive User Association Policies for Wireless Systems with Dynamic Interference', *IEEE/ACM Transactions on Networking*, 19(6), 1690–1703 (2011).
[8] Q. Ye, B. Rong, Y. Chen, M. Al-Shalash, C. Caramanis, and J. G. Andrews, "User Association for Load Balancing in Heterogeneous Cellular Networks", *IEEE Transactions on Wireless Communications*, 12(6), 2706–2716 (2013).
[9] 3GPP TS 36.423, 'X2 Application Protocol (X2AP)', v.12.2.0, June 2014.
[10] 3GPP TS 36.101, 'User Equipment (UE) Radio Transmission and Reception', v.12.4.0, June 2014.
[11] P. Szilagyi, Z. Vincze, and C. Vulkan, 'Enhanced Mobility Load Balancing Optimisation in LTE', Proc. 2012 IEEE 23rd International Symposium on Personal, Indoor and Mobile Radio Communications – (PIMRC), 2012.

16

Smartphone Optimization

Rafael Sanchez-Mejias, Laurent Noël and Harri Holma

16.1 Introduction

Smartphones are the main source of traffic in the mobile broadband networks. This chapter analyses the traffic patterns created by smartphones and focuses on the network optimization for the typical smartphone applications. Section 16.2 explains the main learnings about the traffic patterns from live LTE networks. The analysis covers user plane and control plane and can be used for network dimensioning. Section 16.3 illustrates the optimization of smartphone power consumption with carrier aggregation (CA) and with discontinuous reception. Section 16.4 presents the signalling traffic from smartphones with different operating systems (OSs) and applications. Messaging and streaming application optimization is considered in Sections 16.5 and 16.6. A number of different voice solutions are tested in Section 16.7 in terms of power consumption and bandwidth requirements. The smartphone design aspects are considered in Section 16.8 and the chapter is summarized in Section 16.9.

LTE Small Cell Optimization: 3GPP Evolution to Release 13, First Edition.
Edited by Harri Holma, Antti Toskala and Jussi Reunanen.
© 2016 John Wiley & Sons, Ltd. Published 2016 by John Wiley & Sons, Ltd.

16.2   Smartphone Traffic Analysis in LTE Networks

This section presents the summary of traffic analysis from a number of leading LTE networks. The section covers data volumes, traffic asymmetry, traffic-related signalling, mobility-related signalling and user connectivity. A more detailed analysis can be found from Reference [1].

16.2.1   Data Volumes and Asymmetry

The data volumes are increasing rapidly in the mobile broadband networks because of more users and because each user consumes more data. The user data consumption growth is driven by new applications, like high definition streaming enabled by the attractive smartphones. The data volumes created by the smartphones are typically 1–3 GB/month in advanced LTE networks during 2014. Laptops can create a lot of traffic even beyond 10 GB/month. Typical data volumes are illustrated in Figure 16.1. Laptop-dominated networks are found in the early phase of LTE deployment where the smartphone penetration is low, or in those markets where the fixed broadband availability is low and Universal Serial Bus (USB) modems are used for home connectivity. Data volumes naturally depend also on pricing and data packages.

Most of the mobile broadband traffic today is in the downlink direction. The dominant traffic type is streaming which is practically downlink-only traffic. The downlink data volume can be even 10 times more than the uplink data volume. The asymmetry in LTE networks tends to be higher than in 3G networks even if LTE radio can provide clearly higher uplink data rates. Because of heavy asymmetry, solutions for improving the downlink efficiency are required in LTE networks. The typical asymmetry is shown in Figure 16.2.

The asymmetry turns out differently in the mass events where the uplink traffic is relatively higher, see Figure 16.3 from a smartphone network. The reason is that many people want to share pictures and videos from the mass events. Therefore, mass events have turned out to be uplink limited and solutions for managing uplink interference are required. The mass event optimization is discussed in Chapter 11.

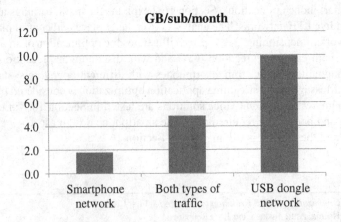

Figure 16.1 Typical data volumes per subscriber per month in advanced networks

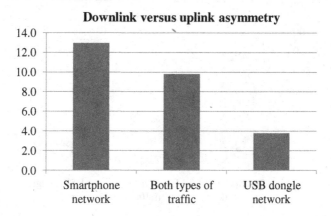

Figure 16.2 Typical asymmetry factor between downlink and uplink

The traffic asymmetry can also be taken into account in the time domain capacity allocation between uplink and downlink in TD-LTE by selecting suitable configuration to match the required capacity split. Configuration 1 provides fairly symmetric split between downlink and uplink, while configuration 2 is better suited for downlink-dominated traffic by using nearly 80% of the time for the downlink direction. The same split needs to be used in the whole coverage area to avoid interference between uplink and downlink.

16.2.2 Traffic-Related Signalling

Smartphone applications create frequent transmission of small packets due to the background activity of the OS and the applications. Even if a person does not actively use the phone, the packet transmissions are still happening in the background. Many mobile broadband networks experience on average even beyond 500 packet calls, that means evolved radio access bearers (eRABs), per subscriber per day. That corresponds to one allocation per subscriber every

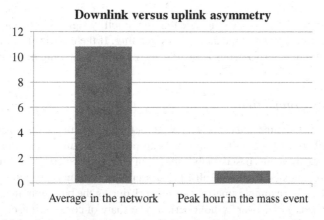

Figure 16.3 Asymmetry between downlink and uplink in the smartphone network

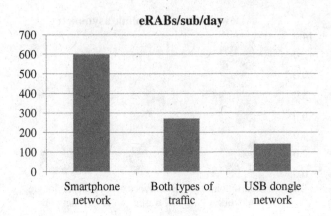

Figure 16.4   Number of LTE packet calls (eRABs) per subscriber per day

two minutes during the busy hour. The typical voice-related activity is just 10 voice calls per subscriber per day which shows that the packet application activity is substantially higher than the voice-related activity. Therefore, the radio network products need to be designed to support high signalling capability in addition to high throughput and high connectivity. Examples of LTE packet call setup volumes per subscriber per day are shown in Figure 16.4 from the network point of view. The frequent signalling creates also challenges for the terminal power consumption.

16.2.3   Mobility-Related Signalling

The amount of mobility signalling has been a concern in the early days of LTE deployment because of flat architecture that makes mobility visible also to the core network. The practical network statistics, however, show that the number of call setups is typically 10 times higher than the handovers. There are two main reasons for this behaviour: the number of packet calls is very high and the packet calls are very short. A handover is unlikely to happen during a short packet call. The relative number of packet calls compared to handovers in three LTE networks is shown in Figure 16.5. We can conclude that the mobility signalling is not a problem in the current networks compared to the call setup signalling. If the cell size becomes considerably smaller in the future heterogeneous networks, the mobility signalling may increase.

16.2.4   User Connectivity

The number of RRC-connected users is relevant for the radio network dimensioning. The share of LTE subscribers that are RRC connected can provide useful information to estimate the radio network requirements based on the amount of LTE subscribers. An example relationship between RRC-connected users and all LTE subscribers is shown in Figure 16.6. The share goes up to 10% in this example smartphone network during the busy hour in the evening, while the average value is 7–8% over 24-hour period. The share of connected users depends heavily on the RRC inactivity timer, which was 10 seconds in this particular network. The share can

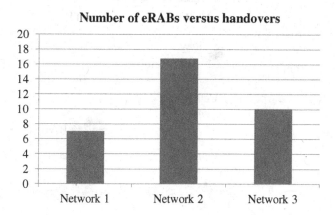

Figure 16.5 Relative number of packet calls compared to number of handovers

be simply explained by the fact that smartphones create a packet call every 2 minutes (every 120 seconds) during busy hour. If the actual packet transmission is short, the typical length of the RRC connection is just slightly more than the inactivity timer, so 10–15 seconds, which explains that 10% of subscribers are RRC connected. In general, a short inactivity timer of 10 seconds is preferred to minimize the smartphone power consumption and the number of connected users. On the other hand, a short inactivity timer leads to high number of packet calls. The share of RRC-connected users tends to be higher in USB dongle networks than in the smartphone networks. Also, if we look the mass events, the share of connected users can be higher. The USB dongle networks can have on average more than 10% of connected users and mass events even more than 20%.

The share of RRC-connected users is clearly higher in HSPA networks since inactive users can stay in Cell_PCH for long periods of time, in the order of 30 minutes. In advanced HSPA networks, Cell_PCH users do not consume any base station resources and a very large number of Cell_PCH users can be supported Radio Network Controller (RNC). Cell_PCH state in HSPA is actually similar to idle state in LTE. Therefore, the number of packet calls (RABs) in

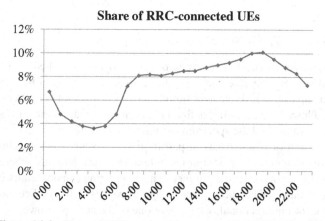

Figure 16.6 Share of RRC-connected users out of all LTE subscribers

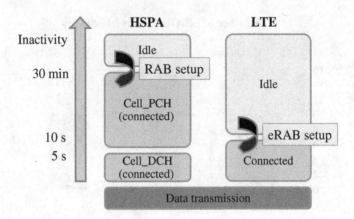

Figure 16.7   Illustration of RRC states in HSPA and in LTE network

HSPA is typically low since the setups only take place between the terminal and RNC while in LTE the setups take place between the terminal and packet core network. Cell_DCH users in HSPA can be considered similar to connected users in LTE. The number of Cell_DCH users in HSPA is normally less than 10% because the inactivity timer on Cell_DCH is usually shorter than LTE RRC inactivity timer. The RRC states in HSPA and in LTE networks are illustrated in Figure 16.7. Optimizing the usage of RRC states is critical for the smartphone performance in terms of network resource consumption and terminal power consumption.

16.3   Smartphone Power Consumption Optimization

Device's battery life is one of the main concerns of the active smartphone users. Since the battery technology is improving only slowly compared to the developments in the digital processing and RF integration, there is a clear need to minimize the terminal power consumption. The following sections present the terminal power consumption measurements and optimization solutions.

16.3.1   Impact of Downlink Carrier Aggregation

Downlink CA is one of the key features of LTE-Advanced, aimed to increase the peak throughput and improve the efficiency of radio resource management by allowing one UE to be scheduled on up to five LTE carriers simultaneously. The exact combinations of intra- and inter-band CA allowed are explicitly defined in 3GPP specifications according to the requests from different operators and the spectrum organization in different areas of the world.

The simplest form of CA consists in pairing two co-located cells serving the same geographical sector, although more complex configurations are possible where more than two carriers of up to 20 MHz could be combined. The cell pairing can optionally be bi-directional, meaning that the same cell can play a role of the Primary (PCell) and Secondary Cell (SCell) at the same time, or unidirectional, where only one of the cells is allowed to behave as PCell and the other as SCell. Both cells can in any case be used as regular LTE carriers for UEs

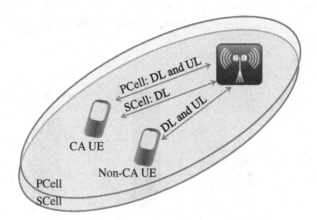

Figure 16.8 Carrier aggregation (CA) principle

which do not support CA, and mobility between them is controlled by same cell reselection and handover settings as in case of unpaired cells. Figure 16.8 illustrates the CA concept.

When additional carriers (SCells) are configured via RRC signalling, the UE may have to periodically monitor the received power level of the SCell(s) causing an increase in the current consumption. Once configured, the SCells may be activated by the network, using an 'Activation/Deactivation MAC Control Element' based on proprietary activity thresholds. Once a SCell is activated, the UE should also start reporting channel state information needed for the packet scheduling.

The CA feature impacts the smartphone battery life because UE has to monitor two frequencies, activate additional RF hardware and increase baseband activity. For large data transfers such as file download, the power consumption with CA increases during the time a file is downloaded. However, the higher throughput makes the download time shorter allowing the UE to go back to RRC Idle state sooner, thus increasing the energy efficiency and saving overall battery life, as can be observed in Figure 16.9. The CA case provides the lowest average power consumption when calculated over the total period of 185 seconds including the active download and idle time.

The instantaneous power consumption measurements for RRC-connected UE are shown in Figure 16.10. The measurements are done without any activity. The CA increases power consumption if it is configured or activated even if no data is transferred. The higher power consumption is explained by the UE monitoring two frequencies. Therefore, for short data transfers, where the RRC connection will spend most of the time waiting for the inactivity timer to expire, the power consumption may increase up to 80% when 10 MHz + 10 MHz CA is configured compared with a 10 MHz single carrier. This result indicates that CA should preferably be configured only when the transferred data volume is large but should not be configured for small background data transmissions. For more details, see Reference [2].

16.3.2 Impact of Discontinuous Reception

In principle, while a UE is in LTE RRC-connected state, it must continuously monitor the PDCCH for new downlink or uplink radio resource allocations even if there is no data to be sent

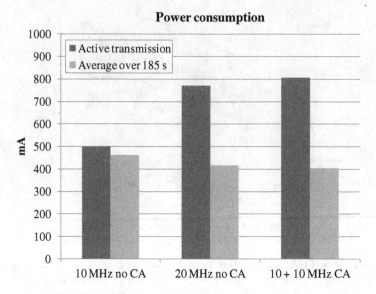

Figure 16.9   UE current consumption during large file download

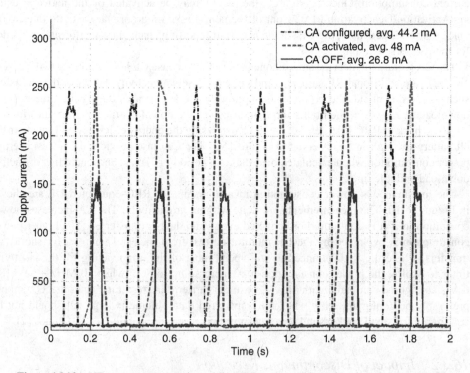

Figure 16.10   UE current consumption while RRC connected but without activity (DRX ON)

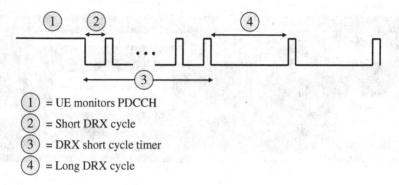

① = UE monitors PDCCH

② = Short DRX cycle

③ = DRX short cycle timer

④ = Long DRX cycle

Figure 16.11 DRX diagram

or received. This continuous PDCCH monitoring requires relatively high current consumption which in the long term can have an impact on the battery life experienced by users. Energy savings are possible in RRC idle state, when UE remains in sleep mode and only wakes up periodically to check if there is any paging indication from the network. Power saving is possible also in connected mode. Connected mode discontinuous reception (cDRX) allows the UE to monitor the PDCCH periodically and enter sleep mode between the monitoring periods while still remaining in RRC-connected state. Thus cDRX reduces UE current consumption in RRC-connected state when no data is transmitted.

DRX concept is illustrated in Figure 16.11. Every time that a packet is sent or received, the DRX inactivity timer is re-started, and the UE keeps monitoring the PDCCH until the timer expires. After this happens, the UE goes to DRX sleep mode for a time period defined by the DRX cycle. The UE comes back to DRX active mode periodically to monitor the PDCCH for few milliseconds, and in case there is no new data for the UE in that period the UE will go back to DRX sleep. This process continues until the RRC Inactivity timer expires and the UE switches to RRC Idle state.

The DRX cycle pattern can be divided into two parts: the long DRX cycle, which can range from 20 ms to 2.5 seconds, and the optional short DRX cycle, which is usually defined between two and four times smaller than the long DRX cycle. Using a combination of short cycle and long cycle enables the usage of a shorter DRX inactivity timer and more frequent monitoring right after the last data is processed, while allowing for higher battery savings when new data takes longer to arrive.

Figure 16.12 shows an example of UE current consumption monitoring for a short one packet transfer, where the power consumption during the inactivity period follows the preconfigured DRX pattern with 50 ms DRX inactivity timer, 80-ms-short DRX cycle and 320-ms-long DRX cycle. It is interesting to note how even if the 'OnDuration' is configured to 10 ms, the actual power consumption on each cycle may expand up to 50 ms due to the time the device requires to switch its radio components on and off. This is of course a factor which depends on the hardware implementation of the device and which is expected to improve with newer generation of chipsets.

From a performance point of view, while having rather large long DRX cycles can be very beneficial for battery savings, it may also increment the latency experienced by users.

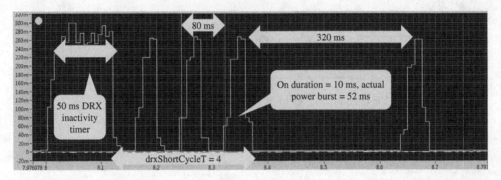

Figure 16.12   DRX impact on current consumption pattern

Furthermore, in some cases it may also force the UE into uplink out of sync state, which means that the UE will have to obtain synchronization again using RACH channel before being able to transfer data in uplink direction.

Choosing the optimal parameter settings for DRX is very important to minimize the current consumption while keeping delay under control and reduce the extra load generated on RACH channel. However, optimal settings also depend very much on the type of services that are used and the traffic pattern that they generate.

The 'tail current' can be defined as the current consumption during the time while there is no data transmission but the RRC inactivity timer is running. We consider a simple traffic pattern consisting of small data transfers (i.e. 'ping'). Figure 16.13 shows the ping latency measurements and Figure 16.14 shows the corresponding power consumption. There is a clear trade-off between low latency and optimized power consumption. Keeping OnDuration small and using mid-range DRX cycle (320–640 ms) seems to provide the best trade-off between latency and battery savings in this case.

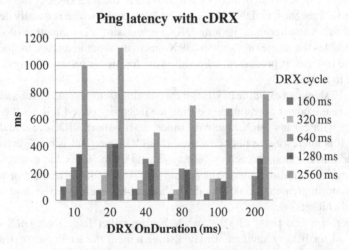

Figure 16.13   Ping latency for different DRX cycles and OnDuration

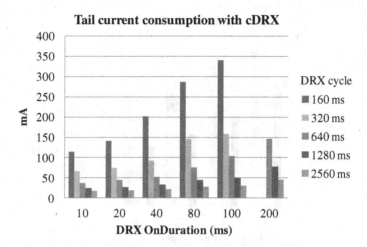

Figure 16.14   Tail current consumption for different DRX cycles and OnDuration

DRX Impact on Standby Performance

Usually smartphones spend most of the day in standby mode, which means that the end user is not actively interacting with the phone, but there is still activity going on in the background between the phone and remote servers, triggered either by the OS or by the installed applications. This continuous background activity is partially responsible for the general perception that modern smartphone's battery life is poor.

Standby behaviour is normally characterized by small packet transfers which are repeated periodically, either to ensure that the service is still available (keep-alive) or to update information that will be presented to the user, like weather updates and messaging status updates. Using a simple model where background activity is represented as a short data transmission (3 seconds), followed by 10 seconds of RRC Inactivity Timer, and using as reference the 'tail current consumption' measured in Figure 16.14, it is possible to estimate how different frequency of background activity events per hour could impact the battery life of the UE. In Figure 16.15, standby battery life is estimated for a 2600 mAh battery considering different frequency of keep-alive messages per hour and different DRX cycle periodicity. Using a 320 ms DRX cycle in this case could increase standby battery life between 21% when the background activity is only four times per hour and up to 50% if there are 12 events per hour.

DRX Impact on Active Profile Performance

When users are actively interacting with smartphone there are other factors which contribute to the battery consumption besides the radio transmission, and it is often the case that the display actually consumes significant amount of battery. Therefore, when trying to estimate the real impact of cDRX during active usage it is important to model a realistic mix of typical services, break intervals and usage of the display.

Figure 16.16 shows an example of the evolution of power consumption during a predefined data session with a mix of different services and periods of inactivity. A similar current

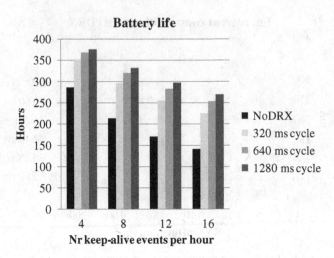

Figure 16.15   Impact of cDRX on standby battery life (2600 mAh capacity, 10 ms OnDuration)

consumption pattern is observed in both cases, driven not only by radio transmission but also by the display usage and processor utilization. However, the current consumption averaged over the whole period still reflects significant savings when cDRX with long DRX cycle is used.

Long DRX cycle adapts well to bursty applications with relatively long silence periods between data transfers. When using short DRX cycle and smaller DRX inactivity timers, it is possible to obtain additional gains by getting into DRX sleep state sooner. Figure 16.17 shows the average current consumption gain when using cDRX with 320-ms-long cycle and a combination of 320-ms-long cycle and 80-ms-short cycle for an extended traffic pattern over 20 minutes duration.

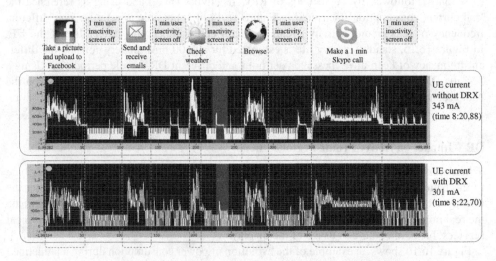

Figure 16.16   Example of current consumption for multiple services

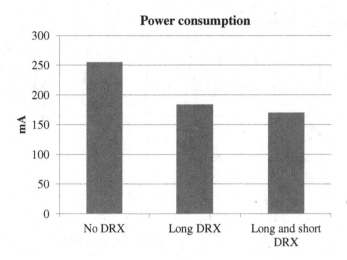

Figure 16.17 Average power consumption savings with cDRX with multi-application profile

16.4 Smartphone Operating Systems

Smartphone OS has an impact on the amount of signalling that one device generates in the network. It not only controls how often the smartphone connects to different servers for basic services, but also may be able to control how installed applications behave towards the network.

With factory default settings and just basic accounts configured most popular OSs only connect to servers sporadically, with a periodicity around 28 minutes or longer. However, when multiple applications are installed, the frequency of background activities increases with intervals as low as 10 minutes for optimized OS with high control over background activities, and even lower for other OSs without strict control of background application activity. Figure 16.18 illustrates the background traffic activity with two different OSs.

In the example illustrated in Figure 16.19, a smartphone creates two eRABs per hour due to the OS without any applications. When some of the popular social media applications (App A, App B and App C) are running in the background, the signalling activity increases up to 16 eRABs per hour. These measurements clearly explain why we can see such a high frequency of eRABs also on the network level.

16.5 Messaging Applications

There are quite a few Instant Messaging (IM) applications (APP) available for smartphones, with different levels of penetration, which include the possibility of sending and receiving text, video and audio messages. Most of these applications have in common that they include some kind of mechanism to determine the availability of the users while others rely on push mechanisms to deliver messages to users whenever they become connected.

Besides users' preferences and usability, from a performance perspective there are some differences in how friendly applications behave towards the cellular network, how much signalling they generate, how much payload they transfer or how much battery they consume.

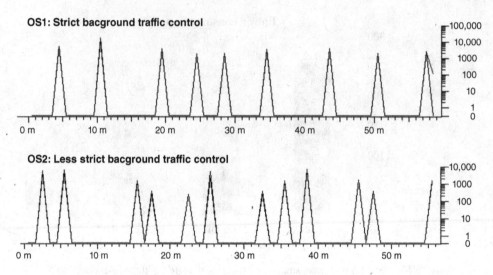

Figure 16.18   Background traffic activity with different operating systems

Probably the two main factors which impact the performance of IM applications are the frequency of keep-alive messages during standby periods and the compression ratio of attachments during active usage of the application.

First, from a signalling point of view, keep-alive intervals between 15 and 30 minutes generate only moderate load to the network, but intervals around 5 minutes or lower can be considered too aggressive and potentially harmful for cell capacity and standby battery life of the UE.

Second, from a payload point of view, there is a significant impact from the strategy used regarding compression of pictures and videos. Some of the most popular applications tend to highly compress the content before transmitting, which makes them very efficient from a network perspective while keeping reasonably good quality when watching the pictures in smartphone's display. However, some applications do very little or no compression in

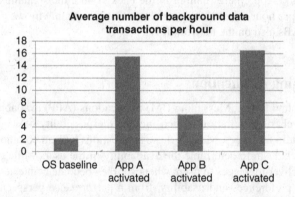

Figure 16.19   Number of LTE packet calls (eRABs) per hour

Table 16.1 Example messaging performance

Type of message	Application	Total delay (seconds)	Power consumption (mA)	Average size of original message sent
Text	IM APP1	6.41	428	~6200 bytes
Text	IM APP2	5.59	323	~750 bytes
Text	IM APP3	4.64	356	~1150 bytes
Picture	IM APP1	16.34	482	~800 kbytes
Picture	IM APP2	12.98	373	~88 kbytes
Picture	IM APP3	11.70	429	~215 kbytes

the UE before uploading the picture to its servers, which creates high uplink traffic and increases the time it takes to upload the picture, the amount of data consumed and the load of the cell.

Compression of attachments in the UE side is therefore very important for efficient messaging applications, and controlling the balance between compression ratio and user satisfaction with the quality of the media received is a key factor to differentiate between friendly and unfriendly applications.

Table 16.1 illustrates the performance of three messaging applications for the transmission of text and picture messages. Application 1 uses clearly more bandwidth and drains also more battery than other two applications. Application 2 takes 20–25% less battery power and uses nearly 10 times less bandwidth than application 1.

16.6 Streaming Applications

Music streaming applications have become very popular in cellular devices, with increased number of users preferring to stream music from remote servers rather than saving it locally in the device. When it comes to how different streaming services interact with the cellular network and smartphones, there are measurable differences in how much signalling, load and battery are used by the application.

Figure 16.20 summarizes the payload generated, battery consumption and radio signalling generated by three different applications during 1 hour of continuous music streaming. The payload reflects the different encoding of the music, where applications with higher quality music will consume more payload than others. Battery consumption on the other hand depends not only on the amount of data transferred, but also on how much processing power is required by the client application to decode and play the music. Finally, the radio signalling indicates how friendly the application is towards the network. Application 3 with significantly lower signalling load than others indicates that it downloads whole songs as a single transfer before playing, while other applications with more signalling download the songs in chunks, causing multiple activation/deactivation of the radio connection for each song. On the other hand, Application 3 uses more power and consumes also more bandwidth than the other two applications.

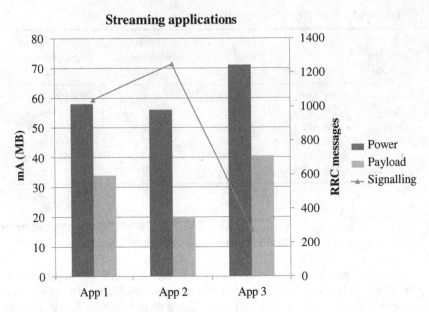

Figure 16.20 Music streaming application performance comparison

16.7 Voice over LTE

LTE networks are designed for packet switched connections and do not support circuit switched voice. During the initial phase of LTE network deployment, operators have relied on solutions such as Circuit Switch Fall Back (CSFB) or dual radio devices connected simultaneously to LTE and 2G/3G voice networks, to provide reliable voice service. However, with the extension of LTE coverage and proliferation of third-party VoIP services, operators need their own solutions for voice over LTE (VoLTE).

At the same time, there are already a number of applications providing over the top (OTT) voice service on smartphones which can be used not only over Wi-Fi connection but also over cellular networks. The voice quality and the mobility performance of OTT voice service are not guaranteed and depend on the instantaneous network conditions.

The aim of this section is to provide a general benchmark between VoLTE and different types of voice clients over LTE network, looking at both user experience and impact on network performance. In addition, legacy circuit switched voice service is used as reference for quality and battery consumption. Different types of VoIP applications can be classified into three categories:

- *Native VoLTE clients:* Embedded VoLTE client which is integrated in the phone software. At the time of testing only a few commercial terminals with native VoLTE clients were available, but more are expected to become available over time.
- *Non-native VoLTE Session Initiation Protocol (SIP) clients:* Third-party applications which can register to IP Multimedia Subsystem (IMS) and establish VoLTE call using Quality of Service (QoS) Class Identifier 1 (QCI1).

- *Over the Top VoIP:* Third-party applications which allow voice or video calls over default non-Guaranteed Bit Rate (non-GBR) bearers, usually for free.

For benchmarking purposes each application was tested individually on two different smartphones with different OSs. All active tests consisted on mobile-to-mobile calls in lab environment in good radio conditions and without additional load. External tools were connected to the UEs and to different network interfaces in order to collect key performance indicators (KPIs) such as battery consumption, voice quality, delay, throughput and network signalling.

16.7.1 VoLTE System Architecture

The VoLTE system enables Voice over IP in a LTE network with QoS. LTE network consists of evolved Node B (eNodeB), a mobility management entity (MME) as well as Serving and Packet Data Network Gateways (SGW and PGW, respectively). The MME and S/PGW form the so-called Evolved Packet Core (EPC). Policy and Charging Resource Function (PCRF) enables QoS differentiation for VoLTE calls.

Core IMS functionality is found in the Call Session Control Function (CSCF). Actual IMS services such as VoLTE and Rich Call Services (RCS) are provided by specific Application Servers (AS). Due to the interworking requirements, there are interfaces from the IMS system and EPC to the CS core and to other relevant control functions such as Media Gateway Control Function (MGCF). Important interworking scenarios are voice calls between VoLTE and CS-only devices as well as call continuity between LTE and 2G/3G coverage areas.

Finally, it is worth noting that subscriptions for both LTE and IMS services are managed in the home subscriber server (HSS). VoLTE system architecture is illustrated in Figure 16.21. More details about VoLTE architecture can be found from References [3] and [4]. VoLTE optimization aspects are considered also in Reference [5].

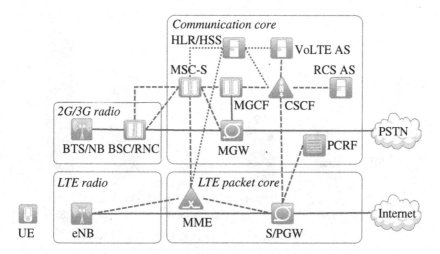

Figure 16.21   Overview of VoLTE system architecture

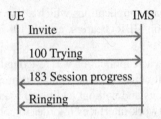

Figure 16.22   SIP call establishment signalling flow

16.7.2   VoLTE Performance Analysis

Call Setup Time

Call setup time is an important KPI, which measures the time it takes for a voice call to be established. In case of VoLTE this KPI can be measured from SIP signalling messages at the UE: from the instant the *INVITE* is sent from the calling party until the *RINGING* indication is received, as described in Figure 16.22.

Figure 16.23 shows how VoLTE can significantly improve the call setup time compared to the legacy circuit switched systems. The total time for the call establishment depends on multiple factors such as whether the terminals were initially on RRC Idle or RRC Connected state when the call was originated. The laboratory measurements show VoLTE call setup time of 0.9–2.2 seconds while the delay in the field is slightly higher depending on the operators' network and transport architecture. The corresponding typical 3G CS call setup time is 4 seconds and with CSFB in both ends approximately 6 seconds.

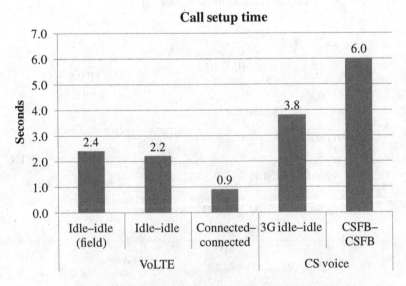

Figure 16.23   Call setup time (VoLTE and CS Voice)

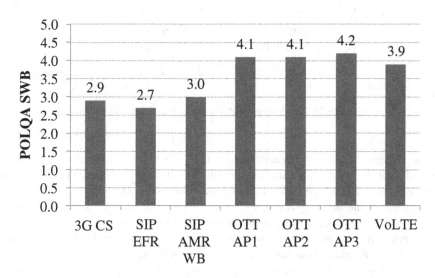

Figure 16.24 Mean opinion score (MOS)

User Experience Benchmark

Different tests were included in the benchmark. First, a 3G circuit switched mobile-to-mobile call using Adaptive Multirate Narrowband (AMR NB) 12.2 kbps codec was used as reference. One native client with AMR Wideband (AMR WB) 23.85 kbps codec was used to show the performance of VoLTE solution. AMR WB is widely known has High Definition (HD) voice. One third-party SIP client was used as representation of non-native clients. That client used two different codecs with Enhanced Full Rate Narrowband (EFR NB) and with AMR WB with variable rate. Finally, three different OTT applications with their own proprietary wideband codec implementation completed the list.

Four metrics are shown in Figures 16.24, 16.26, 16.27 and 16.28 which include the following:

- *Voice quality* is a subjective metric which depends on multiple factors ranging from the speakers' voice tone to the language used in the communication. Traditionally, mean opinion score (MOS) surveys or standardized metrics have been used to measure voice quality, with the latter being more convenient in most cases.
- The *Perceptual Evaluation of Speech Quality* (PESQ, ITU-T P.862) algorithm has been used for narrowband telephony since 2000. However, this methodology was designed for narrowband codecs, limited to 3.4 kHz, and is not well suited for evaluating the quality of new wideband codec used in VoLTE and OTT VoIP applications, which capture a larger range of audio frequencies.
- A new algorithm, Perceptual Objective Listening Quality Assessment (POLQA, or ITU-T P.863) addresses PESQ shortcomings and provides a new narrowband, wideband and super wideband (SWB) that can go up to 14 kHz audio frequency [6].
- *Mouth to ear delay* is a measure of the delay in the speech from the speaker to the listener. It depends on several factors including the size of the buffers to compensate packet jitter and combined delays in radio and transport interfaces. POLQA scores are relatively independent

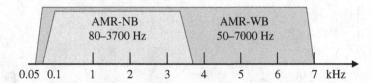

Figure 16.25   Audio bandwidth of narrowband and wideband AMR

of the average voice delay, but according to ITU-T G.114 recommendations, delays lower than 200 ms are needed for excellent quality, 200–300 ms for 'satisfied users' and higher than 400 ms denote a high probability of dissatisfied users [7].

- *Throughput* is the one-way average bit rate measured at IP level during a 2-minute call using a predefined speech sample which combines periods of voice activity and silence. This speech pattern provides approximately 25% of talking time, 25% of listening time and 50% of time where both parties are in silence. Additionally, for VoLTE services PDCP layer may implement robust header compression (ROHC) algorithms which may reduce the actual throughput in the radio interface for VoLTE further.
- *Current consumption* is the average current in milliamperes consumed during a 2-minute call using the predefined voice pattern described in the previous paragraph.

Figure 16.24 presents the average MOS for each of the cases. The voice quality depends heavily on the voice codec sampling rate and the resulting audio bandwidth. AMR NB codec provides audio bandwidth of 80–3700 Hz while AMR WB extends the audio bandwidth to 50–7000 Hz. Furthermore, handset acoustics may delimit the maximum bandwidth provided by the speech codecs. The audio bandwidth is illustrated in Figure 16.25. The CS connections can use either AMR NB or AMR WB while VoLTE in practice always uses AMR WB. The AMR WB data rate for CS connection ranges from 6.6 to 12.65 kbps while the VoLTE connection can use data rates up to 23.85 kbps enhancing the quality of the connection compared to HD voice in CS networks.

The reference 3G CS narrowband call provided a score of 2.9 in the POLQA SWB scale in good radio conditions, while the non-native SIP client with EFR NB codec scored slightly lower with 2.7. The same SIP client with an AMR WB codec configuration provided a score of 3.0. The score of this and other third-party SIP clients could be increased to 3.4 or 3.6, when tweaking some optional functionality such as deactivating voice activity detection (VAD). However, this caused an increase in the power consumption and throughput requirements as the application would transmit a constant data stream regardless of whether the speaker was talking or in silence. It was also observed that some of these clients either had the VAD as an optional feature or included non-standard implementations which did not transmit comfort silence frames with sufficient frequency, which in standard VoLTE should be sent every 160 ms.

Native VoLTE client scored 3.9 with AMR WB 23.85 kbps codec on the same scale, using both VAD and standardized implementation of comfort silence frames. The OTT VoIP applications scored between 4.1 and 4.2 in the POLQA SWB scale using proprietary codecs, quite close to native VoLTE client. The future VoLTE voice quality can be further improved with the new super wideband (SWB) and fullband (FB) codecs, which will be able to cover the whole voice and audio bandwidths. 3GPP has defined a SWB/FB codec in Release 12 called

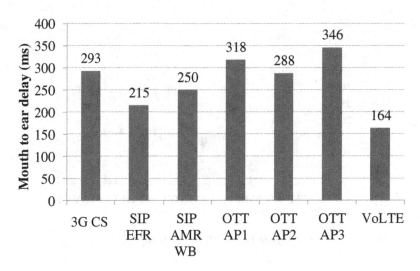

Figure 16.26 Mouth to ear delay

Enhanced Voice Services (EVS) codec. That codec will make it possible for VoLTE to match and beat the voice quality of all OTT clients.

In terms of *mouth to ear delay* in Figure 16.26, VoLTE client provided the lowest value of 164 ms, well below the target of 200 ms for excellent quality. This was even significantly lower than 3G CS call which resulted only slightly under the 300 ms target. The OTT VoIP applications generated higher latency, with two of them exceeding the 300 ms target. However, it is important to note that in these tests the routing of the VoIP streams was done directly between the two calling parties' IP addresses, while in some other implementations a gateway may be used to route the packets adding additional delays to the voice transfer. Additionally, in a live network transport delays may increase the overall delay further, depending on the network architecture.

Figure 16.27 presents the average user throughput and Figure 16.28 the current consumption over a 2-minute call. In this case native VoLTE provided an average of 10.2 kbps, considering the rate during the transmission of voice (40 kbps) and silence frames (3 kbps) and the voice activity of 25% of talking, 25% of listening and 50% of silence. The third-party SIP application with AMR WB codec averaged as low as 8 kbps when using VAD. The lower average throughput in this case was a consequence of not transmitting practically anything during silence periods, which on the other hand contributed to a significant reduction of the MOS. This same application with EFR NB codec generated up to 17.3 kbps, driven by less efficient VAD.

The three OTT VoIP applications ranged between 17.6 and 42.8 kbps average throughput, depending on the codec they used and the particular implementation of features such as of codec rate adaptation or VAD.

Furthermore, at PDCP layer and radio interface, VoLTE and SIP applications using dedicated bearer with QCI1 QoS will further reduce the throughput when ROHC is used.

The lowest power consumption was provided by 3G CS call with 153 mA because of the integration of the voice application on the radio chip. VoLTE power consumption was 232 mA

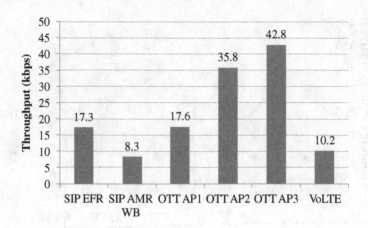

Figure 16.27 Measured throughput

and OTT VoIP power consumption more than 300 mA. OTT VoIP applications were clearly the least efficient, since they used the default non-GBR LTE radio bearer and could not make use of optimized discontinuous reception (DRX) settings for VoIP traffic pattern. These applications required the UE to remain constantly active during the whole duration of the voice call. In the case of native VoLTE and SIP clients, they used dedicated radio bearer with guaranteed bit rate QoS (QCI1) and optimized DRX settings for VoIP traffic pattern. We will show later in this chapter that the integrated VoLTE client can further minimize the power consumption. This kind of optimization will not be available for third-party SIP clients or OTT VoIP which would still need to run on top of the OS and using smartphone's application processor.

Figure 16.29 shows a more detailed analysis of the throughput during the duration of the call for some of the cases. It can be observed how VoLTE provides a clear correlation between the throughput and the voice pattern used for the test, with a peak IP throughput of 40 kbps and a minimum of 3 kbps during silence periods. The SIP AMR WB application shows a similar

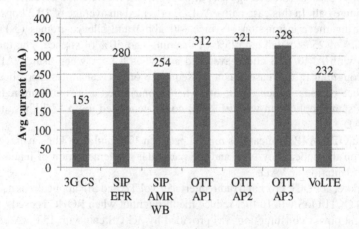

Figure 16.28 Current consumption

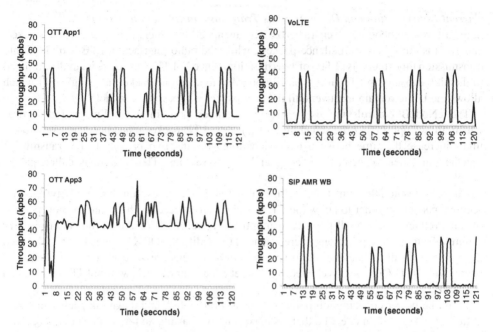

Figure 16.29   Instantaneous throughput over 2-minute call

pattern, but with variable bit rate during speech (25–45 kbps) and practically zero during silence periods. OTT application 1 produced a very similar throughput pattern as VoLTE; however, the peak throughput rose up to 45 kbps, and the low throughput during silence periods did not go below 10 kbps. On the other hand, OTT App 3 did not quite show the same pattern and transmitted around 42 kbps most of the time with some peaks of up to 60 kbps.

Knowing the average throughput figures for different VoIP applications is important for users, considering that today's cellular contracts do have some kind of monthly quota. Taking the numbers from the examples above, it is possible to estimate how many voice minutes could one user afford considering a certain application and a fixed data plan. Table 16.2 shows how in case of VoLTE more than 112 hours of talk time could be provided with a 1 GB data plan, while OTT application 3 could only achieve up to 27 hours with the same plan.

Table 16.2   VoIP minutes with 1 GB data plan using 50% activity voice pattern

	Average throughput (kbps)	Cost of 2-minute call (Kbytes) downlink + uplink	VoIP minutes (hours) with 1 GB data plan
SIP EFR	17.3	519.0	3946 (66)
SIP AMR WB	8.3	249.0	8225 (137)
SIP AMR WB NoVAD	40.6	1218.0	1681 (28)
OTT application 1	17.6	528.0	3879 (65)
OTT application 2	35.8	1074.0	1907 (32)
OTT application 3	42.8	1284.0	1595 (27)
VoLTE	10.2	306.0	6693 (112)

Current Consumption and Discontinuous Transmission/Reception (DRX)

Standard voice encoding is characterized by using 20-ms interval between voice packets and 160 ms during comfort silence packets, while the radio interface in LTE is divided into transmission time intervals (TTIs) of 1 ms duration. Since VoLTE packet size is small compared to the high capacity in LTE, it is typically possible to send one packet in a single TTI which allows the UE to remain inactive until the next packet will become available.

The eNodeB controls the use of radio resources both in the uplink and downlink, which allows optimizing the scheduling of packets. LTE packet scheduler may even sometimes increase the time between two allocations, forcing two packets to be transmitting together, in order to increase capacity and power savings. That feature is called packet bundling.

Connected state DRX functionality in LTE network tries to exploit the gaps between consecutive packets in order to allow the UE to switch off its receive chain during short periods of time, reducing its overall power consumption. However, as VoLTE traffic pattern is very different from other applications, there is a need to modify the DRX settings for the UE when a new VoLTE call is established; so specific DRX settings are sent when a new dedicated bearer with QCI1 QoS class is created. The target behaviour of DRX with VoLTE is illustrated in Figure 16.30.

From the power consumption point of view, if both uplink and downlink packets are sent at the same time within one TTI, there is maximum time available for the power-saving DRX sleep mode. However, when downlink and uplink packets are not synchronized, the probability of obtaining gains is significantly reduced. Silence periods also increment the probability of battery savings, since the interval between packets increases to 160 ms.

Figure 16.31 shows a comparison between the average current consumed by a SIP VoIP client using AMR WB codec with a predefined voice pattern looped between the two UEs in the call. The voice pattern is shown in Figure 16.32. The reference current consumption is obtained for the case where the VoIP client did not activate VAD, which means that each UE was transmitting constantly at a rate of one packet every 20 ms regardless of the voice activity pattern. This kind of behaviour results in no practical opportunities for DRX sleep mode and higher power consumption. In the following cases, connected mode DRX is activated with different settings. DRX long cycle is fixed at 40 ms, but short cycle is changed between 20 and

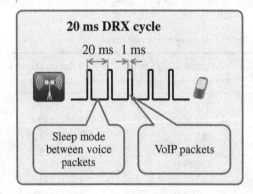
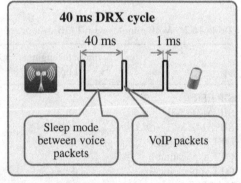

Figure 16.30   Overview of VoLTE transmission and DRX

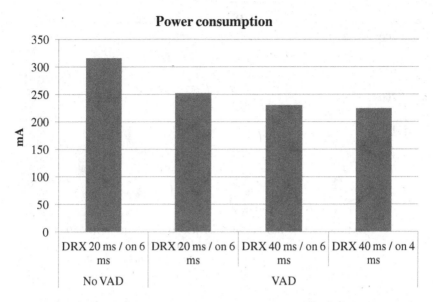

Figure 16.31   DRX impact on VoLTE current consumption

40 ms. Additionally, the OnDuration, which is the time the UE remains active after waking up on every cycle, was changed between 6 and 4 ms.

In the first case, with 20-ms short cycle and 6 ms on duration, the average current consumption was reduced by 20% compared to the reference case. Increasing the short cycle to 40 ms provided additional gains up to 27%. Finally reducing the OnDuration to 4 ms only provided small additional gain. During all the tests, the voice quality in terms of MOS was monitored, but the increase of the packet delay jitter caused by having 40 ms DRX cycles instead of 20 ms DRX cycles did not reflect on any appreciable MOS degradation during the call.

Power Consumption with Integrated VoLTE Client
Power consumption with VoLTE and other VoIP services depends highly on the UE and how efficiently the voice is handled by hardware and software implementation. While the expectation is that VoLTE should provide similar or better power consumption than legacy 3G circuit switched voice to offer a good user experience, some terminals and early implementations of VoLTE may not achieve the required level of efficiency when using solutions which rely on UE application processors to run the VoLTE stack. However, a UE with a chipset-integrated VoLTE implementation and optimized DRX settings can achieve a clearly lower power consumption. Figure 16.33 shows the power consumption of integrated VoLTE implementation with different DRX settings illustrating that VoLTE can obtain the same power consumption as 3G CS voice. OTT VoIP without DRX may consume up to two times more current.

TALK	Silence	TALK	Silence	LISTEN	Silence	LISTEN	Silence
2 s	2 s	3 s	4 s	2 s	2 s	3 s	4 s

Figure 16.32   Voice pattern in power consumption testing

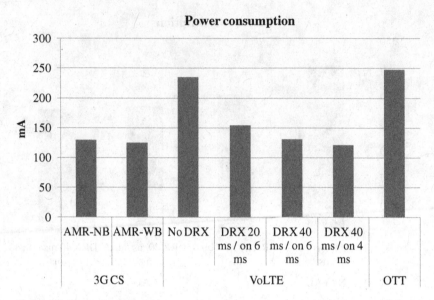

Figure 16.33 VoLTE power consumption measurements with integrated VoLTE client

16.7.3 Standby Performance

Another important aspect of VoLTE performance is the behaviour during standby periods when the application is in background without performing any calls. During this period, different applications usually do some kind of activity to ensure connectivity is maintained either with IMS system or similar proprietary solution in case of OTT VoIP.

Figure 16.34 shows the number of radio signalling messages generated when using different VoIP applications. Native VoLTE service was the one generating the least number of messages,

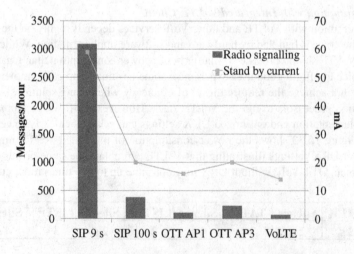

Figure 16.34 Radio signalling and current consumption during standby

which were pretty much the minimum load that the smartphone generated anyway with basic configuration. OTT VoIP applications generated between 50% and 200% more radio signalling than VoLTE. The third-party SIP clients tested included a configurable timer for sending keep-alive messages, which by default were configured between 9 and 600 seconds. The cases with 9 and 100 seconds keep-alive interval are presented in the chart, showing the high impact in both the number of messages generated during 1 hour standby time and the average current consumption, which in the worst case grew up to four times higher than in VoLTE case.

In conclusion, native VoLTE clients are very well optimized to minimize standby impact on UE's battery life and network load, while some of the OTT VoIP and third-party SIP clients may generate significantly higher load and negatively impact the battery consumption and network capacity, especially if very aggressive keep-alive settings are used by default.

16.7.4   Impact of Network Loading and Radio Quality

As shown in previous sections, OTT VoIP applications, although less efficient in terms of data and battery consumption than VoLTE, can still provide very good voice quality under good radio conditions and low load. However, in a real network, the radio conditions can get bad and other users will generate additional traffic in the cell.

One essential difference between OTT VoIP and VoLTE is that the latter gets a dedicated QCI1 bearer to transport the voice packets, which receives high priority in the radio interface and transport, while OTT VoIP generally will be transferred using the default data bearer mixed with other low-priority traffic.

The effect of this difference in how voice packets are delivered can be clearly seen in Figures 16.35 and 16.36, where MOS and mouth to ear delay for one OTT VoIP application and VoLTE are shown for good and bad radio conditions and different levels of background data load. In this case, load was generated with 19 UEs doing continuous uplink and downlink file transfer protocol (FTP) transfers. Higher effective loads were achieved by increasing the

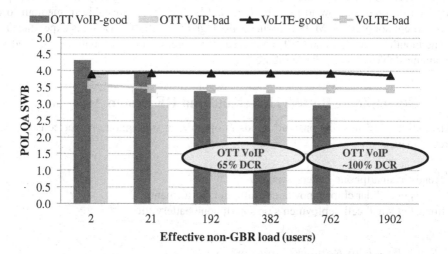

Figure 16.35 MOS (POLQA SWB) versus non-GBR load for VoLTE and OTT VoIP in good and bad radio conditions

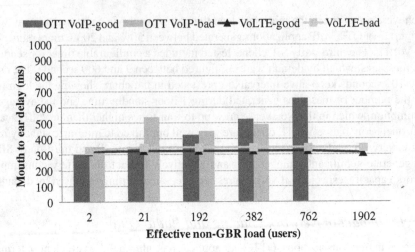

Figure 16.36 Mouth to ear delay versus non-GBR load for VoLTE and OTT VoIP in good and bad radio conditions

relative weight of background data compared to the OTT VoIP users. For example, 19 UEs with a 10 to 1 scheduling ratio generate the radio load equivalent to 190 UEs with a 1 to 1 ratio.

In Figure 16.35 it can be observed that VoLTE is not affected at all by any background load level, since its packets get always prioritized over FTP. Only increased load of other VoLTE users may eventually have an impact on the quality. On the other hand, there is an impact of bad radio conditions on the MOS, which in this case was translated in a drop from 3.9 to 3.5 POLQA score. This reduction is caused by additional jitter and delay due to packet retransmissions at the radio interface. It is important to note that 3.5 is still very good quality, actually better level than AMR NB in good radio conditions.

OTT VoIP, while delivering very good MOS in low load, quickly experiences quality reduction in relatively low to medium load and becomes very unstable in medium to high loads, where not only the MOS is reduced, but high drop call rate (DCR) is experienced.

The mouth to ear delay in OTT VoIP case also increases proportionally to the load while remaining stable for VoLTE; see Figure 16.36.

16.8 Smartphone Battery, Baseband and RF Design Aspects

This section presents solutions and trends in optimizing smartphone power consumption in three domains:

- Trends in smartphone battery capacity
- Trends in cellular chipset power consumption performance
- Impact of small cell deployments on smartphone battery life

16.8.1 Trends in Battery Capacity

Smartphone's battery life is perhaps one of the KPIs when it comes to measuring user experience. In many instances, battery life is still published for the traditional voice talk time

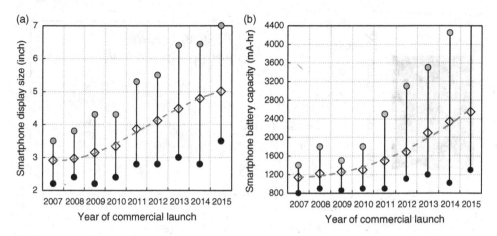

Figure 16.37 (a) Trend in screen size. (b) Battery capacity trend in smartphones. Data extracted from 388 smartphones

and smartphone standby time. These metrics, inherited from the era of the voice-centric feature phones, no longer reflect the end-user experience. The introduction of smartphones has shifted the customer's habits towards data-centric activity profiles. Recent smartphone activity statistics have shown that only 25% of the usage time is spent making voice calls, while the remaining 75% is split between a variety of use cases ranging from radio/video streaming, web browsing to online gaming or social networking. With the exception of radio streaming, all of these activities imply an active and heavy touch-screen activity. And improving touch-screen user experience, such as web browsing, has pushed OEMs to use larger, higher resolution screens. Figure 16.37a shows that the screen size has grown on average from 3.0 inches in 2007 to 5 inches in 2014. This trend, together with ever-increasing screen resolutions, has pushed chipset vendors to deliver always more powerful graphics processors and application engines [8]. The combination of these two factors has dramatically changed the power supply budget of a chipset: in feature phones, the cellular subsystem dominates the power consumption over screen activities, the power amplifier (PA) being the largest contributor at high output powers. In smartphones, the consumption distribution is reversed [8].

Figure 16.38 illustrates these impacts by comparing a smartphone power consumption using animated wallpaper (a) versus a static picture (b). Both measurements are performed with identical backlight levels (50%) and all radio subsystems being switched off, that is, UE is in flight mode. With 1.7 W consumption, the animated wallpaper more than doubles the UE battery power consumption.

This explains why the modern superphone, with its powerful application engines, high-resolution displays, requires a battery capacity which has nearly doubled in 7 years, reaching an average of 2500 mA-h in 2014 (Figure 16.37b). The emergence of phablets can be clearly seen with battery capacities in excess of 4000 mA-h and screen sizes greater than 6 inches.

Had the energy efficiency of lithium-ion (Li-Ion) batteries been constant over this time span, the need for doubling battery capacity would have led to doubling weight. This would translate into a severe penalty in the smartphone weight budget. In reality, battery energy efficiency keeps on improving at a rate of approximately 5–7% every year [9]. For example,

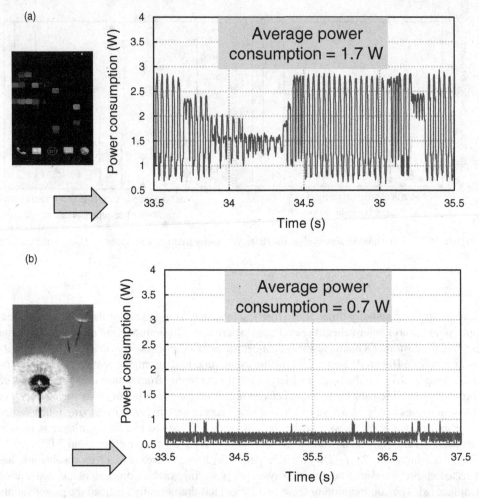

Figure 16.38 UE power consumption in flight mode. (a) animated wallpaper; (b) static picture

in 2014, an advanced 2100 mA-h battery weighs 35 g. This represents an energy density of approximately 230 Wh/Kg. The same battery built with the average 190 Wh/Kg density of the 2009–2010 timeframe would have weighed 42 g. Alternatively, a 2009 35 g battery pack would have delivered a capacity of 1750 mA-h. These examples are only a simplistic summary of improvements in Li-Ion batteries because energy density does not depend purely on weight/volume of the cell, but also on its thickness [9]. These slow improvements explains why device manufacturers have had no other choice but to increase battery weight and/or volume and to put a lot of pressure on chipset vendors to always deliver more power-efficient cellular subsystems. At the other end of the food chain, advanced 3GPP features give telecom operators plenty of tools which can further reduce smartphone power consumption. For example, activating discontinuous reception (DRX) and transmission (DTX) is one option. Deploying small cells is another.

Section 16.8.2 presents the efforts accomplished in successive generations of LTE chipsets to improve power efficiency. The benefits of small cell deployments are presented in Section 16.8.3.

16.8.2   Trends in Cellular Chipset Power Consumption

Assessing the contribution of an entire smartphone, whether it be in idle or in 'active/busy' states, is a rather complex task. When idle, numerous OS-related and application 'keep-alive' specific tasks wake up several hardware blocks, leading to frequent, small packet radio transmissions even when screen is in its 'off' state. In active state, such as radio streaming, battery consumption depends on screen activity, streaming protocol, client payload on application processor, as well as the amount of data transferred. We propose to reduce this complexity by

- Focusing solely on the cellular subsystem-intrinsic contribution, that is, digital baseband (BB) modem, RF transceiver, RF front-end and associated power management circuitry, for example, DC–DC converters and envelope tracking supply modulators
- Restricting measurements to the LTE FDD standard in conducted test conditions, that is, loading the RF front-end with an ideal 50 Ohm load
- Investigating operation in LTE band 4 and 20 MHz cell bandwidth only

The UE power consumption modelling presented in this section is an extension of the work presented in Reference [10] in which the LTE cellular subsystem model can be defined as

$$P_{\text{cellular}} = m_{\text{con}} \times P_{\text{con}} + m_{\text{idle}} \times P_{\text{idle}} + m_{\text{DRX}} \times P_{\text{PDRX}}(\text{W}) \qquad (16.1)$$

where 'm' is a binary variable describing whether the UE is in RRC_connected (con), RRC_idle (idle) or DRX mode. The associated P value describes the power consumption in the given RRC state/mode of operation. For the sake of simplicity, only RRC-connected state measurements are presented. Beyond enabling a detailed benchmark of different chipsets, 'P_{con}' is also used in Section 16.8.3 to illustrate the savings that can be offered by small cell deployments.

From a power consumption characterization perspective, one essential difference between WCDMA and LTE is that the LTE-flexible air interface introduces a much greater number of variables. Assessing the impact of each one on 'P_{con}' can be approximated by defining two categories of transfer functions: either uplink (TX) or downlink (RX) related. Each category is then further split into a baseband and a RF subcategory. The resulting four transfer functions (P_{RxBB}, P_{RxRF}, P_{TxBB}, P_{TxRF}) are described in Figure 16.39. A test case (TC) is dedicated to assess each variable for each function. For example, TC 1 investigates the impact of modulation and coding scheme (MCS) on P_{RxBB} while all other variables are held constant.

Figure 16.39 also shows the experimental setup. Smartphones are tested in a faraday cage and are connected to an eNodeB emulator (Anritsu 8820c) via a pair of coaxial cables. Both supply voltage and current consumption are acquired over at least 30 seconds per point by an Agilent N6705B power supply. The measurements accuracy is estimated at ±10 mW and are performed with screen maintained in its 'Off' state.

Figure 16.40 presents the characteristics of the benchmarked smartphones. With the exception of UE 3, all UEs are high-end flagship devices. The selection is representative of two

Link	Category	Sub-category (impact of)	Transfer function	Test case	Downlink parameters			Uplink parameters		
					MCS	PRB	S_{Rx}	MCS	PRB	S_{Tx}
DL	Baseband	MCS / PRBs in SISO and MIMO	P_{RxBB}	1	[0,28]	100	−25	6	100	−40
				(2)	0	[0,100]	−25	6	100	−40
	RF power	SISO/MIMO	P_{RxRF}	3	0	100	[−25,−90]	6	100	−40
UL	Baseband	MCS	P_{TxBB}	4	0	3	−25	6	[0,100]	−40
		PRBs		5	0	3	−25	[0,23]	100	−40
	RF power	QPSK / 16 QAM	P_{TxRF}	6	0	3	−25	6	100	[−40,23]

Figure 16.39 LTE chipset power consumption test cases and experimental setup. Test case 2 is voluntarily omitted for the sake of clarity. Variables are shown in brackets

generations of baseband modems and three generations of RF transceivers. The devices also allows for comparing two PA control schemes: Average Power Tracking/Gain Switching (APT/GS) in UEs 1, 2 and 3, versus the recently introduced Envelope Tracking (ET) in UEs 4 and 5. The test results shown in Figure 16.41 reveal several remarkable findings.

Starting with P_{RxBB} performance in Figures 16.41a and 16.41b (TC1), it can be observed that baseband decoding energy efficiency has significantly improved in UEs 3 and 4 versus UEs 1 and 2. In 2×2 MIMO operation, the average power consumption slope is nearly halved, from 1 mW per Mbps, to approximately 0.4 mW per Mbps. For example, at 80 Mbps, UE 4 power consumption is 33% better than that of UE 1. This means that operating UEs 3 and 4 (and by extension UE 5) at its maximum downlink throughput whenever possible is beneficial from an energy efficiency point of view. These measurements also show that single input single output (SISO) mode saves on average 100 mW over 2×2 MIMO operation. Note that the earlier measurements in Figure 16.9 showed a clear difference in the power consumption between 10 and 20 MHz bandwidths while Figure 16.41 shows that the number of received PRBs with the given bandwidth has only a minor impact of the power consumption.

An even more pronounced pattern can be observed in Figure 16.41c (TC4) which shows that transmitter baseband performance is nearly independent of the number of transmitted physical resource blocks (PRB). Figure 16.41d shows that, with the exception of UE 1, using 16QAM has no impact on transmitter power consumption. The consumption step in UE 1 is believed

	UE 1	UE 2	UE 3	UE 4	UE 5
Launch date	June 2012	April 2013	July 2014	August 2014	October 2014
Operating system	Android 4.0.4	Android 4.1.2	Android 4.4	Android 4.4	Android 4.4
Modem and CPU	Part #A	Part #B	Part #C	Part #D	Part #D
Modem CMOS node	28 nm LP	28 nm LP	28 nm LP	28 nm HPM	28 nm HPM
RF transceiver	Part #E	Part #F	Part #G	Part #F	Part #F
Transceiver CMOS node	65 nm	65 nm	65 nm	65 nm	65 nm
Band 4 PA control	Gain switching (GS) + average power tracking (APT)			GS + APT + envelope tracking (ET)	
LTE bands	4, 17	1, 2, 4, 5, 17	2, 4, 7, 17	4, 7, 17	1, 2, 3, 4, 5, 7, 8, 12, 17
UE category	3	3	4	4	4

Figure 16.40 Smartphones under test (CPU = central processing unit, LP = lower power, HPM = high-performance mobile)

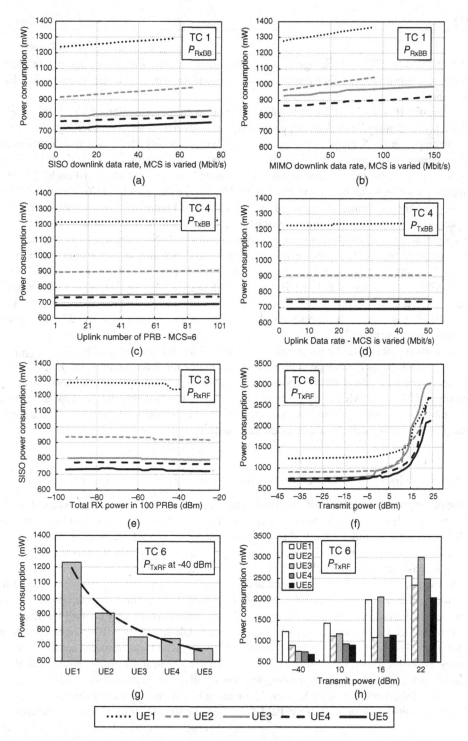

Figure 16.41 Band 4, 20 MHz cell bandwidth, LTE subsystem consumption benchmark (UE 5 results are restricted to SISO operation due to lack of RF Rx diversity test port)

to be related to a PA bias change to accommodate the larger peak to average power ratio in 16QAM transmissions. This adjustment is PA dependent and is no longer required in UEs 2, 3, 4 and 5.

In Figure 16.41e (TC3) it can be observed that, with the exception of UE 1, the receiver chain power consumption does not change significantly across the entire UE dynamic range. For example, in UE 5, the power consumed at cell edge is barely 1% higher than that measured at maximum input power, that is, under the eNodeB. Even with UE 1, the increase is less than 3.5% at cell edge, a value below the estimated accuracy of the test bench. The most impressive savings can be seen in Figures 16.41f, 16.41g and 16.41h (TC6). Figure 16.41g is a zoom on the performance measured at −40 dBm, and Figure 16.41h eases the comparison at key transmit powers. Significant savings are delivered generations after generations. For example, UE 5 consumption is 45% less than that of UE 1. Not only does UE 5 achieve best in class performance, but it also exhibits a near flat consumption over a greater output power range than any other devices. It is also interesting to compare the performance of UE 3 which uses APT versus that of UE 4 or UE 5 which both use ET.[1] While all three UEs have similar performance up to 0 dBm, above 10 dBm UE 4 exhibits an impressive 47% and 17% gains over UE 3 at 16 and 22 dBm, respectively. UE 5 provides gains of 44% and 32% at the same power ratings over UE 3.

In summary, the LTE chipset power consumption can be approximated by using P_{RxBB} and P_{TxRF}, that is, the performance is directly related to downlink throughput and uplink transmit power. Since LTE power control loops aim at maintaining the target uplink SINR for a given number of PRBs and MCS, increasing uplink throughput indirectly leads to increasing the UE transmit power, and therefore to increasing the subsystem power consumption. These aspects are presented using drive test data in the next section.

16.8.3   Impact of Small Cells on Smartphone Power Consumption

One of the primary objectives of small cell deployments is to deliver a higher average downlink RF power to the UE, that is, to ensure UE experiences a lower pathloss than in macro cells. Common sense would then say that UE transmit power should on average be lower. Section 16.8.2 has shown that UE power consumption is heavily dependent on its transmit power. So it is reasonable to expect that small cells deployments should help improving UE battery life. The aim of this section is to estimate consumption savings by establishing a relationship between reference signal received power (RSRP) levels and UE power consumption using the previously defined P_{CON} transfer functions. These estimations can only be used for illustration purposes since UE transmit activity toggles between cDRX/DTX in RRC-connected state and idle states. The estimations are then compared with field data measurements across the six use cases: radio streaming, two distinct video streaming services/applications, application download from Google Playstore, Skype two-way video call and a file upload use case. The analysis is accomplished in three steps: Section 'Estimating UE Transmit Power Profiles' assesses the LTE transmit power profiles from field data, Section 'Establishing Empirical Relationship

---

[1] This comparison does not pretend to draw general conclusions as UE 3 is a mid-range (200–300$), cost-optimized, smartphone, while UE4/5 are high-end iconic devices (600$ price range). It is therefore likely that the selection criteria for UE 3 power amplifier rely on different cost/performance trade-off criteria than those of UE 4/5.

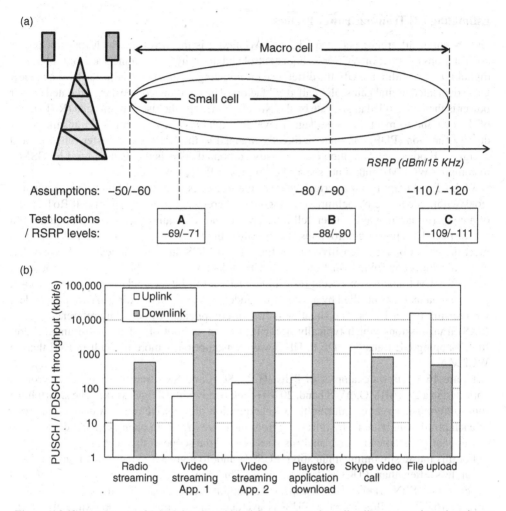

Figure 16.42 (a) Range of RSRP levels assumed for macro and small cells versus selected static test locations. (b) Drive route average physical layer uplink (PUSCH) and downlink (PDSCH) throughput for six use cases. All measurements are performed on UE 4

Between RSRP and UE Transmit Power' establishes an empirical relationship between RSRP versus UE transmit power levels and Section 'Establishing Relationship Between RSRP and UE Power Consumption' compares the resulting P_{con} versus RSRP with field performance. Figure 16.42a shows the range of RSRP levels assumed to define an average macro cell and small cell. Three static test locations, labelled A, B and C, are used to collect UE power consumption at the boundaries of each cell type. All field data measurements are performed using UE 4 in near free space test conditions, that is, with UE laid on a wooden table and with no surrounding objects and no user interaction to ensure that the smartphone-radiated performance is not impacted by hand or head effects. Figure 16.42b shows the average uplink and downlink physical layer throughput measured across several drive tests.

Estimating UE Transmit Power Profiles

One essential difference between LTE and WCDMA is the ways in which transmit power control loops operate. In WCDMA, both control and data physical channels are multiplexed at digital IQ level within the UE, the difference in spreading factors being adjusted by the NodeB using dedicated digital gains, also called 'β' factors. The resulting composite modulated carrier power is then controlled at slot rate by the NodeB to manage the rise over thermal (RoT) noise of the cell and to meet the target signal level link quality. The resulting UE transmit power density function (PDF) can then be directly mapped to the UE P_{TxRF} transfer function, and average UE power consumption can be easily computed. This is the method used by GSMA to compute WCDMA talk time, see also examples in Reference [11].

With LTE, the uplink air interface changes the role of power control loops. First, transmissions within a cell are by definition orthogonal. Therefore, controlling intra-cell RoT is less of a concern than managing inter-cell interference and maintaining adequate link quality. In this respect, LTE power control aims at controlling the UE transmit power spectral density to meet the eNodeB target sensitivity level for a given MCS and PRB allocation. Second, UE either transmits control information on PUCCH or data on PUSCH, but never simultaneously. Since PUCCH transmissions occur on a single PRB, at low bit rate and high coding gain, each physical channel is controlled by a completely independent loop, resulting in two independent UE transmit power PDFs. The situation is further complicated by sounding reference signals (SRS) transmissions which typically occur in the last symbol of a 1 ms sub-frame. Therefore, mapping UE transmit PDF to UE power consumption is more difficult in LTE than in WCDMA.

Figure 16.43 shows examples of PUCCH, PUSCH and SRS transmission power profiles collected in a 2×2 MIMO, AWS band, 20 MHz cell bandwidth LTE network.[2] The smartphone runs radio/music streaming during the whole duration of the drive route. A detailed analysis of each profile can be done by retrieving from the drive logs both open-loop and closed-loop power control parameters. This analysis goes beyond the scope of this section.

For this music streaming application, PUSCH and, to a greater extent, SRS transmissions, occur less often than PUCCH transmissions. This can be clearly seen in both PDF and power versus SFN graphs. This is expected as uplink traffic is small compared to downlink (Figure 16.42b). In this example, PUSCH average transmit power is approximately 18 dB higher than that of PUCCH. Extracting the open-loop P_o nominal parameters from SIB2 and RRC connection setup messages for PUSCH and PUCCH explains that this power offset cannot be less than 14 dB in this network: P_{o_PUSCH} is set to −106 dBm, P_{o_PUCCH} to −120 dBm and alpha 1.0. See Chapter 11 for more details about the power control parameters.

Establishing Empirical Relationship Between RSRP and UE Transmit Power

Figures 16.44a, 16.44c and 16.44e use this set of drive test logs to plot constellations of PUSCH, PUCCH and SRS transmit power versus RSRP levels for a set of applications restricted to radio streaming, Skype video calling and a large file upload to a cloud server.

---

[2] Detailed chipset parameters are captured at a TTI time resolution for each drive test using the Accuver XCAL software suite, and postprocessed using the Accuver XCAP tool.

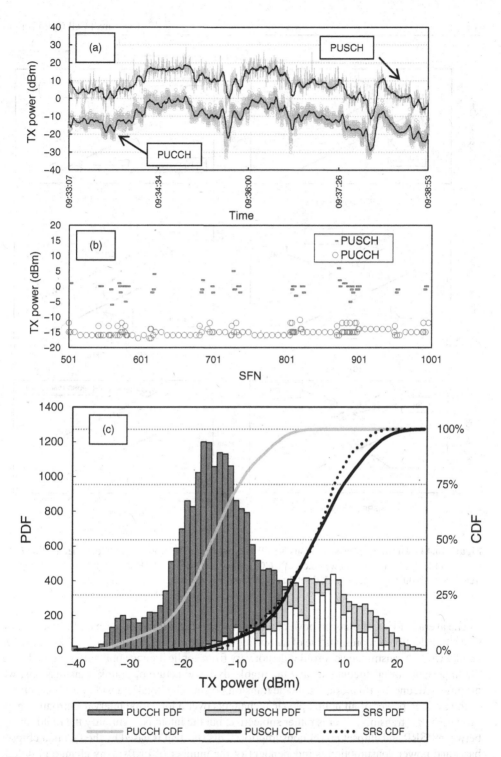

Figure 16.43 Example of PUCCH, PUSCH and SRS transmit power profile in the drive route for a radio streaming application. (a) Transmit power versus time; (b) transmit power versus serial frame number; (c) transmit power CDF and PDF

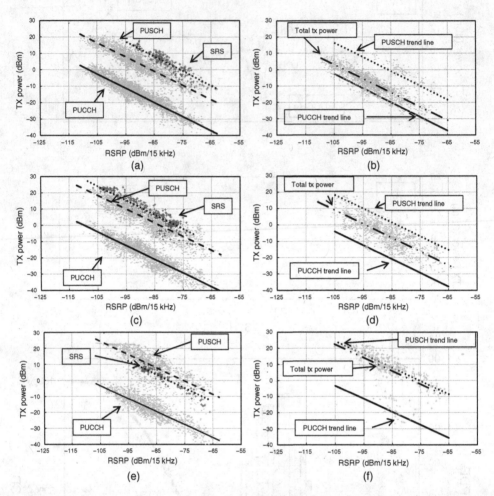

Figure 16.44 UE transmit power versus RSRP in radio streaming (top row), Skype video calling (mid row) and file upload (bottom row). (a,c,e) PUCCH, PUSCH, SRS versus RSRP clouds and approximate trend lines; (b,d,f): approximate 'UE total tx power' versus RSRP cloud and trend lines

The average PUSCH transmit power increases across these three use cases. This is expected since the selection is representative of increasing uplink traffic (cf. Figure 16.42b). On the other hand, PUCCH transmit power profiles do not vary from application to application. Considering the important power fluctuations and the highly dynamic nature of uplink transmissions, we propose to define a virtual 'total' average transmit power. The 'total' transmit power is defined as the average power of all uplink physical channels over a long time interval. Introducing an 'average total' transmit power is rather subjective, but the intent is to simplify the relationship between RSRP and chipset power consumption. This shortcut is eased by the fact that chipset baseband power consumption is independent of the number of PRBs. Any change in uplink PRBs or MCS is accounted for by transmit power control loop adjustments. Therefore, averaging each physical channel transmit power should provide a reasonably accurate estimation and should reflect that UE tx power should increase as uplink throughput is increased. A

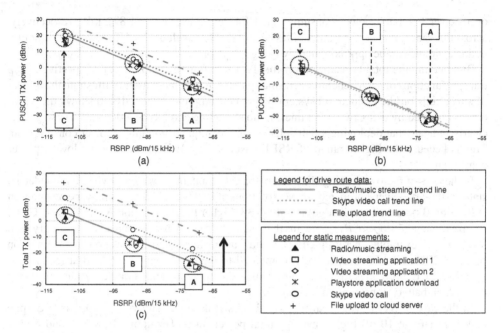

Figure 16.45 PUSCH (a), PUCCH (b) and total transmit power (c) versus RSRP trend lines versus static locations "A", "B" and "C" for six smartphone applications (music streaming, video streaming, application download, skype video call and file upload)

detailed analysis and description of LTE power control loops can be found in Chapter 11. In Figures 16.44b, 16.44d and 16.44f, a 2-second averaging window is used to compute the equivalent 'total' power. For each constellation, an approximate trend line is extracted.

Figure 16.45 overlays each trend line with measurements performed in static locations A, B and C for the six use cases presented in Figure 16.42b.

Measurements performed in static locations globally match within a few decibels the general trend lines extracted from drive route test conditions. Interestingly, the trend line captured for radio streaming is a good match for applications with low uplink traffic. For example, Figures 16.45a and c show that the two video streaming applications and the Playstore download use cases closely follow the PUSCH and total transmit power trend lines of radio streaming. This is somehow expected as the uplink average throughput for this selection of use cases ranges from 12 to 210 kbps for radio and Playstore download use cases, respectively (cf. Figure 16.42b). Total transmit power is on average 10 dB and between 20 and 24 dB higher than that of radio streaming when performing a Skype video call and a file upload, respectively. The impact on battery life is therefore greater for these use cases. Figure 16.45b shows that all PUCCH transmissions are independent of use cases and test conditions.

Establishing Relationship Between RSRP and UE Power Consumption

Section 16.8.2 concluded that P_{con} can be approximated as a function of P_{RxBB} and P_{TxRF}. Yet, the example of Figure 16.42b indicates that the downlink average throughput experienced by UE4 spans from 570 kbps for radio streaming to a little more than 30 Mbps in the case of application download. Figure 16.41b shows that this range of throughput has very little impact on receiver baseband power consumption: UE4 consumes 865 and 872 mW, respectively, across this range, that is, a variation of less than 1%. We conclude that UE power consumption in this example can be approximated by using P_{TxRF} only.

By injecting the total transmit power trend lines of Figure 16.45 into the respective P_{TxRF} transfer functions of each benchmarked smartphones, it becomes easier to illustrate the impact of small cells on the cellular subsystem power consumption. Figure 16.46 shows the set of curves obtained by using two trend lines: radio streaming and Skype video call. These graphs clearly indicate that deploying small cells is one technique to minimize power consumption. This is because the higher range of RSRP ensures the UE is operated in its lowest power consumption state.

In the case of radio streaming (Figure 16.46a), this statement is clearly visible in the latest, state-of-the-art, power-optimized chipset solutions of UEs 3, 4 and 5 where power consumption at location B (small cell, cell edge) is only 2–3% higher than in location A. At location C, the sharp exponential slope of P_{TxRF} means that small cell battery life savings may reach very high levels. In this example, operating UEs 3, 4 and 5 at this location increases power consumption by 25%, 20% and 13%, respectively, compared to the performance obtained at location B. Relatively, UE 5 provides lower savings. This is due to the extended range of transmit power over which battery consumption is minimal.

In the case of Skype video calling, small cells still deliver superior performance despite operating UEs at 10 dB higher average total power. In this example, UEs 3, 4 and 5 only consume 4–7% more power at location B than at location A. At location C one can see the benefits of Envelope Tracking in UEs 4 and 5 over UE 3. Operating UEs at location B versus location C could bring savings in the range of 30–45%. This is due to the near exponential rise in power consumption as UE is operated closer to its maximum output power. However, in practice these gains are moderated by two factors:

- cDRX and transitions to idle state lower the consumption.
- Screen activity and application processor may dominate and flatten these theoretical savings.

Figure 16.47 illustrates the actual performance of UE 4 at the three static locations for the case of radio streaming, two different video streaming applications and Skype video calling. Figure 16.47a shows that listening to radio with screen off at location B delivers a near 20% better power consumption than at location C. Turning 'on' the screen halves the gain. The impact of screen and application engine is more visible in Figure 16.47b. Video application 2 drains less battery than application 1. But battery savings achieved with both video streaming applications are in the same order of magnitude in small cells than those experienced with music/radio streaming, ie approximately 10% savings. On the contrary, Skype two-way video calling power consumption profile flattens the cellular subsystem savings.[3] Note that at location C, a small change in UE position may induce a significant increase in power consumption, even though the RSRP level may only drop by a few decibels (say 3–5 dB). This is because UEs are operated in their near exponential operating regions. For example, during one trial, RSRP levels dropped from −110 dBm to approximately −115 dBm by moving the UE by a few centimetres. This resulted in a 12% power consumption increase during a Skype video call.

---

[3] Power consumption with video calling varies according to the motion of the subject being filmed. Consumption is lower when subject is stationary/still.

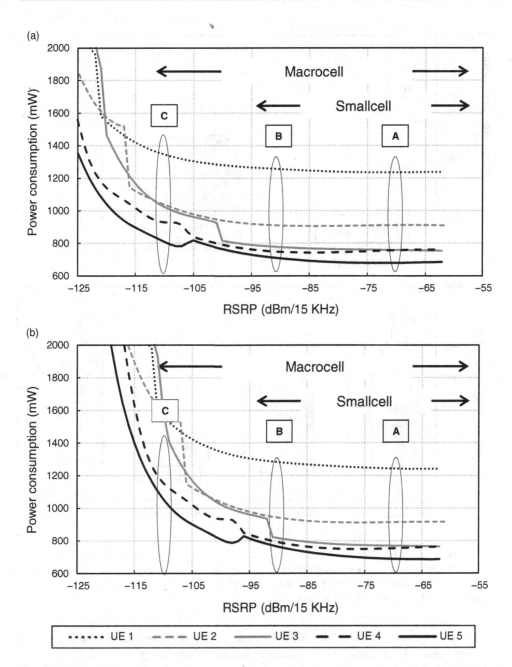

Figure 16.46 Cellular subsystem power consumption versus RSRP versus UE: (a) 'radio streaming profile'; (b) Skype video call profile

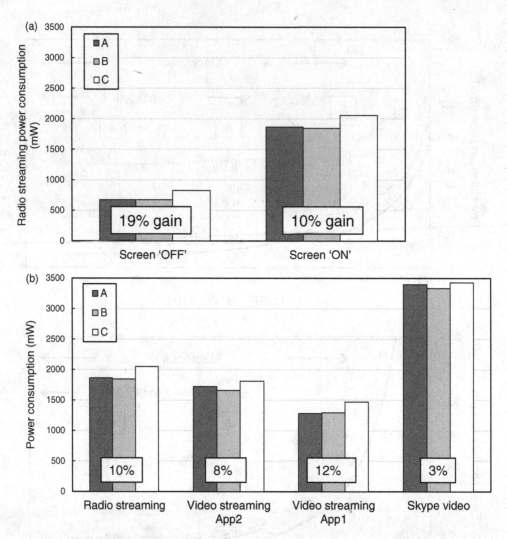

Figure 16.47 Average smartphone (UE4) power consumption measured in field locations "A", "B" and "C". (a) Radio streaming power consumption vs. screen state, (b) Application power consumption with screen in 'active state'. Measurements accuracy is estimated at ±2% and are performed over several 2-minute-long logs using 100 μs time resolution

In summary, several efforts are made to continuously improve smartphone battery life. Chipset vendors have managed to significantly reduce the intrinsic power consumption of the cellular subsystem over a short lapse of time. The use of advanced PA control schemes also helps improving power consumption at high output powers. The deployment of small cells is a promising solution to ensure the contribution of the cellular subsystem is minimal. Yet, the consumption savings accomplished in small cells can be partially, and in certain cases completely, masked by other contributors, such as application engine, encoding and decoding as well as screen activities.

16.9 Summary

Smartphones are the dominant source of data traffic in mobile broadband networks. This chapter showed that the typical LTE smartphone user consumes more than 1 GB data per month and most of the traffic is in downlink direction. One exception is mass events that typically turn to be uplink limited. The traffic-related signalling traffic is high up to 500 packet calls per subscriber per day or one packet call every 2 minutes during the busy hour. That leads to 10% of LTE subscribers being RRC connected simultaneously.

Different OSs and applications such as Instant Messaging or music streaming have different impact on smartphone's performance and cellular network depending on specific implementation decisions. In general, longer keep-alive periods and data compression are important factors which help applications to behave more friendly and improve user's perception.

Battery life for smartphones is one of the optimization areas. The battery life depends on how services and applications are implemented, but there are also functionalities in the cellular network such as cDRX which when optimized can significantly improve current consumption. Other functionalities, such as Carrier Aggregation, can improve battery consumption for large data transfers, but affect negatively with short data transfers.

VoLTE and OTT VoIP solutions are very important since the traditional circuit switched voice service is not available in LTE networks. While various VoIP solutions can provide very good voice quality and user experience in certain conditions, it takes a fully integrated VoLTE solution to achieve high standards of voice quality consistently, maintain similar battery life as legacy circuit switched voice, maximize network voice capacity and ensure seamless mobility towards legacy systems.

The power consumption of the cellular subsystem in the smartphones has improved considerably during the last few years with the new RF optimization solutions. The small cell solutions can also reduce the smartphone power consumption by using lower UE transmit power levels.

References

[1] 'Smartphone Traffic Analysis and Solutions', Nokia white paper, 2014.
[2] R. Sanchez-Mejias, M. Lauridsen, Y. Guo, L. Ángel Maestro Ruiz de Temiño, and P. Mogensen, 'Current Consumption Measurements with a Carrier Aggregation Smartphone', IEEE VTS Vehicular Technology Conference. Proceedings, 2014.
[3] 'IMS Profile for Voice and SMS Version 8.0', Groupe Speciale Mobile Association (GSMA), 2014.
[4] 'From Voice Over IP to Voice Over LTE', Nokia Siemens Networks white paper, November 2013.
[5] 'Voice Over LTE (VoLTE) Optimization', Nokia white paper, 2014.
[6] 'ITU-T Recommendation P.863: Perceptual Objective Listening Quality Assessment', 2014.
[7] 'ITU-T Recommendation G.114', 2009.
[8] H. Holma, A. Toskala, and P. Tapia, 'HSPA + Evolution to Release 12: Performance and Optimization', ISBN: 978–1–118–50321–8, John Wiley & Sons, September 2014.
[9] A. Keates, 'Challenges for Higher Energy Density in Li-Ion cells', Intel Corporation, IWPC Workshop, CA, USA, March 2014.
[10] M. Lauridsen, L. Noël, T. B. Sørensen, and P. Mogensen, 'An Empirical LTE Smartphone Power Model with a View to Energy Efficiency Evolution', *Intel Technology Journal*, 18(1), 172–193 (2014).
[11] H. Holma and A. Toskala. *WCDMA for UMTS: HSPA Evolution and LTE*, 5th ed., John Wiley & Sons (2010), ISBN: 978–0–470–68646–1.

17

Further Outlook for LTE Evolution and 5G

Antti Toskala and Karri Ranta-aho

17.1 Introduction

This chapter covers the expected evolution in Release 14 and beyond from LTE-Advanced evolution point of view. Also the 5G evolution aspects are looked at, including consideration of the relationship between LTE and 5G; the expected timeline in 3GPP for addressing 5G in response to the IMT-2020 process going on in ITU-R is presented.

17.2 Further LTE-Advanced Beyond Release 13

The Release 13 work programme, with the items mostly covered in Chapters 3, 10 and 11, sets the direction of LTE-Advanced evolution, with a new name also being defined in 3GPP to highlight the significant increase in capabilities compared to earlier LTE releases. There is

LTE Small Cell Optimization: 3GPP Evolution to Release 13, First Edition.
Edited by Harri Holma, Antti Toskala and Jussi Reunanen.
© 2016 John Wiley & Sons, Ltd. Published 2016 by John Wiley & Sons, Ltd.

going to be a demand for higher capability due to increased traffic as well as addressing better different use cases from the traditional mobile broadband use cases, reaching to areas such as machine-to-machine (M2M) communication. The existing Release 13 items are presented in Figure 17.1.

Some of the items in the table for Release 13 are rather large and potential and some of the aspects from the on-going study items are likely to materialize (at least part of them) only as Release 14 work item. Further there are several items that have been raised but not considered as urgent due to the work load or other reasons in TSG RAN domain.

The following items are envisaged for Release 14 LTE-Advanced further evolution work, expected to start in the 1H/2016, while some of the items may still end up being reflected in Release 13:

- Enhancements for LTE-Unlicensed (LAA for LTE). While Release 13 work is already on-going as described in Chapter 11, there is a clear need to add further functionality in Release 14, especially for the area of uplink operation with LTE-Unlicensed with the Release 13 covering the basic functionality in downlink side and necessary co-existence methods with Wi-Fi.
- LTE-WLAN carrier aggregation is being specified for Release 13, with the work item started in March 2015. Thus it is unlikely that all items desired to the specified will be completed by the end of 2015.
- Improvement for eMBMS has been done over several releases; still there are items not addressed so far, such as support for dedicated eMBMS carrier that would be beneficial if moving more of the broadcast type of service to be carrier on top of eMBMS.
- Addressing requirements for 5G, as addressed in the next section. For example, aiming for short latency would require shorter TTI and shorter eNodeB and UE decoding times for the received TTI.
- Wider LTE bandwidth to address band allocations with more than 20 MHz of continuous spectrum for an operator. This may be alternatively covered by 5G.
- Use of LTE for communications from/towards vehicles (V2X), as has been addressed in 3GPP with requirements study on-going before actual radio standardizations studies start. Respectively, the device-to-device (D2D) solution, introduced in Release 12 as covered in Chapter 3, as being extended further in Release 13, may be enhanced to cover the vehicle-to-vehicle (V2V) use case also, as is currently being investigated with a specific V2X study item.
- Improvements in mobility, as there is going to be increased signalling with larger number of carriers and use of dual connectivity unless some enhancements are introduced. The proposed approaches include autonomous selection of the SCell by the UE to reduce handover signalling.
- Work for further narrowband LTE solution for M2M or Narrowband Internet of Things (NB-IoT). There has been interest raised for very low cost and narrow bandwidth solution, fitting basically to a single 200 kHz GSM carrier, aiming for cases with low data rate transmission needs. 3GPP is addressing this aspect still for Release 13 in addition to the on-going GSM and LTE M2M work, with the work expected to be finalized potentially slightly later than other Release 13 items around mid 2016.

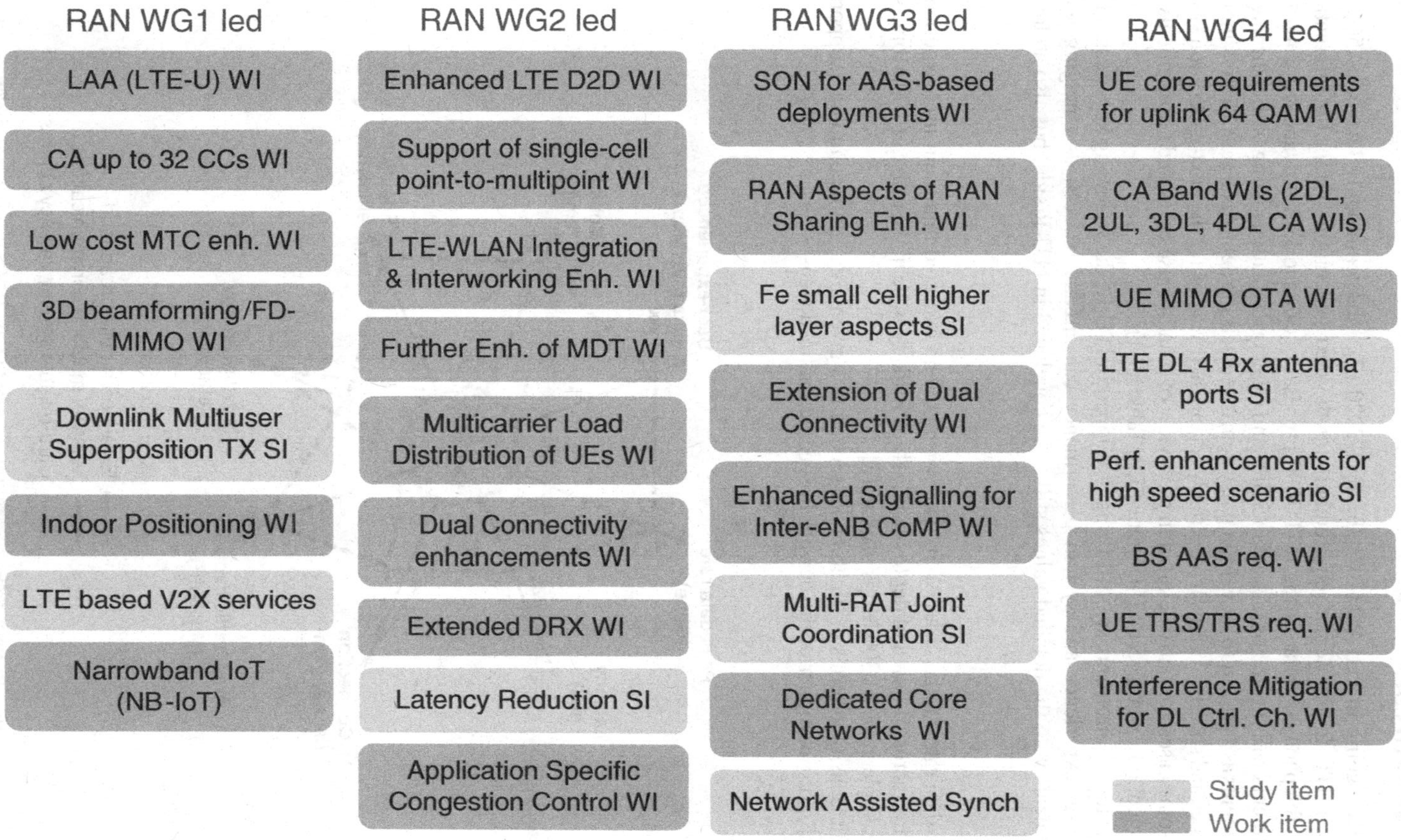

Figure 17.1 Release 13 topics for LTE-advanced evolution

RAN WG1 led
- LAA (LTE-U) WI
- CA up to 32 CCs WI
- Low cost MTC enh. WI
- 3D beamforming/FD-MIMO WI
- Downlink Multiuser Superposition TX SI
- Indoor Positioning WI
- LTE based V2X services
- Narrowband IoT (NB-IoT)

RAN WG2 led
- Enhanced LTE D2D WI
- Support of single-cell point-to-multipoint WI
- LTE-WLAN Integration & Interworking Enh. WI
- Further Enh. of MDT WI
- Multicarrier Load Distribution of UEs WI
- Dual Connectivity enhancements WI
- Extended DRX WI
- Latency Reduction SI
- Application Specific Congestion Control WI

RAN WG3 led
- SON for AAS-based deployments WI
- RAN Aspects of RAN Sharing Enh. WI
- Fe small cell higher layer aspects SI
- Extension of Dual Connectivity WI
- Enhanced Signalling for Inter-eNB CoMP WI
- Multi-RAT Joint Coordination SI
- Dedicated Core Networks WI
- Network Assisted Synch

RAN WG4 led
- UE core requirements for uplink 64 QAM WI
- CA Band WIs (2DL, 2UL, 3DL, 4DL CA WIs)
- UE MIMO OTA WI
- LTE DL 4 Rx antenna ports SI
- Perf. enhancements for high speed scenario SI
- BS AAS req. WI
- UE TRS/TRS req. WI
- Interference Mitigation for DL Ctrl. Ch. WI

Study item
Work item

17.3 Towards 5G

Work is on-going in ITU-R [1] for the IMT-2020 [2], more commonly known as 5G. The process in ITU-R will define requirements for the next-generation system for International Mobile Telephony, similar to the IMT-2000 process for 3G and IMT-Advanced for 4G. The requirements will then be processed in 3GPP and other organizations along with their internal requirements as a pretext for developing a system that meets the defined targets. Key technology companies and specific forums have already set out their visions for the next-generation system and in some cases are fairly detailed in their view of what the use cases and requirements are, as reflected, for example, in [3] (Figure 17.2; Table 17.1).

While the requirements are still being developed in ITU-R, the following key areas related to the user experience and system performance can already be seen as emerging as the requirements with the detailed values still to be concluded in ITU-R side, but also addressed in [4]:

- Peak data rates in the order of 10 Gbps or more and expected minimum user-experienced data rate across the area of deployment, which is likely to be in the order of 100 Mbps, with an example value of 300 Mbps used in [3].
- Traffic density in terms of gigabits per second in a given area, with technology supporting 1 Tbps/km^2 or higher traffic density in specific confined areas.

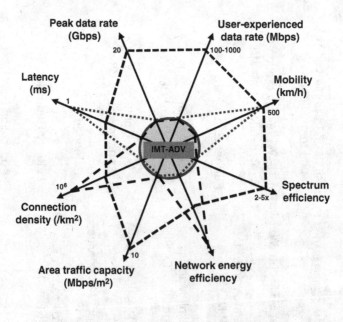

Figure 17.2   IMT-2020 key capabilities and their relation to key use case families

Table 17.1 IMT-2020 use case families and their relation to the IMT-2020 requirements

Requirement	Use case family			Requirement at a different importance level		
	A	B	C	High (H)	Medium (M)	Low (L)
Peak data rate	H	L	L	20 Gbps	~ IMT-A	Best effort
User-experienced data rate	H	L	L	100–1000 Mbps	Several Mbps	Best effort
Mobility	H	L	H	500	Moderate mobility without service degradation	Low mobility
Latency	M	L	H	1 ms	Several ms	Somewhat larger latency acceptable
Connection density	M	H	L	10^6/km^2	10^4–10^6/km^2	< 10^4/km^2
Network energy efficiency	H	M	L	100 × IMT-A	Important	N/A
Spectrum efficiency	H	L	L	3 × IMT-A	~ IMT-A	N/A
Area traffic capacity	H	L	L	10 Mbps/m^2	Several hundred Gbps/km^2	N/A

A, extended mobile broadband; B, massive machine communication; C, ultra-reliable and low latency communication.

- Spectrum efficiency, with the obvious target to have clearly higher efficiency than LTE-Advanced.
- Latency (end-to-end delay) as low as 1 ms. Typically mentioned use cases included tactile Internet type of use cases, but several industry automation use cases require very low latency.
- Higher density of connected devices, with support for 1 million connections per square kilometre.
- Higher mobility, up to 500 kmph.
- Reliability of the connection, which is relevant especially for approaches such as industrial automation.
- Energy efficiency, to achieve lower energy consumption.

17.4 5G Spectrum

Another important aspect for handling more traffic is to have enough spectrum available for 5G. The upcoming WRC-15 is going to address spectrum below 6 GHz which is essential for having good coverage, especially when considering rural areas. It is foreseen in the next phase with WRC-19 to address spectrum above 6 GHz, even up to 100 GHz for more local user and denser network deployments.

The spectrum below 6 GHz and closer to 1 GHz will be important from the coverage point of view especially when one considers use cases such as M2M. The frequency bands above

Table 17.2 Spectrum usage models

	Licensed model	Unlicensed model	Complementary licensed model
Spectrum usage	Exclusively allocated to one operator	Anyone can use the spectrum	Incumbent use has the first-take right – when/where not in use, cell operator(s) may use the spectrum
Interference	No unexpected interference	No control over interference, other access point or user may use the same spectrum	No unexpected interference
Quality	Predictable	Unpredictable	Predictable, when use allowed

6 GHz are then more suited to delivering very large data volumes and data rates for small cell and local connectivity scenarios.

The existing bands with legacy technologies may eventually be refarmed to 5G, with the expectation first aiming to refarm bands used by 2G and 3G technologies.

The mainstream approach for spectrum allocation for mobile communication is the exclusive spectrum licensing, where an operator owns the rights to use a particular spectrum band, and thus the system deployed is protected from external interference as no one else is allowed to use that particular band. The complement to this is the unlicensed spectrum usage, most well known being the Wi-Fi mode of operation – no one owns the spectrum and controls the deployment, thus the coverage and service quality cannot be guaranteed, but there is no associated cost of spectrum licensing. License-Assisted Access for LTE is moving the mobile technologies to benefit from the unlicensed spectrum in addition to the licensed band. There the service is always guaranteed over the licensed band, and additional bandwidth of unguaranteed quality can be aggregated with that from the unlicensed spectrum.

Between the two models, new complementary spectrum access approaches are being considered, typically referred to as Licensed Shared Access and Authorized Shared Access. The intention is to unlock underutilized spectrum being used by other systems for mobile communication use, when and where possible, but protect the incumbent user (e.g. radar, broadcast or military user) from the mobile communication interference when and where it may be harmful. The spectrum-sharing models are of course not directly linked to a particular technology: using all three approaches with LTE as well as 5G is likely in the future (Table 17.2).

17.5 Key 5G Radio Technologies

Different research organizations and public research programs such as the EU-funded METIS and 5G-PPP are conducting radio-related research on multitude of different new technological components that could become 5G building blocks. The well-proven OFDMA with good MIMO compatibility as well as the DFT-spread single-carrier FDMA known from the LTE uplink are the benchmarks in the signal waveform. In the other domain, massive MIMO, native CoMP support, dynamic TDD, full-duplex TDD and solutions like in-band self-backhauling are being looked into.

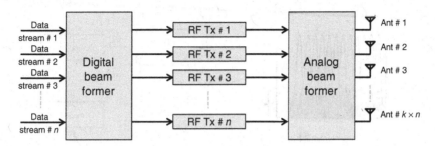

Figure 17.3   A block diagram of a hybrid digital/analog beamformer for massive MIMO

Massive MIMO, where, for example, 64, 128 or 256 antenna elements are used, becomes more commercially attractive when the frequencies increase. The size of one antenna element becomes smaller when frequencies increase, and it becomes possible to integrate an antenna array directly in the chip itself. The large number of MIMO antennas is capable of directing transmitted energy towards the intended user, and thus obtaining higher link efficiency as well as reducing interference. Massive MIMO arrays and high spatial selectivity of the narrower beams are also convenient for multiuser MIMO operation. Fully digital beamforming, for example, for a 256-antenna transmitter would require 256 power amplifiers, one per antenna. This would lead to high energy consumption and high cost. Thus a hybrid digital/analog beamforming is likely more attractive with Massive MIMO (Figure 17.3).

A dynamic TDD structure allows for better overall spectrum utilization than FDD or static UL/DL TDD time allocation. With dynamic TDD the data part of any frame can be allocated either to uplink or to downlink based on the traffic needs. The downside of the dynamic allocation is increased inter-cell interference: at the same time when one cell's UE may be transmitting in the uplink, another UE close may be receiving from another cell, and thus experiences this inter-cell interference to the fullest. Thus dynamic TDD is more promising in relatively small cells. This cell-edge phenomenon can be mitigated with advanced receiver designs and inter-cell interference coordination. The dynamic TDD frame structure can be designed to be compatible with full-duplex TDD, where the transmission and reception take place at the same time, and the transmitted signal is cancelled from the received signal (Figure 17.4). In this case some frames would be scheduled to have both DL and UL transmission at the same time. Due to the massive power difference of the transmitted and received signals, even after the cancellation there is some residual interference injected to the received

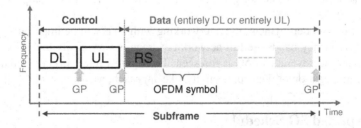

Figure 17.4   An example of a dynamic TDD-compatible frame structure

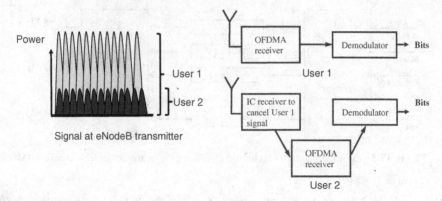

Figure 17.5   NOMA principle

signal leading to somewhat reduced link performance when compared to just transmitting or receiving. The cost and complexity of the self-cancellation related to full-duplex operation may make it more practical for the base stations than terminals. In this set-up the base station would receive a transmission from one UE and at the same time transmit to another one.

Notably the dynamic TDD frame structure with sub-frame duration of a fraction of a millisecond is suitable for low-cost in-band self-backhauling access points. The low-cost design approach allows for using the same RF transceiver for both access and backhaul links in an opportunistically time-divided manner, greatly simplifying the AP design, as only one transceiver operating at one frequency band is required. In a more sophisticated case with two transceivers, a full-duplex operation could be used, where both the access link and the backhaul link operate at the same time on the same band – here it would be possible to constrain the AP operation so that it would transmit at both the links or receive at both the links at the same time, but never transmit and receive simultaneously, removing the need for self-interference cancellation, or alternatively transmitting always on one link only and receiving on another link, eliminating the need for two transceivers, but leading to more challenging self-interference environment.

In the area of multiple access there are foreseen discussions on slightly modified variants of OFDM technology, such as Non-Orthogonal Multiple Access (NOMA), as is foreseen to be studied during 2015 on the LTE-Advanced evolution track, with the example principle shown in Figure 17.5. In general, the expected development of the processing capability will allow more advanced receivers, but if the bandwidths and on the other hand processing times become shorter (to achieve lower latencies) then all the developments for the computing power will definitely be needed as well.

As part of the 5G deployment, the interworking with LTE will be important. As especially with higher frequency bands the full network coverage is very costly to achieve, use of LTE to complement the coverage of 5G is foreseen. The tightest interworking would be achieved with carrier aggregation or dual connectivity-like approach, as also noted in [3].

17.6   Expected 5G Schedule

ITU-R Working Party 5D decided on the process and timeline for the development of the IMT-2020 requirements and certification of a system as IMT-2020 compliant in 2014. The

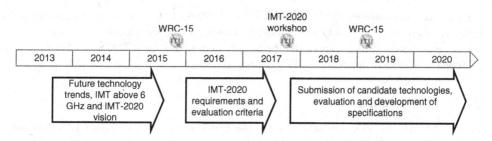

Figure 17.6 Outline of the ITU-R WP5D process and timeline for development and certification of IMT-2020 technologies

standard defining organizations such as 3GPP will then work on developing the system and, in accordance to the process and timeline, submit required evidence to the ITU-R on the developed system meeting the set targets. The ITU-R IMT-2020 timeline is shown in Figure 17.6.

In 3GPP 5G-related activities are expected to start by the end of 2015, first with collection of targets and requirements [5], as illustrated in Figure 17.7. Successful 5G technology studies require some enablers, such as channel model for use of spectrum above 6 GHz [6]. At the same time the 3GPP will keep on working on the LTE-Advanced evolution track, and as seen in the past with HSPA, which developed the capabilities needed to qualify as a 4G technology, it is likely that the LTE-Advanced will in the future meet most of the 5G requirements as well. The actual start of the 5G work in 3GPP was the September 2015 TSG RAN 5G workshop, where visions and technologies were discussed first time and the direction for actual work in Release 14 and onwards was identified.

The potential approach in 3GPP is to consider phasing the 5G introduction in such a way that, first, the 3GPP work would cover radio aspects of 5G only, aiming for addressing the need for extended mobile broadband use case, as reflected in [7] and [8]. This would be answering the need for some of the markets, especially in Asia and US, where there is a desire for early introduction of 5G by 2020 or even before. This would require early phase work item to be

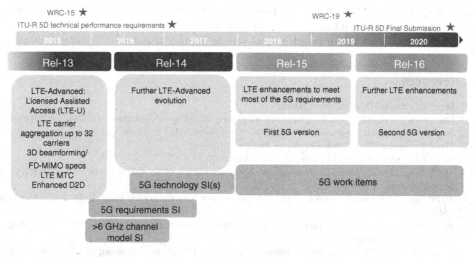

Figure 17.7 Expected 3GPP timeline for LTE-A evolution and the introduction of 5G

started in parallel to the study phase aiming for such 'Phase 1 5G'. Further needs and use cases reflected in Figure 17.2 would then be addressed later, including evolution of the core network for those use cases. This is expected to be the key item in the 3GPP TSG RAN 5G workshop in September 2015 to define the 3GPP roadmap on 5G [9].

17.7 Conclusions

In the previous sections, we have covered the expectations for the further LTE-Advanced evolution beyond 3GPP Rel-12 as well as the first stages and expected timeline of the 5G systems. The process and timeline for 5G is already in place, and the work for detailing out the vision, use cases and requirements is well on the way, but how the 5G system will finally look like will be known with certainty only in the years to come. However, it is already possible to say a few things about it.

5G is needed to answer the ever-growing data demands and to enable mobile communications on higher frequency bands in which LTE is not very suited. The big focus on peak data rates and extending the band access may lead to confusion that 5G is about small cells on high-frequency bands only, when that is just one, albeit important aspect of 5G. 5G will unlock the potential of the higher frequency bands above 6 GHz, but at the same time for wide area coverage and M2M use, 5G will find itself to the more traditional lower cellular frequency bands as well.

The access scheme for 5G is still under research, but dynamic TDD based on OFDMA (on bands below 6 GHz and up to a few tens of gigahertzs) and DFT-generated single carrier waveforms (on several tens of gigahertzs) coupled with massive MIMO, pencil beamforming and enhanced spectrum flexibility emerge as strong candidates for the new access schemes.

It is also important to keep in mind that whatever technological components 5G system will introduce, if suitable, those will be adapted in the LTE-Advanced system as well, making the LTE-Advanced eventually meeting most of the 5G requirements as well, and thus the LTE-Advanced and its evolution with Releases 13 and 14 is typically characterized as one important component of 5G. 3GPP will continue the strong evolution of LTE technology with the new features under standardization work for Release 13 and beyond, enabling evolved LTE-Advanced technology to be the cutting-edge radio solution for addressing the upcoming traffic and service needs for mobile broadband as well as for several other use cases, including M2M, public safety, LTE for broadcast and vehicle communications.

References

[1] http://www.itu.int/ITU-R (accessed June 2015)
[2] http://www.itu.int/en/ITU-R/study-groups/rsg5/rwp5d/imt-2020/Pages/default.aspx (accessed June 2015)
[3] Next Generation Mobile Networks (NGMN), '5G White Paper', v.1.0, February 2015.
[4] ITU-R Document 5D-918, Nokia Networks, January 2015.
[5] 3GPP Tdoc, SP-150149, '5G Timeline in 3GPP', March 2015.
[6] 3GPP Tdoc 151606, 'New SID Proposal: Study on channel model for frequency spectrum above 6 GHz', September 2015.
[7] 3GPP Tdoc RWS-150010, 'NOKIA Vision & Priorities for Next Generation Radio Technology', September 2015.
[8] 3GPP Tdoc RWS-150036, 'Industry Vision and Schedulefor the New Radio Part of the Next Generation Radio Technology', September 2015.
[9] 3GPP Tdoc RWS-150073, 'Chairman's summary regarding 3GPP TSG RAN workshop on 5G', September 2015.

Index

LTE Small Cell Optimization: 3GPP Evolution to Release 13, First Edition.
Edited by Harri Holma, Antti Toskala and Jussi Reunanen.
© 2016 John Wiley & Sons, Ltd. Published 2016 by John Wiley & Sons, Ltd.

Printed in the United States
By Bookmasters